T0140602

Advances in Intelligent Systems and Computing

Volume 744

Series editor

Janusz Kacprzyk, Polish Academy of Sciences, Warsaw, Poland
e-mail: kacprzyk@ibspan.waw.pl

The series “Advances in Intelligent Systems and Computing” contains publications on theory, applications, and design methods of Intelligent Systems and Intelligent Computing. Virtually all disciplines such as engineering, natural sciences, computer and information science, ICT, economics, business, e-commerce, environment, healthcare, life science are covered. The list of topics spans all the areas of modern intelligent systems and computing such as: computational intelligence, soft computing including neural networks, fuzzy systems, evolutionary computing and the fusion of these paradigms, social intelligence, ambient intelligence, computational neuroscience, artificial life, virtual worlds and society, cognitive science and systems, Perception and Vision, DNA and immune based systems, self-organizing and adaptive systems, e-Learning and teaching, human-centered and human-centric computing, recommender systems, intelligent control, robotics and mechatronics including human-machine teaming, knowledge-based paradigms, learning paradigms, machine ethics, intelligent data analysis, knowledge management, intelligent agents, intelligent decision making and support, intelligent network security, trust management, interactive entertainment, Web intelligence and multimedia.

The publications within “Advances in Intelligent Systems and Computing” are primarily proceedings of important conferences, symposia and congresses. They cover significant recent developments in the field, both of a foundational and applicable character. An important characteristic feature of the series is the short publication time and world-wide distribution. This permits a rapid and broad dissemination of research results.

More information about this series at http://www.springer.com/series/11156

Thi Thi Zin · Jerry Chun-Wei Lin
Editors

Big Data Analysis and Deep Learning Applications

Proceedings of the First International Conference on Big Data Analysis and Deep Learning

Editors
Thi Thi Zin
Faculty of Engineering
University of Miyazaki
Miyazaki
Japan

Jerry Chun-Wei Lin
Department of Computing, Mathematics, and Physics
Western Norway University of Applied Sciences (HVL)
Bergen
Norway

ISSN 2194-5357 ISSN 2194-5365 (electronic)
Advances in Intelligent Systems and Computing
ISBN 978-981-13-0868-0 ISBN 978-981-13-0869-7 (eBook)
https://doi.org/10.1007/978-981-13-0869-7

Library of Congress Control Number: 2018944427

Printed on acid-free paper

This Springer imprint is published by the registered company Springer Nature Singapore Pte Ltd.
part of Springer Nature
The registered company address is: 152 Beach Road, #21-01/04 Gateway East, Singapore 189721, Singapore

Preface

This volume composes the proceedings of the first International Conference on Big Data Analysis and Deep Learning (ICBDL 2018), which is jointly organized by University of Miyazaki, Japan, and Myanmar Institute of Information Technology, Myanmar. ICBDL 2018 took place in Miyazaki, Japan, on May 14–15, 2018. ICBDL 2018 is technically co-sponsored by Springer; University of Miyazaki, Japan; Myanmar Institute of Information Technology, Myanmar; and Harbin Institute of Technology, Shenzhen, China.

The focus of ICBDL 2018 is on the frontier topics in data science, engineering, and computer science subjects. Especially, big data analysis, deep learning, information communication, and imaging technologies are the main themes of the conference.

All submitted papers have gone through the peer review process. Forty-five excellent papers were accepted for the final proceeding. We would like to express our sincere appreciation to the reviewers and the International Technical Program Committee members for making this conference successful. We also would like to thank all authors for their high-quality contributions.

We would like to express our sincere gratitude to Prof. Dr. Tsuyomu Ikenoue, the President of the University of Miyazaki who has made the conference possible. Finally, our sincere thanks must go to the host of the conference, the University of Miyazaki, Japan.

March 2018

Thi Thi Zin
Conference Program Committee Chair

Organizing Committee

General Chair

Tsuyomu Ikenoue	University of Miyazaki, Japan

General Co-chairs

Win Aye	Myanmar Institute of Information Technology, Myanmar
Masahito Suiko	University of Miyazaki, Japan
Toshiaki Itami	University of Miyazaki, Japan

Advisory Committee Chairs

Mitsuhiro Yokota	University of Miyazaki, Japan
Masugi Maruyama	University of Miyazaki, Japan
KRV Raja Subramanian	International Institute of Information Technology, Bangalore, India
Pyke Tin	University of Miyazaki, Japan
Hiromitsu Hama	Osaka City University, Japan

Program Committee Chair

Thi Thi Zin	University of Miyazaki, Japan

Program Committee Co-chair

Mie Mie Khin	Myanmar Institute of Information Technology, Myanmar

Publication Chairs

Thi Thi Zin	University of Miyazaki, Japan
Jerry Chun-Wei Lin	Western Norway University of Applied Sciences (HVL), Norway

Invited Session Chairs

Soe Soe Khaing	University of Technology, Yatanarpon Cyber City, Myanmar
Myint Myint Sein	University of Computer Studies, Yangon, Myanmar

International Technical Program Committee Members

Moe Pwint	University of Computer Studies, Mandalay, Myanmar
Win Zaw	Yangon Institute of Technology, Myanmar
Aung Win	University of Technology, Yatanarbon Cyber City, Myanmar
Thi Thi Soe Nyunt	University of Computer Studies, Yangon, Myanmar
Khin Thida Lynn	University of Computer Studies, Mandalay, Myanmar
Myat Myat Min	University of Computer Studies, Mandalay, Myanmar
Than Nwe Aung	University of Computer Studies, Mandalay, Myanmar
Mie Mie Tin	Myanmar Institute of Information Technology, Myanmar
Hnin Aye Thant	University of Technology, Yatanarbon Cyber City, Myanmar
Naw Saw Kalayar	Computer University (Taunggyi), Myanmar
Myint Myint Khaing	Computer University (Pinlon), Myanmar
Hiroshi Kamada	Kanazawa Institute of Technology, Japan
Tomohiro Hase	Ryukoku University, Japan

Contents

Deep Learning and its Applications

Data Mining and its Applications

Big Data Analysis

Data-Driven Constrained Evolutionary Scheme for Predicting Price of Individual Stock in Dynamic Market Environment

Henry S. Y. Tang(✉) and Jean Hok Yin Lai(✉)

Hong Kong Baptist University, Kowloon Tong, Hong Kong
henry.tang.22303@gmail.com, jeanlai@comp.hkbu.edu.hk

Abstract. Predicting stock price is a challenging problem as the market involve multi-agent activities with constantly changing environment. We propose a method of constrained evolutionary (CE) scheme that based on Genetic Algorithm (GA) and Artificial Neural Network (ANN) for stock price prediction. Stock market continuously subject to influences from government policy, investor activity, cooperation activity and many other hidden factors. Due to dynamic and non-linear nature of the market, individual stock price movement are usually hard to predict. Investment strategies used by regular investor usually require constant modification, remain secrecy and sometimes abandoned. One reason for such behavior is due to dynamic structure of the efficient market, where all revealed information will reflect upon the stock price, leads to dynamic behavior of the market and unprofitability of the static strategies. The CE scheme contains mechanisms which are temporal and environmental sensitive that triggers evolutionary changes of the model to create a dynamic response towards external factors.

Keywords: Genetic Algorithm · Artificial neural network · Data-driven Evolutionary · Stock · Prediction

1 Introduction

Stock market is often seen as a dynamic structure with significant changes over time [1], this nature leads challenges in predicting the individual stock price within the market.

Due to the statistical basis and advancement of automatic trading, technical analysis has gained popularity over time. Attempts have been made using different approaches based on human behavior to create a model in predicting the price movement of the stock [2]. Timeseries analysis and machine learning algorithms are commonly used models for the task [3], however, most of the studies are not concerning the dynamic nature of the market where the parameters of the models are obtained and fixed through measuring the statistical confidence or learning algorithm from historical data. This approach might suffer from poor performance in long run due to market structure shift.

T. T. Zin and J. C.-W. Lin (Eds.): ICBDL 2018, AISC 744, pp. 3–8, 2019.
https://doi.org/10.1007/978-981-13-0869-7_1

1.1 Artificial Neural Network

Inspired by human brain's ability of non-linear, parallel and complex computation power, artificial neural networks are proven to be a universal function approximator given enough neurons within the hidden layer of the network. [4] Different attempts using artificial neural network to predict stock trends were presented in [5], where it indicates the possible feasibility of such approach with curtain classes of neural networks.

1.2 Genetic Algorithms (GA)

John Holland proposed the idea of GA in 1970 inspired by the process of biological evolution. [6] This method provided us with a learning method to find an optimal solution given an optimization function (fitness function). GA search the solution by encodes the solution to a chromosome which represents a potential solution to the problem. The chromosome is then tested by the fitness function which will return a fitness value representing the survival abilities. Fitness function is specifically designed for a particular problem, it provides an evaluation of the goodness of the chromosome to the problem. The chromosome with higher fitness value indicating a better solution towards the specific problem. Theses chromosome will produce offspring based on the fitness value through crossover operation, the chromosome with higher fitness value will be selected out and have more offspring. All offspring will then experience a mutation operation, where the chromosome might mutate under predefined probability. Given enough generation, the GA will find a near-optimum solution to the problem. In terms of financial applications, in [7], an ensemble system based on GA were proposed to forecast stock market. In [8], a framework based on Web robot, GA and Support Vector Machine was proposed for data analysis and prediction.

2 Problem Scenario and Assumptions

The problem of stock price movement prediction can be seen as a binary classification problem. The output y of our model has the property of $y \in [0, 1]$, where 0 representing a down-trend prediction and 1 representing an up-trend prediction.

In real-world, stock price of individual company was affected by multi-factors including temporal events, competitors, collaborators and other factors simultaneously. Since the weighting of the combination of factors very likely to change over time, making any non-dynamic model might not be suitable for stock price movement prediction in long run.

We assume that for any given time t there exist a $f(.,.)$ decision boundary such that the expected error of the decision boundary is smaller than a critical value ε, where

$$E(error(f(x,t), true\ label)) \leq \varepsilon \quad (1)$$

Unfortunately, the function f is far too complex to model and almost impossible to observe. However, we can approximate the function at a given time where the approximate function $f^*(x)_t$ *at timet* $\sim f(x, t)$. We further assume that the approximated function $f^*(x)_t$ decision boundary will shift continuously against time, where

$$f^*(x)_t \sim f^*(x)_{t+\Delta t} \tag{2}$$

Based on the assumptions, for creating an effective approximation function, the scheme should solve below problem.

2.1 Shift Detection (Trigger)

For the scheme to be functional against market structure shift, we require a mechanism to detect the shifting signal and start triggering the evolution process. This mechanism might only consider the presence of the shift but not consider the reason of the shift and the reaction against it.

2.2 Shift Direction

Once we have detected the presence of the shift, we need more information on what type of shift is occurring, for example there is a new player introduced into the market or new regulation announced. Different external environment influence will cause the market and investor behavior changes drastically. Therefore, we need to understand what kind of shift we are experiencing currently.

2.3 Shift Degree

After we know the goal of the evolution, it is sensible for us to consider how fast the model should evolve and from which path. Scheduling of the evolution will allow us progressively reaching the goal. During the evolution, result from problem 2 need to revise for more accurate estimation.

3 Methodology

In this section, we introduce a scheme to solve the problem indicated in previous section. The architecture of the scheme consists of three major structure. A detection function, evolution function and model base. In terms of operation, the detection function continuously monitoring the new data feed to the system, once the function detected the market shift, it will start triggering the evolution process and request the model base to support its operation. Finally, the new model produced will replace the current model and stored in the model base.

3.1 Shift Detection Function

In phase one of the scheme, most recent historical data is used, e.g. 4 months to the past from now, and separate into different sections $U(i)$ chronologically, and for each section, we run the current model $f^*(x)_t$ and compute the error rate, where

$$section\ error\ (i) = \frac{1}{no.\ of\ data\ point\ in\ U(i)} \sum\nolimits_{x \in U(i)} \left(f^*(x)_t - y(x)\right)^2 \tag{3}$$

If observed some progressive error increase against i, then the scheme determines that there exists a shift in the market structure and trigger the evolutionary process.

3.2 Shift Direction and Shift Degree

In the proposed model, we used the word constrained to describe the evolutionary process being bounded by current and previous models stored in model base.

The chromosome of each model is the weight and activation function of the model. Figure 1 show that the construction of a chromosome of a neural network, where each layer of the network can be represented as a m × n matrix $M^L_{m,n}$. Each row i of matrix $M^L_{m,n}$ is the weight of all connection of neuron i at layer L to its descendant. Each model will contain one or more layer of matrix $M^L_{m,n}$ and a vector of activation functions attached to each layer.

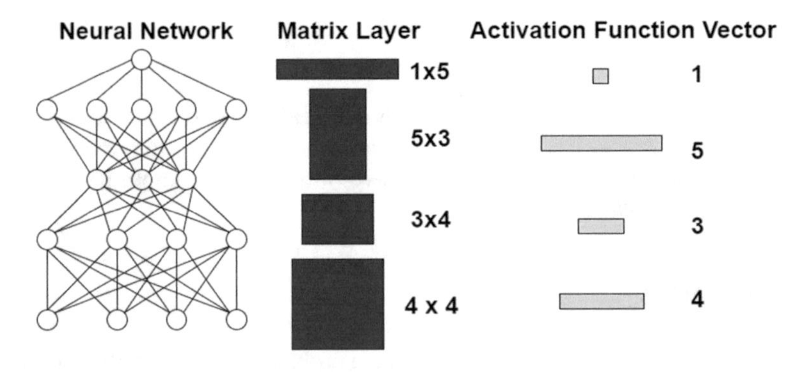

Fig. 1. Illustration of an ANN model and its respective chromosome representation.

Two types of crossover operation exist within the evolutionary scheme. First is the neuron swap operation, where the crossover lines only locate between rows of matrix $M^L_{m,n}$. Due to consistency between different layers, the child layer must be the same dimension as its parents. Notice that for this operation, the activation function vector is following the swap as illustrated in Fig. 2a. Second crossover operation will be the connection swap operation. In this operation, the crossover lines only locate within randomly selected rows of the matrix $M^L_{m,n}$, see Fig. 2b. The activation function vector in this operation will change based on the parent that contributed the most number of weight.

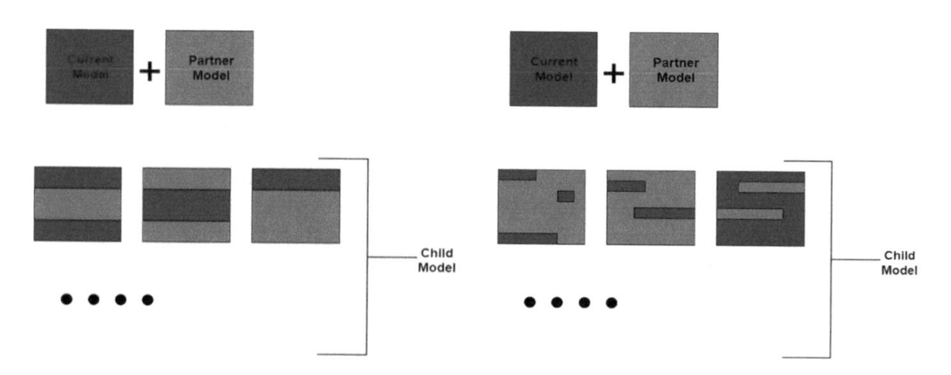

Fig. 2a. Type one crossover operation

Fig. 2b. Type two crossover operation

Mutation will occur with certain probability on both the weight on the matrix set and the set of activation function vector. Therefore, the activation function of a neuron can mutate from sigmoid function to tanh or other possible activation functions or vice versa.

The number of child produced by the current model and a previous model will base on the error rate of the previous model produced with current data feed. More children will be produced by models that having lower error rate.

The reason of choosing the current model as the major partner is because according to assumption 2, the current model will be a good starting point as the parent of the next evolutionary point.

3.3 Fine Tuning

After the crossover and mutation process, the child model will experience a fine tuning and selection stage. At this stage, the data feed will be used to fine tuning the child model by using backpropagation training method with small learning rate. The fitness function is the cross-validation error result of the fine-tuned child model. The next generation will be created based on the fitness function result and the crossover of the fine-tuned child model. The process will terminate until reaching the desire score of fitness or reaching maximum generation.

4 Discussion

Since the possible outcome of the model are based on the number of previous model that stored on the model base, the method for initializing the scheme can be further improved to introduce more constructive model at early stage.

This paper only provided the concept of the proposed model, it is encouraged for further experiment on the real-world data and compared the proposed model with a randomized approach to illustrate the efficiency of the speed of convergence of the GAs and accuracy towards individual stock price prediction for a long-extended period.

References

1. Hamilton, J.D., Lin, G.: Stock market volatility and the business cycle. J. Appl. Econometrics **11**(5), 573–593 (1996). Special Issue: Econometric Forecasting
2. Barberis, N., Thaler, R.: A survey of behavioral finance. In: Handbook of the Economics of Finance, vol. 1, Part B, pp. 1053–1128 (2003). Chap. 18
3. Murphy, J.J.: Technical analysis of the financial markets: a comprehensive guide to trading methods and applications, New York Institute of Finance (1999)
4. Haykin, S.: Neural Networks and Learning Machines, 3rd edn. Pearson, Upper Saddle River (2009)
5. Saad, E.W., Prokhorov, D.V., Wunsch, D.C.: Comparative study of stock trend prediction using time delay, recurrent and probabilistic neural networks. IEEE Trans. Neural Netw. **9**(6), 1456–1470 (1998)
6. Holland, J.H.: Adaptation in natural and artificial systems, p. 183. The University of Michigan Press, Michigan (1975)
7. Gonzalez, R.T., Padilha, C.A., Couto, D.A.: Ensemble system based on genetic algorithm for stock market forecasting. In: IEEE Congress on Evolutionary Computation (CEC) (2015)
8. Wang, C.-T., Lin, Y.-Y.: The prediction system for data analysis of stock market by using genetic algorithm. In: International Conference on Fuzzy System and Knowledge Discovery, pp. 1721–1725 (2015)

Predictive Big Data Analytics Using Multiple Linear Regression Model

Kyi Lai Lai Khine[1] and Thi Thi Soe Nyunt[2(✉)]

[1] Cloud Computing Lab, University of Computer Studies, Yangon, Myanmar
kyilailai67@gmail.com
[2] Head of Software Department,
University of Computer Studies, Yangon, Myanmar
thithi@ucsy.edu.mm

Abstract. Today fast trending technology era, data is growing very fast to become extremely huge collection of data in all around globe. This so-called "Big Data" and analyzing on big data sets to extract valuable information from them has also become one of the most important and complex challenges in data analytics research. The challenges of limiting memory usage, computational hurdles and slower response time are the main contributing factors to consider traditional data analysis on big data. Then, traditional data analysis methods need to adapt in high-performance analytical systems running on distributed environment which provide scalability and flexibility. Multiple Linear Regression which is an empirical, statistical and mathematically mature method in data analysis is needed to adapt in distributed massive data processing because it may be poorly suited for massive datasets. In this paper, we propose MapReduce based Multiple Linear Regression Model which is suitable for parallel and distributed processing with the purpose of predictive analytics on massive datasets. The proposed model will be based on "QR Decomposition" in decomposing big matrix training data to extract model coefficients from large amounts of matrix data on MapReduce Framework with large scale. Experimental results show that the implementation of our proposed model can efficiently handle massive data with a satisfying good performance in parallel and distributed environment providing scalability and flexibility.

Keywords: Big data · Multiple linear regression · Predictive analytics
MapReduce · QR decomposition

1 Introduction

Nowadays, the Internet represents a big storage where great amounts of information are generated every second. The IBM Big Data Flood Infographic describes 2.7 Zettabytes of data exist in the today digital universe. Moreover, according to this study from Facebook there are 100 Terabytes updated daily and an estimate of 35 Zettabytes of data generated leading to a lot of activities on social networks annually by 2020. Amir and Murtaza expressed that big data moves around 7 Vs: volume, velocity, variety, value and veracity, variability and visibility. Storing huge volume of data available in various formats which is increasing with high velocity to gain values out it is itself a

T. T. Zin and J. C.-W. Lin (Eds.): ICBDL 2018, AISC 744, pp. 9–19, 2019.
https://doi.org/10.1007/978-981-13-0869-7_2

big deal [3]. Big data analytics can be defined as the combination of traditional analytics and data mining techniques together with any large voluminous amount of structured, semi-structured and unstructured data to create a fundamental platform to analyze, model and predict the behavior of customers, markets, products, services and so on. "Hadoop" has been widely embraced for its ability to economically store and analyze big data sets. Using parallel processing paradigm like MapReduce, Hadoop can minimize long processing times to hours or minutes. There exists three types of big data analytics: descriptive analytics which answer the question: "What has happened?", use data aggregation and data mining techniques to provide insight into the past, predictive analytics which also replies like this "What could happen in future?" applying statistical models like regression and forecasts to understand the future. It comprises a variety of techniques that can predict future outcomes based on historical and current data and the last one, prescriptive analytics for optimization and simulation algorithms to advice on possible outcomes for the question: "What should we do to happen in future?" [7]. Extracting useful features from big data sets also become a big issue because many statistics are difficult to compute by standard traditional algorithms when the dataset is too large to be stored in a primary memory. The memory space in some computing environments can be as large as several terabyte and beyond it. However, the number of observations that can be stored in primary memory is often limited [10].

Therefore, the two challenges for massive data in supervised learning are emerging explained by Moufida Rehab Adjout and Faouzi Boufares. First, the massive data sets will face two severe situations such as limiting memory usage and computational hurdles for the most complicated supervised learning systems. Therefore, loading this massive data in primary memory cannot be possible in reality. Second, analyzing the voluminous data may take unpredictable time to response in targeted analytical results [1]. One of the important major issues in predictive big data analysis is how to apply statistical regression analysis on entire huge data at once because the statistical data analysis methods including regression method have computational limitation to manipulate in these huge data sets. Jun et al. [8] discussed about the sub-sampling technique to overcome the difficulty in efficient memory utilization. They also presented that this approach is useful for regression analysis that only brings the regression parameters or estimators in parts of data and which are less efficient in comparing with the estimators that are derived from the entire data set rather than by parts. However, the desirable regression estimators on entire data set may be impossible to derive [9]. That is why; we propose an approach to lessen the computational burden of statistical analysis for big data applying regression analysis especially multiple linear regression analysis on MapReduce paradigm. The paper is organized as follows. Section 2 presents the concepts and relationships between regression analysis and big data. The background theory of multiple linear regression and its equations and then MapReduce Framework explanations are described in Sect. 3. Our main implementation of the proposed algorithm and respective explanations in detail are presented in Sect. 4. Some performance evaluation results, discussions and final conclusion to illustrate the appropriateness of the proposed approach are given in Sect. 5.

2 Regression Analysis and Big Data

Statistics takes important role in big data because many statistical methods are used for big data analysis. Statistical software provides rich functionality for data analysis and modeling, but it can handle only limited small amounts of data. Regression can be seen in many areas widely used such as business, the social and behavioral sciences, the biological sciences, climate prediction, and so on. Regression analysis is applied in statistical big data analysis because regression model itself is popular in data analysis. There are two approaches for big data analysis using statistical methods like regression. The first approach is that we consider extracting the sample from big data and then analyzing this sample using statistical methods. This is actually the traditional statistical data analysis approach assuming that big data as a population. Jun et al. [8] already expressed that in statistics, a collection of all elements which are included in a data set can be defined as a population in the respective field of study. That is why; the entire population cannot be analyzed indeed according to many factors such as computational load, analyzing time and so on. Due to the development of computing environment for big data and decreasing the cost of data storage facilities, big data which close to the population can be analyzed for some analytical purposes. However, the computational burden still exists as a limitation in analyzing big data using statistical methods. The second approach is that we consider about splitting the whole big data set into several blocks without using big population data. The classical regression approach is applied on each block and then respective regression outcomes from all blocks are combined as final output [6]. This is only a sequential process of reading and storing data in primary memory block by block. Analyzing data in each block separately may be convenient whenever the size of data is small enough for implementing the estimation procedure in various computing environments. However, a question, how to replace sequential processing of several data blocks that can adversely affect in response time still remains as an issue for processing increasing volume of data [12]. Jinlin Zhu, Zhiqiang Ge and et al. proved that MapReduce framework is a sort of resolution to this problem for the replacement of sequential processing with the use of parallel distributed computing that enables distributed algorithms in parallel processing on clusters of machines with varied features.

3 Multiple Linear Regression

Multiple linear regression is a statistical model used to describe a linear relationship between a dependent variable called “explain” and a set of independent or predictor variables called “explanatory” variables. The simplest form of regression, we mean linear regression, uses the formula of a straight line ($yi = \beta iXi + \varepsilon$) and it determines the appropriate value for β and ε to predict the value of y based on the inputs parameters, x. For simple linear regression, meaning only one predictor, the model is:

$$Y = \beta_0 + \beta_1 X_1 + \varepsilon \quad (1)$$

This model includes the assumption that the ε is a sample from a population with mean zero and standard deviation σ. Multiple linear regression, meaning more than one predictor is represented by the following:

$$Y = \beta_0 + \beta_1 X_1 + \beta_2 X_2 + \cdots + \beta_n X_n + \varepsilon \quad (2)$$

where Y is the dependent variable; X_1, X_2,…., X_n are the independent variables measured without error (not random); $\beta_0, \beta_1, \ldots, \beta_n$ are the parameters of the model. This equation defines how the dependent variable Y is connected to the independent variables X [5]. The primary goal of multiple linear regression analysis is to find $\beta_0, \beta_1, \ldots, \beta_n$ so that the sum of squared errors is the smallest (minimum). The most powerful and mathematically mature data analysis method, multiple linear regression is focused on a central approach traditionally where the computation is only done on a set of data stored in a single machine. With an increasing volume of data, the transition to the algorithm in distributed environment is hardly possible to implement. Multiple linear regression, a classical statistical data analysis method, also proves unsuitable to facilitate the scalability of the data processed in the distributed environment due to computing memory and response time. In this work, our contribution is to show the adaptation of classical data analysis algorithms generally and predictive algorithms specifically for multiple linear regression providing a response to the phenomenon of big data. In big data era, it is an essential requirement to solve the transition to the scalability of the algorithms for parallel and distributed massive data processing with the use of MapReduce paradigm seems like a natural solution to this problem.

3.1 MapReduce Framework

Zhu et al. (as defined by [14], p. 2) have discussed about infrastructure, data flow, and processing of MapReduce Framework. MapReduce, a programming platform cooperating with HDFS in Hadoop, which is popular in analyzing huge amount of data. There are two kinds of computational nodes in MapReduce Framework: one master node (NameNode) and several slave nodes (DataNode). This can be known as master-slave architecture and all the computational nodes and their respective operations are in the form of massively parallel and distributed data processing. The master node serves the duty of entire file system and each slave node serves as a worker node. Actually, each slave node performs the two main phases or processes called Map () and Reduce (). The data structure for these both phases exists in the form of <Key, Value> pairs. In the Map phase, each worker node initially organizes <Key, Value> pairs with same key nature and then produces a list of intermediate <Key, Value> pairs as intermediate Map results. Moreover, MapReduce system can also perform another shuffling process in which intermediate results produced from all Map operations by lists of same-key pairs with an implicit set of functions such as sort, copy and merge steps. Then, the shuffled lists of pairs with the specific keys are combined and finally passed down to the Reduce phase. In the Reduce phase, it takes lists of <Key, Value> pairs that are resulted from previous process to compute the desirable final output in <Key, Value> pairs.

4 The Proposed MapReduce Based Multiple Linear Regression Model with QR Decomposition

With the massive volume of data, training multiple linear regression on a single machine is usually very time-consuming task to finish or sometimes cannot be done. Hadoop is an open framework used for big data analytics and its main processing engine is MapReduce, which is one of the most popular big data processing frameworks available. Algorithms that need to be highly parallelizable and distributable across huge data sets can also be executable on MapReduce using a large number of commodity computers. In this paper, a MapReduce based regression model using multiple linear regression will be developed. We focus particularly on the adaptation of multiple linear regression in distributed massive data processing. This work shows an approach that the parallelism of multiple linear regression, a classical statistical learning algorithm that can meet the challenges of big data in parallel and distributed environment like MapReduce paradigm. However, we have still a big problem or issue to solve how to split or decompose the large input matrix in computing the regression model parameter "β" for the multiple linear regression analysis. In resolving the values of "β", we actually need to load the transpose of the input matrix and multiplication with its original matrix and then other subsequent complex matrix operations. It is impossible to process the entire huge input matrix at once. Therefore, matrix decomposition for the proposed regression model is contributed to overcome the limitations and the challenges of multiple linear regression in huge amount of data. We would like to present a new computational approach; the proposed regression model with QR Decomposition which provides computing on the decomposed or factorized matrix with scalability that is much faster than computing on the original matrix immediately without any decomposition.

The fundamental building block of many computational tasks consists of complex matrix operations including matrix decomposition utilized in the fields of scientific computing, machine learning, data mining, statistical applications and others. In most of these fields, there is a need to scale to large matrices in big data sets to obtain higher accurateness and better results. When scaling large matrices, it is important to design efficient parallel algorithms for matrix operations, and using MapReduce is one way to achieve this goal. For example, in computing "β" values from the Eq. (3), inversion of matrix "R" must be calculated. Matrix inversion is difficult to implement in MapReduce because each element in the inverse of a matrix depends on multiple elements in the input matrix, so the computation is not easily splitting as required by the MapReduce programming model [13]. QR Decomposition (also called a QR Factorization) of a given matrix A is a decomposition of matrix X into a product $X = QR$ of an orthogonal matrix Q if $Q^T = Q^{-1}$ or $Q^T Q = I$ and an upper triangular matrix R [11]. It is used to solve the ordinary least squares problem in multiple linear regression and also the standard method for computing QR Factorization of a matrix which has many rows than columns (m > n) causing a common problem arisen in many real-world applications. As we already known that data in a MapReduce processing is represented by a collection of Key-Value pairs. When we apply MapReduce to analyze matrix-form data, a key represents the identity of a row and a value represents the elements in that row [4]. Therefore, the matrix is also a collection of Key-Value pairs assuming that each row has a distinct key for simplicity although sometimes each key may represent a set of rows [2]. To determine multiple linear regression model's coefficient, "*β*" the computational approach QR Decomposition is to simplify the calculation by decomposing the data matrix X into two matrices "Q" and "R" as follows: $\beta = \left(X^T X\right)^{-1} X^T Y$

By Substituting $X = QR$, $\beta = \left(Q^T R^T QR\right)^{-1} Q^T R^T Y$, we obtain

$$\beta = (R)^{-1} Q^T Y \tag{3}$$

4.1 Implementation of the Proposed Model

The computation of the coefficients "$\beta_i[]$" of multiple linear regression using QR Decomposition on MapReduce framework will be three-stage processing or iterations to facilitate parallel and distributed processing of the proposed model in efficient manner. In the following section, we would like to present the algorithm with three-stage MapReduce processing including main or driver function in respective tables.

Algorithm for the Proposed Model

The algorithm for the proposed model takes as parameters block numbers to be used to divide the large training input matrix 'X' and distribute it on several tasks of "Map" functions. The 'Map' function of the first stage takes 'Xi' sub-matrices decomposed from big training data 'X' matrix for all 'noBlock'. The two result matrices 'Qi' and 'Ri' with the respective 'Keyi' are produced for the 'Reduce' function. The main idea here is to highlight that each 'Map' process load into memory at the maximum size of a matrix (BlockSize, n) which significantly overcomes the problem of "out of memory" with big matrix training data. Likewise, the 'Reduce' process will also receive a maximum array with size (n*noBlock, n). Therefore, choosing the number of blocks should be considered according to the size or number of machines in the cluster we applied. The more increasing computing power may be simply adding new machines into the cluster for the purpose of improving the 'MapReduce Framework Parallelism'. The second stage receives input from the result of first stage and the yi of for all blocks 'i'. In the 'Map' function, the vector y is decomposed into several vector yi (number of blocks) and then sent to 'Reduce' function with associated key 'Keyi'. The third or final stage uses the input from second stage including set of vectors of 'Vi' and 'R_{final}'. The 'Map' function constructs a list 'ListRV' combining with all sets of 'Vi' from all blocks 'i' and 'R_{final}' with the associated key 'Key_{final}'. The 'Reduce' function takes the list 'ListRV' and adding the values of all 'Vi' vectors together to get the final vector V. Moreover, 'R_{final}' is applied as inverse matrix and finally multiplying with 'V' to obtain the '$\beta_i[]$' as final output for the proposed model (Tables 1, 2, 3 and 4).

Table 1. Main or driver function for the proposed model

Input: Matrix X and Vector y from Input Dataset
Start
noBlock = Size (X) / BlockSize
n (No of Observations for each Block 'i'): = X /noBlock

X_i: = n
For all row of y
y_i:= y / noBlock
End
Output: X_i; with i: =1...noBlock

Table 2. First stage map/reduce function for the proposed model

Map Function *(Mapper 1)*	Reduce Function *(Reducer 1)*

Input: X_i of X for all blocks
Output: Q_i, R_i
Start
For all block X_i of X
 	
(Q_i, R_i)= Map1 (X_i)
Produce (Key_i, Q_i)
Produce (Key_i, R_i)
End for
Function Map1 (X_i)
Input: X_i of Matrix 'X'
Start
(Q_i, R_i):= QRDecompose (X_i)
Output: Q_i, R_i
End
Function QRDecompose (X_i)
Start
Factorizing X_i into Q_i and R_i for all respective blocks 'i'
End
End
	Input: <key, value> pairs for 'R_i' Matrix from Mapper1.
	Output: Q_i' for all block 'i',
	R_{final} for intermediate result in 'β' calculation(Q_i', R_{final})= Reduce1 ((Key_r, [R_1,R_2,.....,R_i])
	
	For all blocks Q_i' of Q'
	Produce (Key_i, Q_i')
	End for
	Produce (Key_{final},R_{final})
	Function Reduce1 (Key_r,[R_1,R_2,.....,R_i])
	Input: R_{temp}=Matrix[R_1,R_2,.....,R_i]
	Start
	(Q_i', R_{final}):= QRDecompose (R_{temp})
	For all row of Q'
	Q_i':= Decompose (Q', BlockSize)
	End For
	Function QRDecompose (R_{temp})
	Start
	Factorizing R_{temp} into Q' and R_{final} for subsequent decomposition of Matrix 'R_{temp}'
	End
	Function Decompose (Q', BlockSize)
	Start
	Decomposition of Q' into Q_i' for all blocks 'i' applying Q'/ BlockSize
	End
	End

Table 3. Second stage map/reduce function for the proposed model

Map Function *(Mapper 2)*	Reduce Function *(Reducer 1)*
Input: $ListQ_i$: = List [Key_i,(Q_i, Q')] y_i of Vector y for all blocks 'i' Output: (Key_i, y_i) for all blocks 'i' Start y_i = Map2(y_i) Function Map2 (y_i) Input: y_i Start Produce (Key_i, y_i) End End	Input: List of Q_i, Q_i', y_i with the same key 'Key_i' Output: V_i[] for intermediate result in 'β' calculation For all block 'i' V_i[ ]= Reduce2(Key_i, List[Q_i, Q_i', y_i]) End for Function Reduce2 (Key_i, List [Q_i, Q_i', y_i]) Input: List [Q_i, Q_i', y_i] Start Q: = Multiply1 (Q_i, Q_i') Q^T: = Transpose (Q) V_i: = Multiply2 (Q^T, y_i) End Function Multiply1 (Q_i, Q_i') Start Matrix multiplication of two matrices Q_i and Q_i' producing result matrix 'Q' End Function Multiply2 (Q^T, y_i) Start Matrix multiplication of transpose matrix 'Q^T' and y_i resulting 'V_i' arrays for all blocks 'i' End Function Transpose (Q) Start Transpose matrix operation of 'Q' End

Table 4. Third stage map/reduce function for the proposed model

Map Function (*Mapper 3*)	Reduce Function (*Reducer 3*)
Input: V_i[ ] for all blocks 'i' , <key, value> pairs for R_{final} with (Key_{final}, R_{final}) Output: ListRV ([R_{final}, V_1,...., V_i]) Start ListRV: = Map3 (V_i[ ],(Key_{final},R_{final})) Function Map3 (V_i[ ],(Key_{final},R_{final})) Input: V_i[],(Key_{final},R_{final}) Start Mapping Key_{final} with all of V_i[] from all blocks 'i' (Key_{final}, List[R_{final}, V_1,...., V_i]) Produce ListRV ([R_{final}, V_1,...., V_i]) End End	Input: ListRV ([R_{final}, V_1,....,V_i]) Output: β_i[] coefficients of proposed model and final output of three stage MapReduce processing Start β_i[] := Reduce3(ListRV ([R_{final}, V_1,....,V_i])) Function Reduce3 (ListRV ([R_{final}, V_1,....,V_i])) Start $InvR_{final}$:= Inverse (R_{final}) SumV := Sum (V_i) β_i[] := Multiply ($InvR_{final}$, SumV) End Function Inverse (R_{final}) Start Inverse matrix operation of 'R_{final}' End Function Sum (V_i) Start $\sum V_i$ for all blocks 'i' End Function Multiply ($InvR_{final}$, SumV) Start Multiplication of inverse matrix $InvR_{final}$ and the values of SumV for all summation from V_i values of blocks 'i' End End

5 Experimentation, Discussion and Conclusion

5.1 Experimental Setup

We applied Apache Hadoop (Version 2.7.1) Framework which already consists of MapReduce processing engine and Hadoop Distributed File System (HDFS). For cluster setup, there are three machines with the specification of CPU (2.4 GHz 4 Core, RAM 4 GB, HDD 500 GB). One machine serves as master ("NameNode") and the rest two machines act as slaves ("DataNode") to test our proposed model with java implementation. In this experiment, the dataset with 1.5 hundred thousand samples which is applied in the simulation of one way roads or streets navigation for city Yangon, Myanmar. This training matrix form dataset composed of "150000" rows and "225" columns. Then, the performance measures in execution time upon four conditions (Figs. 1 and 2):

1. Single or conventional processing not distributed environment without applying decomposition technique
2. Parallel processing not in distributed environment and also without applying decomposition technique
3. Single or conventional processing not distributed environment but applying decomposition technique
4. Parallel and distributed processing and also applying decomposition technique (the proposed idea) are presented in the following diagrams.

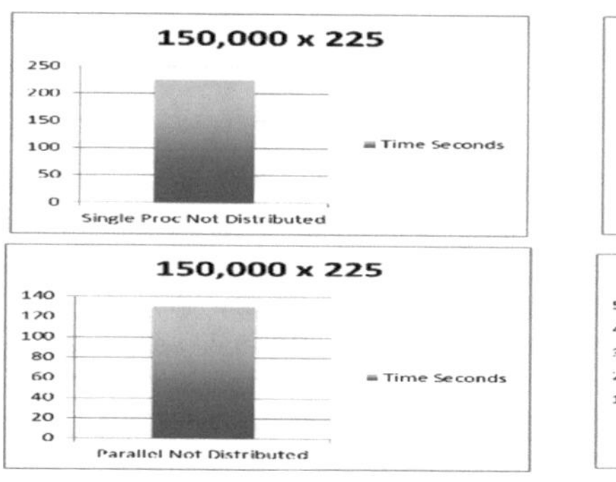

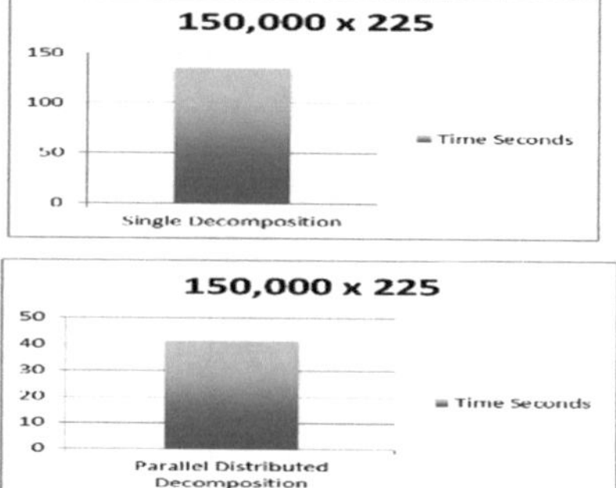

Fig. 1. The performance measures in four conditions

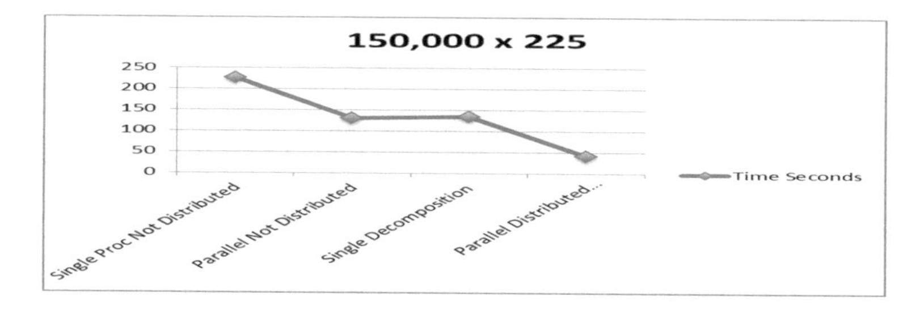

Fig. 2. The overall performance measures between four conditions

5.2 Discussion and Conclusion

According to the experimental results, our proposed work can handle the input massive training matrix (m, n) by distributing the computation on "Map" tasks and then local matrix decomposition function for optimization of the proposed model and finally combine and extract model coefficients "β" on "Reduce" tasks without facing any risk of "out of memory". Moreover, we can also prove that our approach provides more efficient computation and response time in compare with others. In this paper, our contributing idea is to show that the adaptation of classical regression analysis generally and multiple linear regression especially is possible to provide a response to the phenomenon of big data analysis. Therefore, our focus mainly places upon traditional multiple linear regression for parallel and distributed massive data processing with MapReduce paradigm. We intend to increase processing performance by avoiding limited memory utilization on massive data providing scalability and flexibility. Moreover, the proposed model will be provided with the purpose of solving extremely large matrices where the problems of entire matrix would not be able to fit in memory and several reads and writes to the hard disk drive would be required to do. We will consider further improvements in our proposed model by adding a preprocessing step making the input training matrix into tall-and-skinny matrix form (very large number of rows but fewer numbers of columns) which is an important and commonly used in linear regression models for the model. And then, we will present further performance evaluation results and comparative studies for prediction accuracy outcomes obtained from the model.

References

1. Adjout, M.R., Boufares, F.: A massively parallel processing for the multiple linear regression. In: 2014 Tenth International Conference on Signal-Image Technology & Internet-Based Systems (2014)
2. Ahsan, O., Elman, H.: QR Decomposition in a Multicore Environment (2014)
3. Amir, G., Murtaza, H.: Beyond the hype: Big data concepts, methods and analytics. Int. J. Manage. **35**, 137–144 (2014)
4. Benson, A.R., Gleich, D.F, Demmel, J.: Direct QR factorizations for tall-and-skinny matrices in MapReduce architectures. In: 2013 IEEE International Conference on Big Data (2013)
5. Dergisi, T.B., Sayfasi, D.W.: Multivariate multiple regression analysis based on principal component scores to study relationships between some pre- and post-slaughter traits of broilers. J. Agric. Sci. **17**, 77–83 (2011)
6. Fan, T.H., Lin, D.K.J., Cheng, K.F.: Regression analysis for massive datasets. Data Knowl. Eng. **61**, 554–562 (2007)
7. Florina, C., Elena, G.: Perspectives on Big Data and Big Data Analytics (2013)
8. Jun, S., Lee, S.J., Ryu, J.B.: A divided regression analysis for big data. Int. J. Softw. Eng. Appl. **9**, 21–32 (2015)
9. King, M.L., Evans, M.A.: Testing for block effects in regression models based on survey data. J. Am. Stat. Assoc. **63**, 1227–1236 (1986)
10. Li, R., Li, B., Lin, D.K.J.: Statistical inference in massive data sets. Appl. Stochast. Models Bus. Ind. **29**, 399–409 (2013)

11. Nugraha, A.S., Basaruddin, T.: Analysis and comparison of QR decomposition algorithm in some types of matrix. In: 2012 Proceedings of the Federated Conference on Computer Science and Information Systems (2012)
12. Tang, L., Zhou, L., Song, P.X.K.: Method of divide-and-combine in regularized generalized linear models for big data (2016)
13. Xiang, J., Meng, H., Aboulnaga, A.: Scalable matrix inversion using MapReduce (2014)
14. Zhu, J., Ge, Z., Song, Z.: Distributed parallel PCA for modeling and monitoring of large-scale plant-wide processes with big data. IEEE Trans. Industr. Inform. **13**, 1877–1885 (2009)

Evaluation for Teacher's Ability and Forecasting Student's Career Based on Big Data

Zun Hlaing Moe(✉), Thida San, Hlaing May Tin, Nan Yu Hlaing, and Mie Mie Tin

Myanmar Institute of Information Technology, Mandalay, Myanmar
zunhlaing@gmail.com, thidako22@gmail.com,
hlaingmaytin1982@gmail.com, nanyu.man@gmail.com,
miemietin1983@gmail.com

Abstract. This paper attempts to offer the evaluation of teacher's ability and the forecasting of students' career opportunities. Teacher's ability is decided based on the student's feedback, active participation in the class, students' result in the tests and the teacher's competency. Feedback is an essential element in the learning process. Students' feedback is an effective tool for teacher evaluation resulting in teacher development. The career opportunity available for a student is a significant area that determines the ranking of a university. This research will also forecast the student's career based on their individual subject grade. The system analyzes the teacher's ability by using Sentiment Analysis which is known as Opinion Mining technique. Student career forecast is based on predictive analytic. It comprises of a variety of techniques that predict future outcomes based on historical and current data.

Keywords: Big data · Mining · Sentiment Analysis · Predictive analytic Career · Ability · Grade

1 Introduction

The role of teachers and students' performances is vital for the reputation of a university. An effective teacher will normally have the following features: teaching skill, class room management, knowledge of subject, knowledge of curriculum, clear objectives for lessons, engaging personality and teaching style, higher expectation upon students and communication skill.

Evaluation based on student feedback is an important strategy for informing and refining teaching quality. So, this research will analyze the teacher ability by collecting feedback from students for an individual teacher. Purpose of collecting feedback from the student is to know how the teacher's ability can affect on the progress of teaching and learning system of the university. There are a number of factors to analyze the ability of a teacher. The teacher's experience, knowledge and abilities largely affect to the career opportunity of students. Therefore, the system will also analyze the teacher ability through their teaching methods and classroom management, etc.

T. T. Zin and J. C.-W. Lin (Eds.): ICBDL 2018, AISC 744, pp. 20–27, 2019.
https://doi.org/10.1007/978-981-13-0869-7_3

Students and their performance play a key role in a country's social and economic growth. Their creations and innovations enable to improve their university's image. So, students' success is one of the most important things for a university.

This research will also predict the students' career. So, we can know how many students have graduated from the university and who were getting great jobs in which industries. Student's career is forecasted by reviewing their grade upon their each subject.

This paper includes the basic concepts of relating to big data. Big data is the data sets that are voluminous and complex. There are three dimensions of challenges in data management to big data; extreme Volume of data, the wide Variety of data type and the Velocity at which the data must be processed.

A big data volume is relative and varies by factors, such as time and the type of the data. What may be deemed big data today may not meet the threshold in the future because storage capacities will increase, allowing even bigger data sets to be captured. A big data variety refers to the structural heterogeneity in a dataset. Technological advances allow firms to use various types of structured, semi-structured, and unstructured data. A big data velocity refers to the rate at which data are generated and the speed at which it should be analyzed and acted upon [1].

This paper is organized into 7 sections. Related work is presented in Sect. 2 and process of the system in Sect. 3. In Sect. 4, implementation of the system that consists of important factors on analyzing the teacher's ability and forecasting the student's career. Section 5 contains the expected results. Conclusion is in Sect. 6.

2 Related Work

There are many researches which have already analyzed the teacher's ability and predicted the students' grades and performances. This section describes the related work of the system. The paper by [2] compared the results of the student feedback gathered paper-based and web based survey of faculty's teaching. Students are the main stakeholders of institutions or universities and their performance plays a significant role in a country's social and economic growth by producing creative graduates, innovators and entrepreneurs [3].

The paper by [4] is predicting the student's performance based on student's interest, ability and strengths. The research work [5] predicts the final grade of the student from each course by using Pearson's correlation coefficient method. Our research mainly focuses on analyzing the teacher's ability and forecasting the student's career. Our system will use predictive analytic.

Baradwaj and Pal [6] conducted a research to analyze students' performance based on a group of 50 students. They focused during a period of 4 years (2007–2010), with multiple performance indicators, including "Previous Semester Marks", "Class Test Grades", "Seminar Performance", "Assignments", "General Proficiency", "Attendance", "Lab Work", and "End Semester Marks". Multiple factors are theoretically assumed to affect students' performance in higher education [7] and they predicted the students' performance based on related personal and social factors.

In this research paper, the aim is to focus on analyzing the teacher's ability from the student's feedback and forecasting the student's career using the following attributes such as grade for each semester.

3 Processing

The student marks from period of 2015 to 2017 are collected as a dataset, which is used to create a regression to forecast the student's career based on their final grade of each year. This dataset includes features, such as practical marks, lab marks, quiz marks, surprise test marks, assignment marks, class activity marks and mid-term marks and final exam marks, which will be used to forecast the career of students.

3.1 Data Collection

For our research, the data of undergraduate students will be collected from Myanmar Institute of Information Technology during the period of (2015 to 2017). Initially about 360 student records will be collected. In our University, there are two degree programs, Computer Science & Engineering and Electronic & Communication Engineering. In our university, one hundred and twenty students increase every year. So, in the next ten years, student data will be bigger and bigger. And then, if our university expand the program such as (e.g. diploma, master, PhD), so our data will also increase more and more. So, the main goal of research is to forecast the student's career opportunity after graduation.

Training Data. In this research paper, the whole dataset will be split into two, one is training dataset with 80% and 20% is used as testing dataset. In this research, the system will use the period of 2015 to 2017 students' data as the training phase.

Testing Data. In the testing process, we will separate 20% of CSE students from the third year from the original dataset. It will be used to test with the model.

4 Implementation

This system will implement two sections. First is analyzing the teacher's ability and second is forecasting the student's career opportunity.

4.1 Analyzing the Teacher's Ability

There are three main sections to analyze the teacher's ability. One is teacher, another is student and university. To analyze the teacher's ability, the system will first check on to how many students have attended the teacher's class. System will also analyze teacher's ability from the student's feedback and result from the test. When the system collects the feedback from students, it is not necessary to reveal the student's name or id. By collecting feedback, the system can know the students' frank opinion about the course and the teacher's ability. Based on the feedback, teachers will know their strength and weaknesses in a teaching and then can improve their teaching ability. In the feedback, there are five rating scales to analyze the teacher's ability. R1 is unsatisfactory, R2 is fair, R3 is satisfactory, R4 is very good and R5 is excellent. This system will analyze the teacher's ability based on technical skill, management skill and communication skill. The following table shows some facts that deal with above three skills (Table 1).

Table 1. Features for three skills to collect feedback

No	Skill	Question
1	Technical skill	• Does the teacher possess deep knowledge of the subject taught? • Clearly explains the objectives, requirements and grading system of the course • Use words and expressions within the student level of understanding • The teacher discusses topics in detail • Presented subject matter clearly and systematically
2	Management skill	• The teacher is punctual to the class • Use class time effectively • Manages a classroom that allows you to work and learn with few disruptions • Makes class interesting and relevant • Movement during class to check all the students
3	Communication skill	• Is approachable and willing to help you • Encourages cooperation and participation • Provides opportunities for student choice • Is involved and supportive of students within the university • Teacher has patience for all students • Keeps you informed of your progress • Confidence level exhibited

Teacher Ability Based on Three Skills. The system gathered data from a total of hundred students who studied the degree program in Computer Science in the Myanmar Institute of Information Technology (MIIT) during the years 2015 and 2017. The system will show the teacher's ability by bar chart based upon the skills.

The Fig. 1 shows the evaluation results of the teacher's ability based on the questions on technical, management and communication skills of a teacher. The students responded 100% excellent rating on these questions: "The teacher is punctual to the class", "The teacher discusses topics in detail" and "The teacher possesses deep knowledge of the subject taught". All students gave very good rate on the question: "Teacher can thoroughly explain or use tactics another way to understand the topic". 62% of students gave satisfactory rate and 38% of students responded in very good rate on the question: "Teacher has patience for all the students". According to these responses, teacher can be assumed as an excellent teacher.

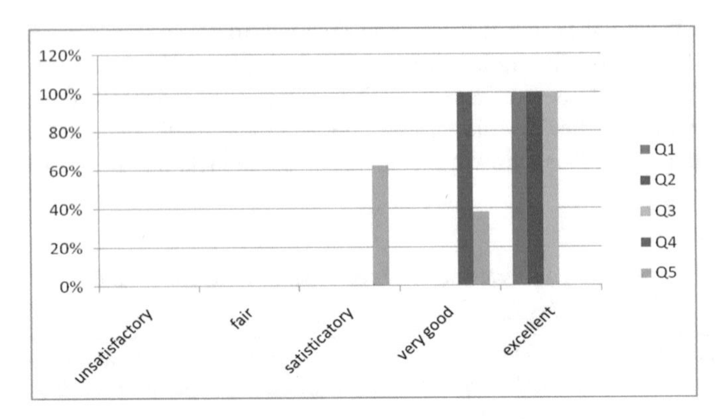

Fig. 1. Student survey for teacher ability

4.2 Forecasting the Student's Career

The important attributes in forecasting student's career are student's grades. Forecasting student's career, the system will mainly use semester grade point average (SGPA) and cumulative grade point average (CGPA). Some papers have used CGPA to predict student's performance [8–10].

The system will forecast the student's career by reviewing their grades. If a student possessed and conserved their best grade till he graduated, it will help them to get good career.

Student's Career Based on Grades. The following graphs are reviewing the student's grade points for each course. For instance, the system reviews the CSE student's grade from first year to second year.

Figure 2(a) and 2(b) represent course's grade point of each subject for each semester of the first year. There are five courses in first semester and seven courses in second semester.

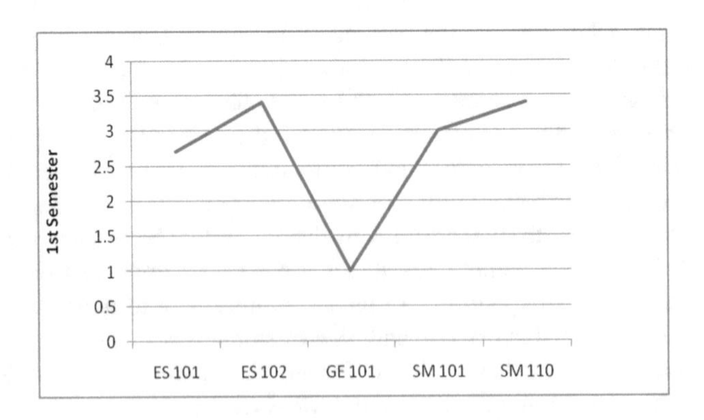

Fig. 2(a). 1st Semester of 1st Year grade point

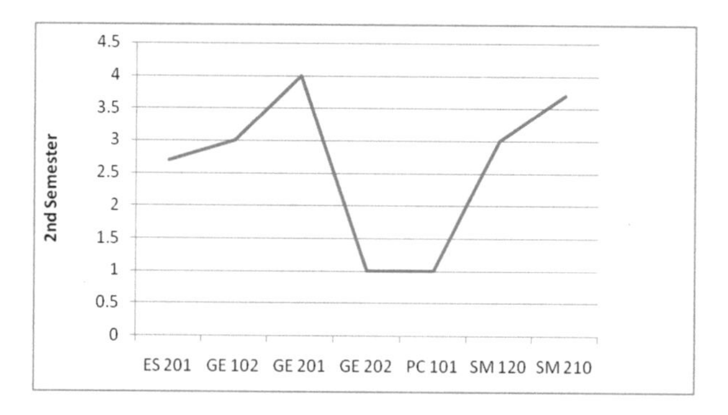

Fig. 2(b). 2nd Semester of 1st Year grade point

Figure 3(a) and 3(b) represent course's grade point of each subject for each semester of the second year. There are seven course and five lab grade points in first semester and six courses in second semester.

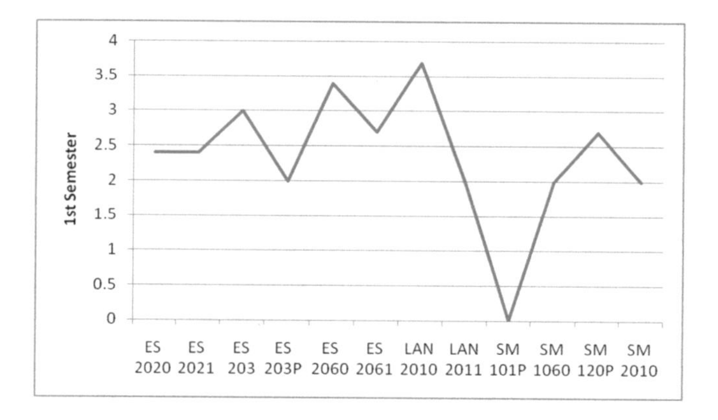

Fig. 3(a). 1st Semester of 2nd Year grade point

Figure 4 shows the student's progression according to the result of his grade. The system will show the student progression by comparing Semester Grade Point Average (SGPA) every year. According to the figure, student career will be unfavorable because his grade decreased year by year.

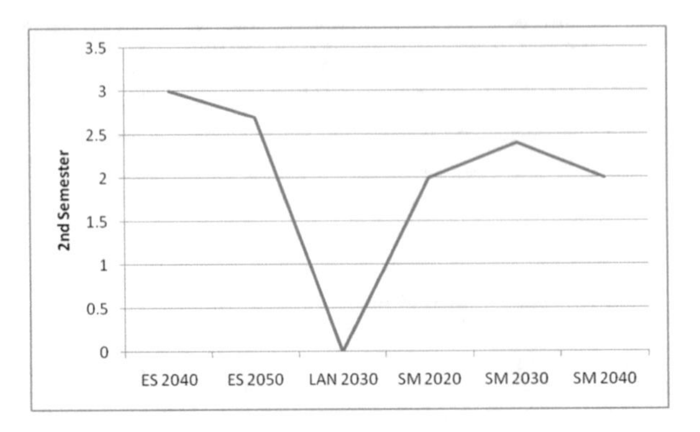

Fig. 3(b). 2nd Semester of 2nd Year grade point

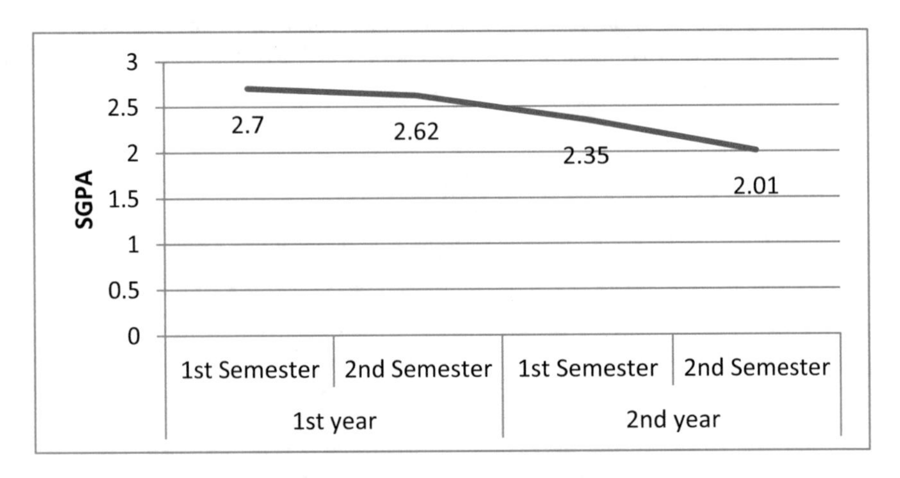

Fig. 4. 1st Year to 2nd Year grade point using SGPA

5 Expected Result

An expected result will show number of qualified teachers in our university, number of students who get greater jobs and how the teacher ability can effect on the students' career. The system will show the performance of the teacher and how many outstanding students have been produced.

6 Conclusion

Our research paper is analyzing the teacher's ability and forecasting the student's career. Analyzing the teacher's ability is useful to help teachers improve their teaching methodology. Feedback is an effective tool for teacher development. This paper has reviewed the feedback from the students. Forecasting the student's career is also very useful to know the product of university and its standard. This paper forecasts the students' career with analytical methods and predictive methods. In conclusion, analysis and forecast on teacher and student performance has motivated us to carry out further research to be applied in our environment. It will help the educational system to monitor the students' performance in a systematic way.

References

1. Osmanbegovi, E., Sulji, M.: Data mining approach for predicting student performance. Econ. Rev. **10**(1) (2012)
2. Ardalan, A., Ardalan, R., Coppage, S., Crouch, W.: A comparison of student feedback obtained through paper-based and web-based surveys of faculty teaching **38**(6) (2007)
3. Yadav, S.K., Bharadwaj, B., Pal, S.: Data mining applications: a comparative study for predicting student's performance. Int. J. Innov. Technol. Creative Eng. **1**(12), 13–19 (2011)
4. Dietz-Uhler, B., Hurn, J.E.: Using learning analytics to predict (and improve) student success: a faculty perspective. J. Interact. Online Learn. **12**, 17–26 (2013)
5. Ephrem, B.G., Balasupramanian, N., Al-Shuaily, H.: Projection of Students' Exam Marks using Predictive Data Analytics
6. Gandomi, A., Haider, M.: Beyond the hype: big data concepts, methods and analytics
7. Saa, A.A.: Educational data mining & students' performance prediction. (IJACSA). International Journal of Advanced Computer Science and Applications **7**(5), 212–220 (2016)
8. Baradwaj, B.K., Pal, S.: Mining educational data to analyze students' performance. (IJACSA) Int. J. Adv. Comput. Sci. Appl. **2**(6), 2011 (2011)
9. Angeline, D.M.D.: Association rule generation for student performance analysis using apriori algorithm. SIJ Trans. Comput. Sci. Eng. Appl. (CSEA) **1**(1), 12–16 (2013)
10. Quadri, M.M., Kalyankar, N.: Drop out feature of student data for academic performance using decision tree techniques. Glob. J. Comput. Sci. Technol. **10**(2) (2010)

Tweets Sentiment Analysis for Healthcare on Big Data Processing and IoT Architecture Using Maximum Entropy Classifier

Hein Htet(✉), Soe Soe Khaing(✉), and Yi Yi Myint(✉)

University of Technology (Yatanarpon Cyber City),
Pyin Oo Lwin, Myanmar
nightstalker.hh.5005@gmail.com, khaingss@gmail.com,
yiyimyint.utycc@gmail.com

Abstract. People are too rare to discuss or talk about their health problems with each other and, it is very poor to notice about their realistic health situation. But nowadays, most of the people friendly used social media and people have started expressing their feelings and activities on it. Focus only on Twitter, users' created tweets composed of news, politics, life conversation which can also be applied for doing a variety of analysis purposes. Therefore, healthcare system is developed to mine about the health state of Twitter user and to provide health authorities to easily check about their continental health behavior based on the Twitter data. Maximum Entropy classifier (MaxEnt) is used to perform sentiment analysis on their tweets to suggest their health condition (good, fair, or bad). It is interacting with Twitter data (big data environment) and so, Internet of Things (IoT) based big data processing framework is built to be efficiently handled large amount of Twitter user' data. The aim of this paper is to propose healthcare system using MaxEnt classifier and Big Data processing using Hadoop framework integrated with Internet of Things architecture.

Keywords: Sentiment analysis · Big data framework · IoT

1 Introduction

Most of the people regard as an unimportant case about sharing the situation of their health problems and most of them are very poor to understand about their realistic health condition. During these days, people use social media (Twitter, Facebook, etc.) and start sharing in the public domain about feelings and activities.

Therefore, health monitoring system is developed by doing social media sentiment analysis to be useful for the people who are very poor to understand about their health state. Moreover, it is used to provide health authorities for checking about the level of their continental health behavior. Among supervised machine learning algorithms, Maximum Entropy classifier is applied in the case of tweets sentiment analysis for data training and accurately classifying to get the positive, negative, and neutral health results. Big Data processing framework is built to efficiently handle for storing steadily increased Twitter users' data. As an overall, this paper purposes fetching tweets by applying the Twitter API, then preprocessed these data on the cloud server and crawled

T. T. Zin and J. C.-W. Lin (Eds.): ICBDL 2018, AISC 744, pp. 28–38, 2019.
https://doi.org/10.1007/978-981-13-0869-7_4

to the big data processing framework to store for further usages. Then, it is analyzed by the MaxEnt classifier and finally produced about the health condition percentage by positive, negative, neutral for the specific testing users or for the specific continents.

This paper is composed with eighth sections. Sections 2 and 3 are about the related work and overall system architecture. Data collection, Data store and processing, data analysis are in the Sects. 4, 5 and 6 consequently. Experimental results are explained in the Sect. 7 and the Sect. 8 is about the conclusion and future work.

2 Related Work

Sentiment analysis research on the Twitter data are doing by most of the researchers in the world. **Syed Akib Anwar** [1] proposed that Public sentiments are the main things to be noticed for collecting the feedback of the product. The twitter is the social media used in this paper for collecting the reviews about any product. The reviews collected are analyzed based on the locations, features and gender. Then, data extraction, data processing is performed and the product analysis using sentiment score is performed.

The next researcher**, Aarathi Patil** [2] proposes that the sentiment analysis can be done on any product or event by using the social media based on the location. There are four major steps involved in this paper. First step is to create Twitter application which is used to mine the twitter4j and analyze the data. Next, the tweets are collected by using the secret tokens from the twitter. These collected tweets are saved in an excel file. Preprocessing of data has been carried out and then, the filtered out tweets are classified by using the Naive Bayes classifier. The sentiment scores are provided as 1 for negative sentiment, 2 for neutral sentiment and 3 for positive sentiments.

3 Overall System Architecture

Overall system architecture described in Fig. 1, it is firstly crawled data from the Twitter via Twitter API to the cloud web server. These raw tweets-data need to be cleaned and so, it is going to the preprocessing stage. Preprocessing stage is very important stage which transforms the raw data into the valuable data for doing analytical process. Firstly, it is needed to convert the raw tweets text into the same format (lowercase). After that, it is needed to remove stop words such as a, an, the, is, etc. Then, tokenization which breaks tweets set into each words is performed and it is needed to pass to the lemmatization stage that is grouping words based on the several grouping format in this system which are nouns, verbs, adjectives and adverbs.

After that, feature sets correspondence with the health data is added to the feature vector and this resultant feature vector will be stored back to the cloud database server. In the local side, IoT based big data distributed processing framework is built using one PC for master node and four Raspberry pi boards for data nodes. Big data processing ecosystem such as Hadoop, MapReduce, and Hive data warehouse is configured within this framework.

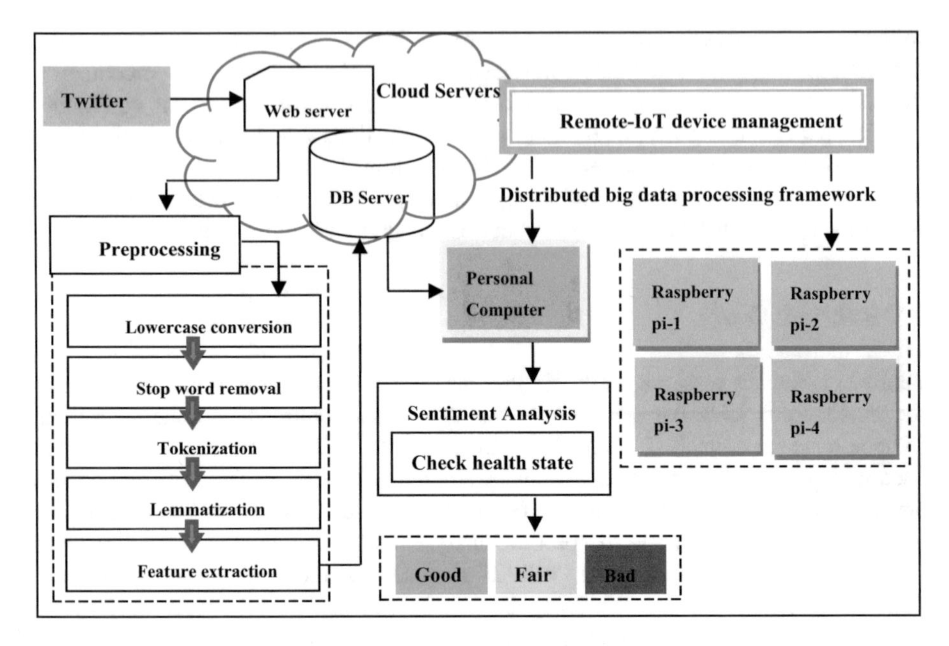

Fig. 1. Overall system architecture

Sentiment analysis for stating the behavior of Twitter users' health is performed at local side framework and the predicted health state (good, fair, or bad) percentage results are shown to the tested Twitter' users.

4 Data Collection

This section is also the first stage of doing data analytics and involves collection of data from several types of data sources, data marts and data warehouses. In this system, data are gathering from the social media- Twitter. Open API is used to fetch data from the Twitter. To get data from the Twitter, it is required to have API, API secret key, access token and access token secret. After fulfill these requirements, it can be obtained Twitter data with the help REST API and Twitter4J. It is returned the data file to the developers by the csv or json format. These are some sample training data concerned with health.

Positive Tweets
"Health risks of light drinking in pregnancy confirms that abstention is the safest approach\u2026 https://t.co/YcfHfoZfcB UK".

Neutral Tweets
"RT @9DashLine: China and Brunei pledge closer cooperation in infrastructure, health and defense https://t.co/mxxzVpTgY5\n\n#".

Negative Tweets
"15-Year-Old Gets Extreme #Cosmetic surgery-https://t.co/2Yk3XN1qPY <<"

5 Data Store and Processing

In Data Store and Processing stage, data is storing into big data' data warehouse (Hive) which is a batch oriented data warehousing layer built on the core elements of Hadoop (HDFS and MapReduce). At shown in Fig. 2, the resultant feature words are imported to the Hadoop Distributed File System (HDFS) with the aid of FLUME which is used for the purpose of import and export unstructured data. The imported data from the HDFS must be send to map reduce paradigm which is also a primary usage tools in solving big data problems.

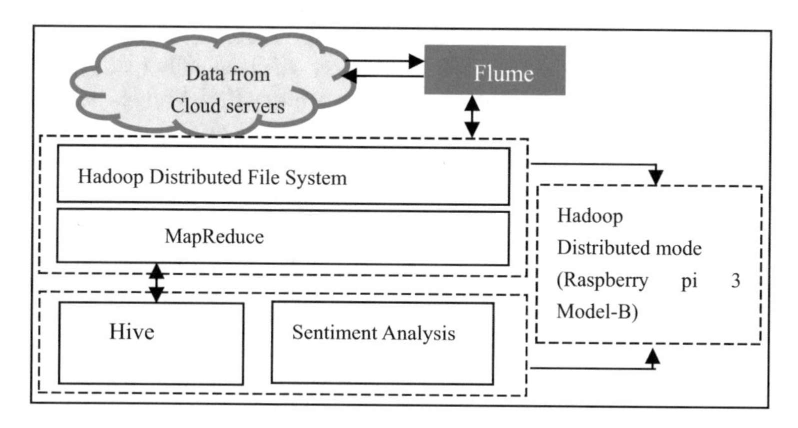

Fig. 2. Work flow between Hadoop distributed file system, MapReduce, and HIVE

Twitter API can response up to 3600 tweets per user and maximum 200 tweets once at a time. A tweet will have maximum 140 characters and so, there will be the data size 3.6 Mb for each Twitter user to predict their health state. The more the tested users, the more storage space is required to persistence their data. Therefore, big data processing framework, Hadoop ecosystem is applied.

5.1 Hadoop

Hadoop is an open source platform based on the Java that provides implementations of both MapReduce and Google file system technologies. Moreover, Hadoop is an open source framework in the case of processing large scale data sets across clusters of low-cost commodity hardware. Hadoop ecosystem, one of the big data solution techniques, is used for handling large amount of users' testing data that are used to predict their health state and two primary components of Hadoop mainly used in this proposed system are [3]:

1. **Hadoop Distributed File System (HDFS)**

The Hadoop Distributed File System is a versatile, resilient, clustered approach to managing files in a big data environment. It is a data service that offers a unique set of capabilities needed when data volumes and velocity are high. There are two types of service in HDFS- NameNode and Data nodes.

2. **MapReduce (MR)**

MapReduce (MR) is a data processing paradigm and this process starts with a user request to run a MapReduce program and continues until the results are written back to the HDFS. MR tasks will be performed on the testing data to store efficiently for further processing.

The first step of Map phase is to locate and read the input file containing the raw health data. This is the function of InputFormat and RecordReader. InputFormat decides how the file is going to be broken into smaller pieces for processing using a function called InputSplit. The, it is assigned to a RecordReader (RR) to transform the raw data for processing by the map. RR is the class which actually loads the data from the source [4]. It is the class which converts the data into <Key, Value> pairs as sample examples, [<health, 2>, <medical, 1>], [<risk, 3>, <medical, 2>], [<health, 1>, <defense, 1>]. The Mapper will receive one <Key, Value> pair at a time until out split is consumed.

The Reduce phrase is performed by gathering intermediate results from the output of the Map phrase and then, by shuffling, sorting, and combining to get the desired result. The output of reduce task is also a key and a value such as [<health, 3>], [<medical, 3>], [<risk, 3>], and [defense, 1]. Then, Hadoop provides an OutputFormat feature which takes key-value pair and organizes the output for writing to HDFS. Finally, RecordWriter is used to write the user testing data to the HDFS.

5.2 Raspberry Pi Hadoop Cluster

Hadoop has developed into a key enabling technology for all kinds of Big Data analytics. To inspire the prototype of the Hadoop, it can be installed open source Apache Hadoop from scratch on Raspberry pi 3 Model B. Hadoop is designed for operation on commodity hardware so it will do just fine on Raspberry Pi as a Master and Slaves type. At shown in Fig. 3, IoT based big data processing framework is built by using the four Raspberries pi, four cat-5 network cables, and five-port switch.

5.3 Internet of Things (IoT)

The IoT is the network of physical devices, which enable these objects to connect and exchange. With the aid of this technology, distributed big data processing framework can be transformed from local networking system to over internet. There are a lot of IoT platforms and but in this system, Remote-IoT services are used. Remote-IoT uses a secure AWS IoT cloud platform to connect to networked devices from anywhere and all network communications are encrypted by the SSH tunnel as shown in Fig. 4. Moreover, it can be fully controlled the device including monitoring CPU, Memory

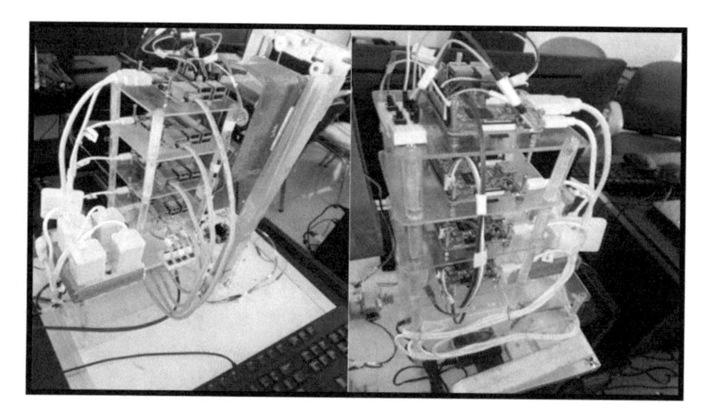

Fig. 3. Hadoop cluster with four raspberry pi3 Model-B

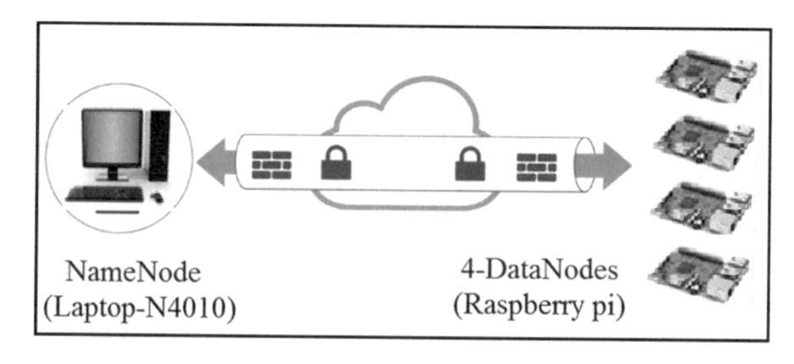

Fig. 4. Secure remote connection (Master and Slave) nodes using Remote-IoT

and Network usage, and run batch jobs on devices. Then, it doesn't required dedicated VPN/Firewall and can be used by installing the Remote-IoT service on the devices with a TCP/IP stack. In this IoT based big data processing architecture, network interfaces for each device are set in accordance with the Table 1.

Before building IoT based architecture, simple distributed cluster topology is built using NameNode and DataNodes. In this cluster, Java environment, SSH server/clients, Hadoop are configured inside each node and static IP addresses are firstly used which are described in Table 1' Internal IP column but the master (NameNode) interface is subdivided into two- DHCP and Static. After successfully finished, IoT based cluster is

Table 1. Network interfaces table

Devices	Network interfaces	
	Internal IP	External IP
NameNode	DHCP/192.168.56.100	103.52.12.61:15865
DataNode01	192.168.56.101	103.52.12.61:24244
DataNode02	192.168.56.102	103.52.12.61:18845
DataNode03	192.168.56.103	103.52.12.61:31654
DataNode04	192.168.56.104	103.52.12.61:20018

transformed by setting proxy addresses and ports into each network interface which are shown in Table 1' External IP column with the aid of Remote-IoT.

6 Data Analysis

Data analytics refers to methods and tools for analyzing large data sets to support and improve decision making. A number of organizations used sentiment analysis in order to collect feedback from the user. In this proposed system, it analyzes the data to focus the health state of social media users. It can also be analyzed health rate according to each continent. As shown in Fig. 5, the sentiment analysis steps consist of two phases: Training phase and testing phase. In both phases, raw input texts need to be preprocessed and extract features which are described details in the next sections.

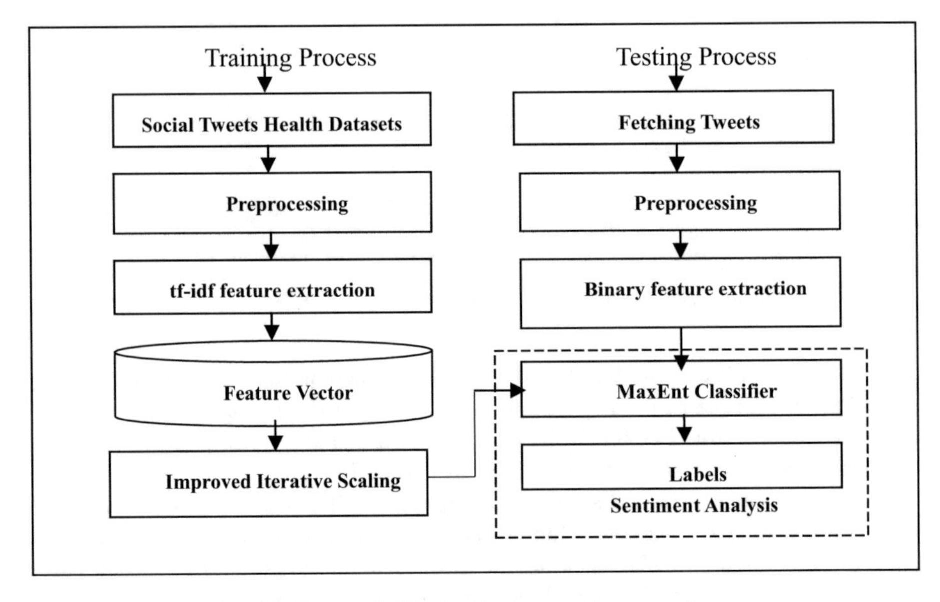

Fig. 5. Schematic block of tweets sentiment analyzer

6.1 Preprocessing Text Data

The preprocessing is really important because there are some words or expressions in the review that don't return any meaning and by the presence of those words that cannot get the correct sentiment analysis. So by doing pre-processing it can also get higher accurate results.

In preprocessing stage, the following steps contained: Lowercase conversion, stop word Removal, Text Segmentation and then, Normalization. The following is the resultant data of the sample positive tweets stated in Sect. 4 after doing these steps.

Positive Tweets-'health', 'risk', 'light', 'drinking', 'pregnancy', 'confirms', 'abstention', 'safe', 'approach', 'experts', 'urge', 'action', ' poison', 'Asia'.

6.2 Feature Extraction: (tf-idf)

After preprocessing stage is done, the cleaned words which relate with health data are getting. Then, the system is trained by the Maximum Entropy Classifier on TF-IDF [5] weighted word frequency features. It is also perform like that frequently occurring words present in all files of corpus irrespective of the sentiment, like in this case, 'health', 'fever', 'wellness', 'happy', 'sick', etc. will be selected as features words.

Term Frequency increases the weight of the terms (words) that occur more frequently in the document.

$$tf(t, d) = \log(F(t, d)), \tag{1}$$

where $F(t, d)$ = number of occurrences of term 't' in document 'd'.

Inverse Document Frequency diminishes the weight of the terms that occur in all the documents of corpus and similarly increases the weight of the terms that occur in rare documents across the corpus.

$$idf(t, D) = \log\left(\frac{N}{N_{t \in d}}\right) \tag{2}$$

6.3 Maximum Entropy Classifier (MaxEnt)

Supervised machine learning algorithm (MaxEnt) classifier is effectively used in a number of natural language processing applications. Sometimes, it outperforms [6] Naïve Bayes at standard text classification. Its estimate Of P(c | d) takes the exponential form as described in Eq. (3),

$$P_{ME}(c/d) = \frac{1}{Z(d)} \exp\left(\sum\nolimits_i \lambda_{i,c} F_{i,c}(d, c)\right) \tag{3}$$

Where, Z(d) is a normalization function. Fi, c is a feature/class function for feature fi and class c, as in Eq. (4),

$$F_{i,c}(d, c') = \begin{matrix} 1, \ if \ (n_i(d) > 0 \, and \, c' = c) \\ 0, \ \text{otherwise} \end{matrix} \tag{4}$$

6.4 Classifier Trainer Algorithms

Two classifier trainer algorithms are introduced in this paper.

1. **Improved Iterated Scaling (IIS)**

Firstly, IIS is used for calculating the parameters of a maximum entropy classifier given a set of constraints. IIS performs by collecting labeled documents ‘D’ which is also training data and a set of features function fi. For every feature fi, estimate its expected value on the training documents. Then, initialize all the λ‘s to be zero and iterate the convergence. At each step, IIS must find an incrementally more likely set of parameters. If it is guaranteed that IIS succeeds in improving the likelihood, then it is known that it will converge to the globally optimal set of parameters. The log likelihood of an exponential model can be calculated by using Eq. 5.

$$\begin{aligned} l(\Lambda/D) &= log \prod\nolimits_{d \epsilon D} P_{\Lambda}(c(d)/d) \\ &= \sum\nolimits_{d \epsilon D} \sum\nolimits_{i} \lambda_i f_i(d, c(d)) - \sum\nolimits_{d \epsilon D} log \sum\nolimits_{c} exp \sum\nolimits_{i} \lambda_i f_i(d, c) \end{aligned} \quad (5)$$

2. **Generalized Iterative Scaling (GIS)**

GIS is a method that searches the exponential family of a Maximum Entropy solution of the form:

$$P^{(0)}(x) = \prod\nolimits_i \mu^{(0) f_i(x)} \quad (6)$$

The next iteration of each is intended to create an estimate that will match the constraints better than the last one. Each iteration ‘j’ follows the steps:

1. Compute the expectations of all the f_i’s under the current estimate function, i.e.,

$$\sum\nolimits_x P^{(j)}(x) f_i(x) \quad (7)$$

2. Compare the present values with the desired ones, updating the

$$\mu_i^{(j+1)} = \mu_i^{(j)} \cdot \frac{K_i}{E_{p^{(j)}} f_i} \quad (8)$$

3. Set the new estimate function:

$$P^{(j+1)} = \prod\nolimits_i \mu_i^{(j+1) f_i(x)} \quad (9)$$

4. If convergence or near-convergence is reached stop; otherwise go back to step1.

7 Experimental Results

All experimental results of proposed research are showing that it is on the 8768 training data sets. By applying the Maximum Entropy Classifier with the use of Improved Iterated Scaling (IIS) algorithm, this system is iterated with the number of 100

Table 2. Performance comparison between IIS and GIS

Process status	Health data sets (8768)	
	IIS	GIS
Training time	3-hr & 32 min	3 h & 14 min
Testing time	1.55-min	1.05-min
Positive precision	0.944250	0.899910
Positive recall	0.957763	0.905531
Neutral precision	0.901954	0.912234
Neutral recall	0.912263	0.779985
Negative precision	0.881472	0.801172
Negative recall	0.857440	0.784542
Average precision	0.909225	0.871105
Average recall	0.909155	0.823352
Average accuracy	90.91%	84.72%

iterations. In Table 2, the evaluation results are discussed based on the experimental results between two classifier trainer algorithms, IIS and GIS. For performance comparison, the following two factors are considered:

- **Precision** = fraction of the returned result that are relevant to the information need
- **Recall** = fraction of the relevant documents in the collection that were returned by the system

The analysis result of health state can be shown by using the pie chart and it can also be tested the health situation for the seven continental areas and for each twitter user. The pie chart shown in Fig. 6 is also the practical analysis result of health state for Asia. The system is predicted that positive (50.5%), negative (16.2%), and neutral (33.3%) based on the testing datasets 300 tweets. Then, it is generally concluded as Asia health state is good situation upon 300 testing tweets.

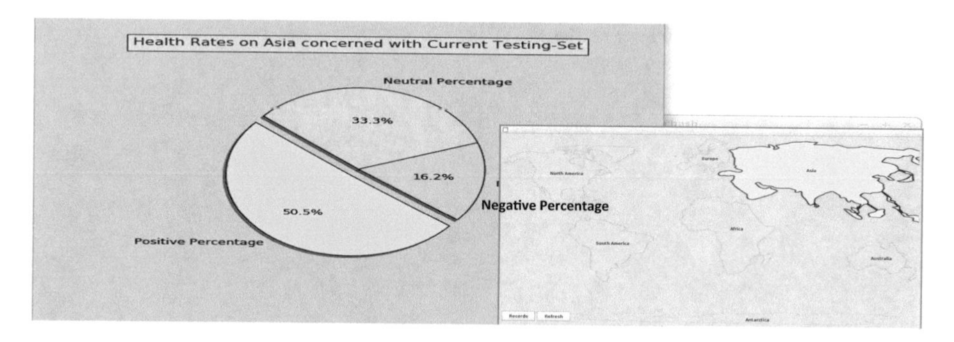

Fig. 6. Asia' health analysis result

8 Conclusion and Future Work

There are some kinds of work for further extending on this proposed system to get more accurately results relate with health. The first thing is that Hadoop MapReduce can be replaced by Hadoop YARN which is more convenient in I/O transaction. Then, Raspberry pi boards should be replaced more powerful desktop computers for real-time big data processing. Moreover, it is needed to be added meta-data analysis in the case of sentiment health state analysis. And so, it can also be produced more accurate health states/ conditions for the users.

This proposed system is building the sentiment analyzer for solving this problem and notifying about their health situation based on the data from the social media that they always use. But, the one significant thing is that social media is not a small place to play and so, there are existing big data challenges. So, it consists of two main parts:

- Building social media mining sentiment analyzer
- Examining big data solutions on this analytics case

This system will be developed by using Python programming language, Open API, open source big data platforms and cloud computing technology.

References

1. Anwar, S.A.: Localized Twitter Opinion Mining Using Sentiment Analysis, India
2. Patil, A.: Location Based Sentiment Analysis of Products or Events over Social Media, India (2014)
3. Pyo, P.S.: Introduction of Big Data and Hadoop Ecosystem. Head of Korea-Myanmar E-learning Center, Korea, March 2016
4. Shinde, G., Deshmukh, S.N.: Sentiment TFIDF feature selection approach. Int. J. Comput. Commun. Eng. (2016)
5. Mehra, N., Khandelwal, S., Priyank, P.: Sentiment identification using maximum entropy analysis of movie reviews. Int. J. (0975-8887) (2016)
6. Nigam, K., Lafferty, J.: Using Maximum Entropy for Text Classification. Andrew McCallum, Carnegie Mellon University, Pittsburgh (2013)

A Survey on Influence and Information Diffusion in Twitter Using Big Data Analytics

Radia El Bacha(✉) and Thi Thi Zin(✉)

Graduate School of Engineering, University of Miyazaki, Miyazaki, Japan
radia.elbacha@gmail.com, thithi@cc.miyazaki-u.ac.jp

Abstract. By now, even if we are still geographically situated, we're able to reach, connect and know about each other through social networks like never before. Among all popular Social Networks, Twitter is considered as the most open social media platform used by celebrities, politicians, journalists and recently attracted a lot of attention among researcher mainly because of its unique potential to reach this large number of diverse people and for its interesting fast-moving timeline where lots of latent information can be mined such as finding influencers or understanding influence diffusion process. This studies have a significant value to various applications, e.g., understanding customer behavior, predicting flu trends, event detection and more. The purpose of this paper is to investigate the most recent research methods related to this topic and to compare them to each other. Finally, we hope that this summarized literature gives directions to other researchers for future studies on this topic.

Keywords: Influence · Social networks · Information cascade · Twitter
Big data analytics

1 Introduction

With the advent of Web 2.0, user experience in the Web shifted from a monologue-oriented towards a dialogue-oriented environment where the user become able to interact and share information through services that are now inevitable in our daily lives such as Social Network Services (SNS). As such, the growth of generated data has been phenomenal. In 2013, the size of the 'digital universe' reached 4.4 zettabytes and it is estimated to grow up to 44 zettabytes by 2020 [1]. All in a model that is often not structured. But most of all is the velocity of data generation. For instance, every second there are more than 9,100 tweets sent on Twitter [2]. This is what bring as to a new digital era, the era of Big Data.

Currently, one of the hottest topic related to Big Data is typically SNS. The large amount of data exchanged in SNS attracted many researcher, companies and data scientist to deep dive into those data in order to understand information diffusion models and get the hidden value from it.

In this paper, we investigate research results related to information diffusion and influence in SNS, particularly on Twitter since it is a popular and fast growing platform with 330 million active users [2], with public profiles rather than protected ones which make it a good source of data targeted by researchers on this topic.

T. T. Zin and J. C.-W. Lin (Eds.): ICBDL 2018, AISC 744, pp. 39–47, 2019.
https://doi.org/10.1007/978-981-13-0869-7_5

By investigating a Twitter dataset, lots of hidden information can be mined. Many research effort has been put into this topic but overall the related literature can be classified into three approaches: Network Topology based approach, Twitter metrics based approach which investigate on user actions on Twitter or another approach that combine both of them.

This paper is organized as follows. Section 2 describes Twitter Network properties and APIs (Application Programing Interface). In Sect. 3, we describe and compare influence models used in literature. Some applications to other field are presented in Sect. 4, and finally, we finish this work with some conclusions in Sect. 5.

2 Twitter Network

Nowadays, Twitter is one of the main SNS in the world. Launched in 2006 but rapidly become one of the most popular micro-blogging sites where people can exchange small elements of content such as short sentences of 140 characters [3], individual images, URL or video links. Twitter users follow others or are followed but reciprocity is not necessary. It is important to know that being a follower on Twitter means that the user receives all the messages (called tweets) from those the user follows.

2.1 Twitter Metrics

In Twitter they are four actions that a user can make to interact within the network: Tweet, Retweet, Mention and Reply. An interesting thing about twitter is its Markup culture such as the *symbol* '@' followed by a Twitter username which is used to mention a user address in a tweet, it's a link to that Twitter profile and the *symbol* '#' followed by a keyword which is called a *hashtag*. It can be placed anywhere in the tweet. Searching for a hashtag on Twitter shows all other tweets marked with that hashtag. Often trending topics are hashtags that become very popular. The notation **RT** or sometimes **R/T** in a tweet message stands for retweet. This means that this is a retweet of someone else tweet.

Twitter user's actions are widely used as metrics or measure to define influential users or influence propagation. The details of these different metrics are summarized in Table 1.

Table 1. Twitter metrics used in research literature

Twitter metrics	Explanation
Tweets (T)	Number of messages posted by a user
Retweets (RT)	Number of times a user's tweet had been republished by other users
Mentions (M)	Number of times a user was mentioned by other users
Replies (R)	Number of replies to another user tweets
Followers (F)	Number of users who follow a user (Indegree)
Time (TI)	Activeness of the user or time difference between his tweets and the first time they get retweeted by another user to evaluate the speed of getting reaction to his tweets in the network

2.2 Twitter API

To compute the metrics above, Twitter data must be collected. For this purpose, Twitter API can be used. Twitter provide two public APIs for developers. First one is called Rest API, it is suitable for searching historic of tweets or reading user profile information. The other one is the streaming API, it gives access to a sample of tweets as they are published in twitter. On average, about 9,100 tweets per second are posted on twitter and developers can crawl a small portion of it (<=1%).

Usually the output rendered by either of the APIs is in JSON (JavaScript Object Notation) format which consists of key-value pairs. A tweet metadata can have over 150 attributes in addition to the text content itself. The complete list of attributes for a tweet can be found on [4].

2.3 Data Collection

Due to privacy concerns, it is quite difficult to find a public Twitter dataset therefore more often researchers collect their own dataset from Twitter APIs.

Considering the limitations of the Twitter Rest API, more often researcher choose to use Twitter streaming API to collect the datasets to test their method. Twitter provides a list of possible request and their associated tokens [5]. For instance, to get the user tweets we need to request 'GET lists/statuses' which has 900 tokens.

Some researchers mentioned using a software system called TweetScope [6–19] in order to collect a dataset. This system can collect tweets streams by connecting to Twitter API.

After testing the proposed method on the collected Twitter dataset, usually Experimental results are compared to a survey conducted by the researchers to verify the accuracy of the results [7, 8].

3 Influence Models in Social Networks

One of the key aspects of information diffusion in social networks is Influence. Therefore, it is considered as one of the most important points that cannot be ignored when reviewing information diffusion research.

3.1 Definition of Influence

Many research was conducted on Twitter network data to investigate influence propagation. But, even if we focus only in one specific SNS, namely Twitter, still we can't find a global definition for influence, because it depends on the used influence metrics which usually vary from one researcher to another.

Some researchers consider a more influential user as someone who has the potential to lead others who are connected with him/her to act in a certain way considering Indegree and 2 other user activities on Twitter: retweets and mentions [8]. In a slightly different way, other researcher [9] suggest that influence is measured by the 'excitation' a user causes in the network by receiving attention from other users. 'Excitation' was defined as how much interest or attention a user receives on the network and it was estimated considering three mechanisms of interaction: retweets, replies and mentions. In some other research methods, categories of influence were introduced [10] usually determined based on the specific role a user is playing in conversations instead of his static followership network. Most of all, retweet network is the most common method used to determine influence.

3.2 Information Diffusion in Twitter

Besides the influence of a node, it is important to know how information can propagate from a user to another distant user on the network. Offline, this is known as Word of mouth diffusion model, a traditional way in which information propagate from person to person in our daily life. Similarly, the same phenomenon can be seen on SNS, Jansen et al. [11] define this as the electronic word of mouth while Bakshy et al. [12] call this a cascade and suggest that the largest cascades tend to be generated by users who have been influential in the past and who have a large number of followers. For instance, in Fig. 1, let's consider node ***a*** as a Twitter user, suppose that he made a Tweet, all his followers can immediately see his tweet in their feeds among them user ***b***. Then ***b*** also decides to retweet this tweet to let his followers know about it. At t_i, ***c*** a follower of ***b*** will decide to retweet ***b*** then at moment T, ***d*** a follower of ***c*** will also make a retweet. This is how Tweets can propagate from ***a*** to ***d*** even if they are not directly connected in a cascade model. It is a very popular way of information diffusion on SNS to spread ideas, campaign, trends, fashions, etc.

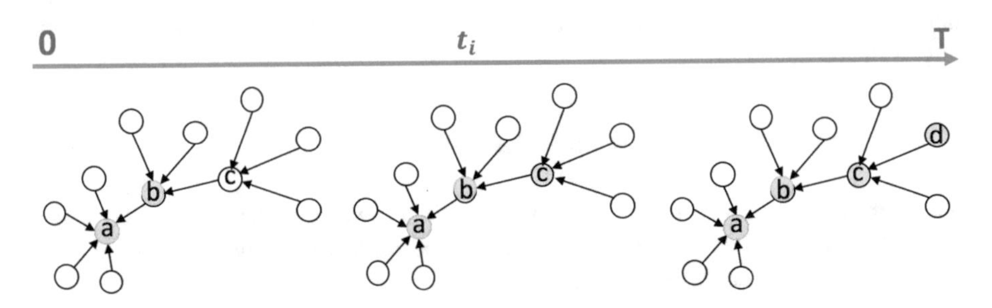

Fig. 1. A graph representing cascade diffusion model in a Twitter network sample

Galuba et al. [13] use this mechanism of information diffusion to predict the spreading of URLs on Twitter. In their study they insist on the importance of Retweets as a strong indication of the direction of information flow. The proposed model takes into consideration the influence of users on each other and time dimension as they introduce a new metric called diffusion delay which is the duration between a user tweet and the first retweet of his tweet.

To the best of our knowledge, most of the research works about Information Diffusion focus on Cascade Models perceived from a producer perspective rather than a consumer perspective.

In fact, a Twitter user can have a dual role, content producer or a consumer of content. A user is producer if he tweet/retweet and a consumer when he sees the content produced by user he is following.

Rotabi et al. [14] proposed a method to measure the effect of cascade on the audience by considering views and engagement with the tweets in a user timeline. Surprisingly, they found that users show a larger engagement for cascades than for non-retweeted content on their timeline. The proposed method was also based on a new Twitter metric which is tweet view. It is determined by calculating if the tweet stayed enough time in the mobile screen of the user.

Table 2 gives a summary of reviewed literature about cascade Models. In the second column, we mark as yes (✓) or no (✗) if the cascades on Twitter are studied from a Producer Perspective (PP) or a Consumer Perspective (CP).

Table 2. Comparison of cascade models in literature

Reference	Perspective		Applications
	PP	CP	
Jansen et al. [11]	✓	✗	Investigate consumer opinions concerning brands through word of mouth diffusion model
Bakshy et al. [12]	✓	✗	Investigate events diffusion on Twitter by tracking Tweets which include URL
Galuba et al. [13]	✓	✗	Predictions of URL mention by users on Twitter
Rotabi et al. [14]	✗	✓	Modeling cascades for audience (Theoretical model)

3.3 User Influence Models in Literature

According to [15] almost half of the existing influence measures are based on the PageRank algorithm [16] or Social Networks Analysis methods [17]. PageRank was created by google founders [18], it is one of the algorithm used by Google to order search engine results. It gives a numerical weight to each web page to represent its importance among all other pages on the web. This weight is based on a link analysis. The weight of a page is defined recursively and depends on the number and incoming link to it. A page that is linked to many pages receives a high rank. Alp and Öğüdücü [7] proposed a Personalized version of PageRank algorithm based on a spread score and other baseline methodologies. Their analysis considers user actions and specific topics to identify topical influencers.

Cha et al. [8] proposed a method for measuring user influence in Twitter. This method investigates the network topology by considering Indegree measure (the number of followers of a user) and 2 user actions: retweets and mentions. They ranked users respectively by Indegree count, retweet count and mention count. Then calculate Spearman's rank correlation coefficient to quantify how a user's rank varies across different measures. The finding of this research suggest that Indegree measure doesn't reveal much about influence of a user and proves that retweet and mentions are better for indicating influential users. Authors also has investigated the validity of this results over time and across a variety of topics. They found that the most influential users hold significant influence across different topics and need concerned efforts and involvement in a topic to maintain their influence on the network all the time.

In Twitter everything happens so fast, therefore immediacy of results has become very important. Cappelletti and Sastry [9] proposed a method to rank influential Twitter users in real-time for large scale events. Their rank was designed based on the concept of information amplification, which takes as influential those users who have a potential to reach a high audience. Two influence measures were presented, Cumulative influence which is achieved by receiving regular attention from other users and Instantaneous influence which can occur if an influential user is interested in an ordinary user, temporary the ordinary user is considered as influencer. The two measures were weighted by an amplification potential that combine two factors: Buzz and Structural advantage.

$$Buzz = \frac{\#Mentions}{\#Event\ Activity} \tag{1}$$

$$Structural\ Advantage = \frac{\#Followers}{\#followers + \#following} \tag{2}$$

#Event Activity is the number of tweets, retweets and mentions or replies of a user, all related to one event.

Structural Advantage is similar to popularity, it measures whether the surrounding network of the user is capable to provide him information or is seeking from him information, this is represented by Eq. (2).

This method shows a good performance for near real-time ranking of the top 4 or 5 influential users. Moreover, results show that it is much faster compared to PageRank algorithm which proves its quality for near real-time ranking. But, considering the accuracy of the results PageRank still win with a much higher accuracy of the ranking.

Another relevant method but slightly different than the previous ones was proposed by Tinati et al. [10]. It's based only on the Twitter user's dynamic communication behaviors and doesn't include any network topology measure. In this method, a classification model based on the work of Edelmam's topology of influence was applied to a Twitter dataset in order to create a network where the interactions between users determine their role and influence against each other. This model has five categories of users based on their communicator roles – amplifier, curator, idea starter and commentator. All those roles were determined using the retweeting feature.

Table 3 gives a summary of methods used for finding influential users on Twitter. We compare them from many aspects. In the fourth column we mark as yes (✓) or no (✗) if the measures are based on Twitter metrics: tweet (T), retweets (RT), mentions (M), replies (RP), Followers (F), Time (TI). All these metrics are explained in Sect. 2.1.

Table 3. Comparison of influence methods

Reference	Research method	Network topology	User actions						Applications
			T	RT	M	R	F	TI	
Alp et al. [7]	– PageRank – Spread score – Baseline methodologies	✓	✓	✓	✗	✗	✗	✓	Identify topical social influencers in a network
Cha et al. [8]	– Correlation between measures of influence: – Spearman's rank	✓	✗	✓	✓	✗	✓	✗	Investigate if influence hold across different topics
Cappelletti and Sastry [9]	– Amplification Rank or IARank	✓	✗	✓	✓	✓	✓	✗	Influence rank in large Scale events: case study of London Fashion Week and London Olympics
Tinati et al. [10]	– Classification Model based on Edelman's Topology of influence	✗	✗	✓	✗	✗	✗	✓	Evaluation performed using different datasets ranging in Topic, size and geographic context

4 Applications

A number of studies has been conducted for identification of Influential users or spread of influence on Twitter in the context of social network analysis, computer science, and sociology. But recently we can find in literature creative applications of this research for different purposes, such as political sciences, business, health, among many others.

Some studies suggest that analyzing emergence of tweets including content related to flu or epidemics can be a productive way to evaluate discourse surrounding health and promote health awareness. Achrekar et al. [19] simulate twitter users as sensors and their tweets that mention key words related to flu as early indicators to track and predict the emergence and spread of flu or influenza-like illness in a population.

Another research conducted by Piccialli and Jung [20] introduces a case study of customer experience diffusion in twitter. This research demonstrate how information shared by companies is distributed and investigate about the main factors that stimulate the diffusion process more quickly and widely for advertisement or sale promotion.

In the same way but for different field, Chung et al. [21] investigate the use of twitter for health promotion, the goal of this study was to evaluate the content of tweets related to Breast Cancer Awareness Month. Understanding the way twitter is used to fight health issues is very important to improve this process.

Twitter can also have implications in political communication such as promotion of presidential election [22], in economics as well to predict stock market fluctuations [23] or even for detection of real-time events, such as earthquakes [25].

Sakaki et al. [25] took advantage from the popularity of Twitter in Japan as Statistics shows [24], Japan is ranked 2nd country on number of Twitter accounts outside U.S with 25.9 million accounts, to use it for detecting earthquakes in real-time.

5 Conclusion

There is a large literature about the various measures and methods used to find influential users and understand information propagation on Twitter. In this survey, we tried to focus more on recent literature and we were able to conclude that most of the research methods refer to either of user actions in Twitter or the network topology. Mostly the oldest research works on this topic (around 2009) focus on network topology using PageRank algorithm [18] whereas more recent ones focus on metrics based on user actions on the network.

Considering influence measures, we would like to emphasize the use of retweets. From all the metrics presented in Table 1, the metrics that is used in all of the methods was Retweets. It is very useful indeed, it enables us to find the source and the direction of propagation of tweets. Despite of this, research results have proven that time dimension is very important, therefore we recommend to consider time dimension in influence metrics.

Finally, we hope that this survey will provide researchers an overview of the variety of influence measures used recently to identify influential entities on Twitter Network.

References

1. The digital universe of opportunities: rich data and the increasing value of the Internet of Things. https://www.emc.com/leadership/digital-universe/2014iview/executive-summary.htm. Accessed 2 Dec 2018
2. Twitter statistics. https://www.statisticbrain.com/twitter-statistics/. Accessed 2 Dec 2018
3. Makice, K.: Twitter API: up and running learn how to build applications with the Twitter API, 1st edn. O'Reilly Media, Sebastopol (2009)
4. Tweet data dictionary. https://developer.twitter.com/en/docs/tweets/data-dictionary/overview/tweet-object. Accessed 2 Dec 2018
5. Rate limits. https://developer.twitter.com/en/docs/basics/rate-limits. Accessed 2 Dec 2018
6. Trung, D.N., Jung, J.: Sentiment analysis based on fuzzy propagation in online social networks: a case study on TweetScope. Comput. Sci. Inf. Syst. **11**(1), 215–228 (2014)
7. Alp, Z.Z., Öğüdücü, S.G.: Topical influencers on twitter based on user behavior and network topology. Knowl. Based Syst. **141**, 211–221 (2018)
8. Cha, M., Haddadi, H., Benevenuto, F., Gummadi, P.K.: Measuring user influence in twitter: the million follower fallacy. In: ICWSM 2010, pp. 10–17 (2010)
9. Cappelletti, R., Sastry, N.: IARank: ranking users on twitter in near real-time, based on their information amplification potential. In: International Conference on Social Informatics 2012, Lausanne, pp. 70–77 (2012)
10. Tinati, R., Carr, L., Hall, W., Bentwood, J.: Identifying communicator roles in Twitter. In: Proceedings of the 21st International Conference on World Wide Web, pp. 1161–1168. ACM, New York (2012)

11. Jansen, B.J., Zhang, M., Sobel, K., Chowdury, A.: Twitter power: tweets as electronic word of mouth. JASIST **60**, 2169–2188 (2009)
12. Bakshy, E., Hofman, J.M., Mason, W., Watts, D.J.: Everyone's an influencer: quantifying influence on Twitter. In: Proceedings of the 4th ACM International Conference on Web Search and Data Mining (WSDM 2011), pp. 65–74 (2011)
13. Galuba, W., Aberer, K., Chakraborty, D., Despotovic, Z., Kellerer, W.: Outtweeting the twitterers - predicting information cascades in microblogs. In: Proceedings of the 3rd Conference on Online Social Networks (WOSN 2010) (2010)
14. Rotabi, R., Kamath, K., Kleinberg, J., Sharma, A.: Cascades: a view from audience. In: Proceedings of the 26th International Conference on World Wide Web, pp. 587–596 (2017)
15. Riquelme, F., González-Cantergiani, P.: Measuring user influence on Twitter: a survey. Inf. Process. Manage. **52**(5), 949–975 (2016)
16. Li, M., Wang, X., Gao, K., Zhang, S.: A survey on information diffusion in online social networks: models and methods. Information **8**, 118 (2017)
17. Wu, X., Zhang, H., Zhao, X., Li, B., Yang, C.: Mining algorithm of microblogging opinion leaders based on user-behavior network. Appl. Res. Comput. **32**, 2678–2683 (2015)
18. Brin, S., Page, L.: The anatomy of a large-scale hypertextual web search engine. In: Proceedings of the Seventh International Conference on World Wide Web 7 (WWW7), Amsterdam, The Netherlands, pp. 107–117 (1998)
19. Achrekar, H., Gandhe, A., Lazarus, R., Yu, S.-H., Liu, B.: Predicting flu trends using twitter data. In: IEEE Conference on Computer Communications Workshops 2011 (INFOCOM WKSHPS), Shanghai, pp. 702–707 (2011)
20. Piccialli, F., Jung, J.E.: Understanding customer experience diffusion on social networking services by big data analytics. Mobile Netw. Appl. **22**, 605–612 (2017)
21. Chung, J.E.: Retweeting in health promotion: analysis of tweets about breast cancer awareness month. Comput. Hum. Behav. **74**, 112–119 (2017)
22. Kreiss, D.: Seizing the moment: the presidential campaigns' use of Twitter during the 2012 electoral cycle. New Media Soc. **18**, 1473–1490 (2014)
23. Bollen, J., Mao, H., Zeng, X.: Twitter mood predicts the stock market. J. Comput. Sci. **2**(1), 1–8 (2011)
24. Twitter by the numbers: stats, demographics & fun facts. https://www.omnicoreagency.com/twitter-statistics/. Accessed 2 Dec 2018
25. Sakaki, T., Okazaki, M., Matsuo, Y.: Earthquake shakes Twitter users: realtime event detection by social sensors. In: Proceedings of the 19th International Conference on World Wide Web, WWW 2010, pp. 851–860. ACM, New York (2010)

Real Time Semantic Events Detection from Social Media Stream

Phyu Phyu Khaing(✉) and Than Nwe Aung(✉)

University of Computer Studies, Mandalay, Myanmar
phyuphyukhaing07@gmail.com, mdytna@gmail.com

Abstract. Real time monitoring of twitter tweet streams for events has popularity in the last decade. This provides effective information for government, business and other organization to know what happening right now. The task comprises many challenges including the processing of large volume of data in real time and high levels of noise. The main objective of this work is timely detection of semantic bursty events which have happened recently and discovery of their evolutionary patterns along the timeline. We present semantic burst detection in adaptive time windows and then retrieve evolutionary patterns of burst over time period. Burst is the task of finding unexpected change of some quantity in real time tweet stream. Moreover burst is highly depending on the sampled time window size and threshold values. Thus we propose how to adjust time windows sizes and threshold values for burst detection in real time. To get accurate burst from real time twitter stream, semantic words and phrase extraction from noise polluted text stream is proposed. Our experimental results show that this semantic burst detection in adaptive time windows is efficient and effectiveness for processing in both real time data stream and offline data stream.

Keywords: Burst detection · Real time events detection
Semantic words extraction · Adaptive sampling in stream mining
Big stream data mining

1 Introduction

Twitter is social information sharing system for twitter users. The features of social media messages contain a large number of irregular and incomplete forms in short and grow fast over time. In the case of Twitter, such documents are called tweets. It is usually related to activities which contain many people all over the world. Tweets cover all imaginable events which are ranging from simple activity updates over news to opinions on arbitrary topics.

By detecting bursty patterns from text streams [3–6], social events can retrieve. However, previous work has mostly focused on traditional text streams such as scientific publications and news articles. There is still a lack of systematic studies for the identifying trends related bursty patterns via social media activities. Moreover their method deals with singular type of social activities.

The detection of the bursts is the critical importance under some circumstances. If the length of the time period or the size of the spatial region when a burst occurs is

T. T. Zin and J. C.-W. Lin (Eds.): ICBDL 2018, AISC 744, pp. 48–57, 2019.
https://doi.org/10.1007/978-981-13-0869-7_6

known in advance, then the detection can easily be done in linear time by keeping a running count of the number of events. However, in many situations, the window size is not known a priori. The size itself may be an interesting subject to be explored. Many methods have been proposed to detect bursts in a fixed predefined time windows. However these methods are not adaptive enough for burst detection real time applications since they used fixed parameter settings for burst detection like time window size and threshold values. The goal of this work is to develop an efficient algorithm for detecting adaptive aggregation semantic bursts in data stream dynamically sized windows. With the help of the statistical characteristics and sketch summary, bursts are detected accurately and efficiently on line.

2 Related Works

Keywords are frequently used in many events as indicators of important information contained in documents. Unlike traditional Natural Language Processing Task (NLP), Twitter data has many noisy data because of tweet size limitation. The number of maximum characters for each tweet is up to 140. Thus, Twitter tweets include many noisy keywords like acronym and other type of abbreviations due to inherent size limitation. To solve this noise of text stream, some author used the corpus of tweets from Twitter annotated with keywords. It is created by using the crowdsourcing method [5]. To provide semantic keywords extraction, Algorithm 1 is proposed to handle the typical usage such as abbreviations, typing and spacing errors.

The events detection from bursty terms are proposed and they used the burst detection scheme presented by Kleinberg. In their system they retrieve events by identifying bursty keywords from digital newspapers and grouping these keywords into clusters to identify bursty events [2, 3]. Their work showed such detection tasks are feasible and it did well in identifying trending events. [9] Paper proposed event detection from Twitter stream on earthquakes in Japan. In [4], they constructed the probabilistic popular event tracker.

3 Semantic Events Detection

There are three steps to perform semantic burst detection and analyzing events is performed as the fourth step. First, tweets are crawled from twitter using Twitter API which is REST APIs provides programmatic access to read and write Twitter data. Second, the keywords are extracted from tweets and solve the keywords variants in parallel. Third step identifies ‘bursty’ keywords, i.e. keywords that abruptly appear in tweets with an abnormally high rate. Then, it groups bursty keywords into events based on their co-occurrences in tweets. The events are identified a set of bursty keywords that occur frequently together in the tweets as fourth step. Bursts detection over adaptive time window is proposed in this study. By using statistical summarization over keywords counts, sampling time window size and threshold will be adjusted. Each of the four steps described above is pictured as a component of the diagram shown (see Fig. 1). To detect bursty events in real time, all of the tasks are processing in parallel.

Furthermore, to be scalable over massive document streams, an approach is required that makes as few passes over the data as possible. So one pass processing for burst detection is used.

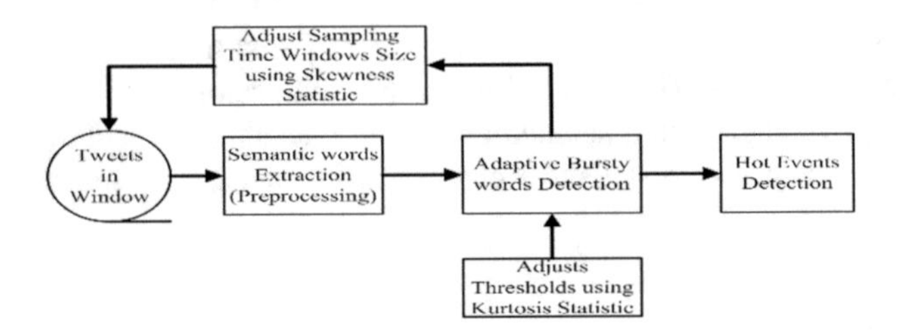

Fig. 1. System overview

4 Semantic Words Extraction

To detect events and bursty keywords, firstly the candidate keywords from tweets are extracted. The various abbreviations, slang words and minor typing errors occur very frequently in tweets. To solve that problem, the 'semantic keyword extraction algorithm (see Algorithm 1) is proposed for semantic filtering words from tweets.

```
Algorithm 1. Semantic Keyword Extraction Algorithm
1. Input: bag of tweets over time  tweetJsonfile,
tweetcount t,
2. Output: Map<String,Integer>  wordCountList, null for
otherwise
3.   If t==0 then
4.    wordCountList=null;
5.   End if
6. If t>0 then
7.          Read line in file until not end of file;
8.           While( line != null )
9.           {
10.               cleaningTweetSign(line);
11.               removeStopWords(line);
12.               NameEntityRecognition( line);
13.               SolveKeywordVariants(line);
14.               wordList[] = line.split;
                  Save word in wordCountList
15.            }// end while
16. End if
17. return wordCountList;
```

The above proposed algorithm returns semantic words and count list for tweet of time window. The Algorithm is developed for data preprocessing. It is described as detail solution in the next section.

4.1 Data Preprocessing

In order to get tweets posted by users in specific country, we defined geographic bounding boxes. Example; bounding box using Latitude and Longitude of Yangon (16.8661° N, 96.1951° E) is [lon−0.5, lat−0.5] [lon+0.5, lat+0.5]. By using tweet content, we save the tweet id, creation time, user id and text in the database. To get effective and accurate keywords from the tweets stream, there are five steps to get meaningful keywords from twitter tweets stream.

Cleaning Tweet Sign. To clean the tweet, remove all the #hashtags, @-mentions, emoticons and URLs (http) and other punctuation marks from tweets. The length of word less than 3 is removed from tweets.

Stop-Words Removal. The stop words are common words such as 'an', 'the' and so on. Before detecting bursty terms stop words are removed. In this approach, English stop word list is used. It is downloaded from Natural Language ToolKit (NLTK) website.

Check Spelling Error. Checking spelling errors by comparing with predefined English dictionary. The dictionary is Brown Corpus that created in Brown University.

Name Entity Tagging. Finding and replacing human name and country name is developed by using predefined corpus. A person's name is typical example. In our experiment dataset, several different formats are used for a person's name. For example, for "Aung San Suu Kyi," the leader of Myanmar, "Suu Kyi," "ASSK," etc., are also used. The text file contains human name and country name. The next section explains keyword variant types and how to solve it in our approach.

Keyword Variants. The keywords variants of this steaming text of social media include abbreviations, typing and spacing errors and word expansion. First variant type is abbreviations/acronyms and we use predefined abbreviation corpus to expand Acronyms. For the keyword "National League for Democracy," we construct "NLD" as its acronym. Somctimes, there may have an ambiguous situation like as "Myanmar" and "Burma" are the same meaning of country name. Since the two keywords are not related, so it will not count as same keywords. To eliminate this ambiguity, we add predefined the synonym words of the same meaning for some words. The third variant type is typing and spacing errors. In the tweet, the typing and spacing errors occur very often. It can be handled by determining the size and character histogram of keywords.

5 Real Time Burst Detection

The streaming data might be transient and are potentially unbounded in size. Data are discarded or archived as soon as an element from a data stream has been processed. Queries from data streaming requires techniques for storing summaries or synopsis information about previously perceived data. There is a tradeoff between the size of summaries and the ability to provide precise answers.

In our approach, discrete time windows are used for sampling. The sampling time window size have a significant impact on the tradeoff between the delay and accuracy of bursty event detection. A shorter window will lead to a smaller delay but may have a poor performance when the post rate is low and there are not many tweets posted during the window. Moreover many spikes or many burst come out and noisy in the burst detection. On the other hand, the sampling time window size is large, there is rare to detect bursts. Besides medium level of words are left. So sub-events of low level burst are left. To handle dynamic time window size and threshold value, summary statistics over rate of tweet arrivals are used. Detailed is shown in Algorithm 2.

```
Algorithm 2. Adaptive Burst Detection Algorithm
1. Input: Streaming Tweets;
2. Output: Map<String, Integer> burstList;
3. Initialization
3.1 winSize=5 minutes
3.2 threshold=30
3.3 Timer timer is set to 10 minutes
4. If timer is expire then
5 Map<String, Integer> wordlist= TweetPreprocessing
(TweetList per time period)
6. for i=0 to wordlist.count
7. If (wordcount > threshold && wordcount > total tweets
count) then
    7.1. burstList[word] = bursty word;
    7.2.     burstList[count] = bursty word count;
8.   End if
9.  End for
10.  Distribution rate is calculated. (Skewness Equation)
11.  Peakness is calculated. (Kurtosis Equation)
12.   Adjust the window sizes and set new winSize value
13.   Adjust the threshold and set new threshold value
14. End if
15.    Next Period for Burst Detection
16. return burstList;
End Algorithm
```

After burst terms are detected, the next time windows sizes are calculated by using distribution rate of tweets in two successive time windows. Thus sampled time window size and threshold values are changed after processing two 10 min time windows (20 min processing). To detect distribution rates of tweets and burst defined threshold value, summary statistics are calculated as in Eqs. 1 and 2. Skewness provides us the information about the asymmetry of the data distribution and Kurtosis gives us an idea about the degree of peakedness of the distribution.

$$skewness = \frac{\sum (x - \bar{x})^3}{ns^3} \quad (1)$$

$$kurtosis = \frac{\sum (x - \bar{x})^4}{ns^4} \quad (2)$$

Where x = data point, s is variance and n = number of keywords over time window. By comparing Skewness of the two continuous time windows, we can get next time window size. Beside by calculating two Kurtosis values of the successive windows we can adjust threshold values for the burst detection [10].

6 Semantic Events Detection

To retrieve semantic events from Twitter stream, bursty list and tweets in that time period are used. We can find correlated bursty terms by finding occurrences of burst terms in same tweets. EventDetection Algorithm assesses co-burst occurrences in same tweets. For this purpose, the related history of tweets is retrieved for each bursty keyword and keywords that are found to co-occur in a number of recent tweets are placed in the same group as in Algorithm 3.

```
Algorithm 3. EventDetction Algorithm
1: Procedure CoBurstyTerms (burstylist, total tweets in
window)
2:   count=0;
3:   do
4:   urstTe  BurstTerm• burstylist. Get (count)
5:         Foreach (String line in totaltweet)
6:       If (Line contains BurstTerm) then
7:              Save word and word count in temporary list
8:            End if
9:       End for
10:        do
11: If (wcount of each words in temporary list >= burst
defined threshold/4) then
12:     Save this word as correlated bursty terms
13:              End if
14:        while (words exit in temporary list)
15:        count = count +1
16:    while (count < burstyListLength);
17:    return CorelatedBurstyTermsList;
```

7 Experimental Results

To measure the effective of our approach of the adaptive time window sampling and adaptive thresholds for burst detection, two offline data sets are used and they are described in Table 1. These data are crawled from Twitter website.

Table 1. Dataset description

Data set name	Date range of data crawling
Myanmar Democracy Government 2016	3/22/2016 to 3/28/2016
Myanmar Earthquake 2016	8/30/2016 to 9/5/2016

Most of the events in this dataset related with the events of that National League Democracy (NLD) won the Myanmar Democracy Government 2016. Examples; Daw Aung San Suu Kyi given Myanmar cabinet position, Myanmar parliament announces nominees of cabinet members including Aung San Suu Kyi and so on. One of the events that is not related with new government news is New Myanmar Stock Exchange trading event. As shown (see Fig. 2), our proposed method can detect more events than the equal window method. Window Number 7 and 8 have no events in equal time window method but our method can detect one event. This is because with the monitoring the arrival rate of tweets, bursty terms and events are detected. The detected events are (1) NLD won Election 2015, (2) Aung San Suu Kyi given Myanmar cabinet position, (3) Aung San Suu Kyi decide on the presidents of Myanmar and (4) Yangon Stock Exchange (YSX).

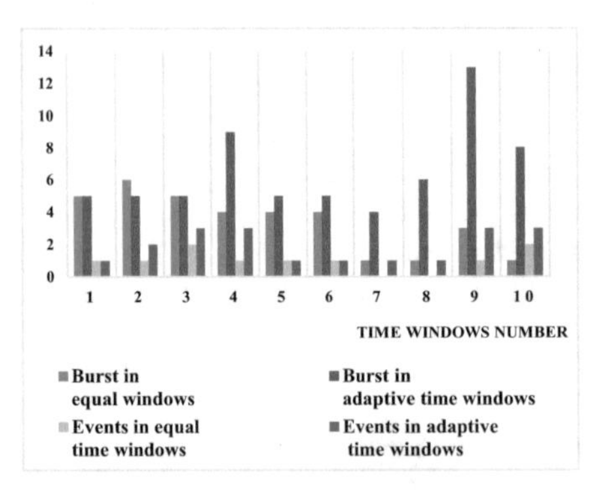

Fig. 2. Burst detection in equal time windows and adaptive time windows for Myanmar Democracy Government 2016.

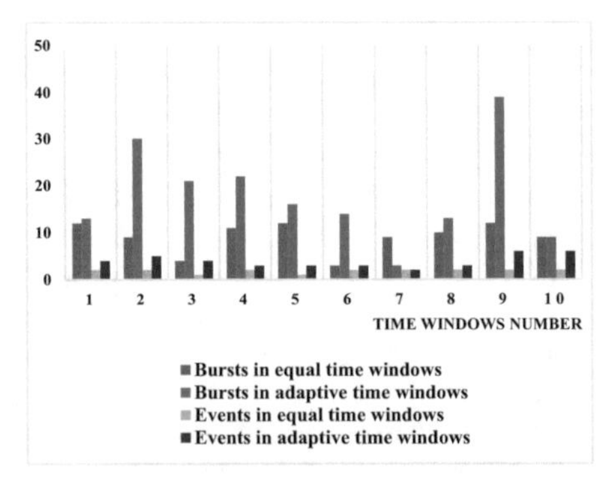

Fig. 3. Burst detection in equal time windows and adaptive time windows for Myanmar Earthquake 2016

The next offline dataset of Myanmar Earthquake 2016 is used to measure effective of our approach. During this time, Events of Italy Earthquake also occurred. Figure 3 describes burst detection for this offline dataset of Myanmar Earthquake 2016. The results, 39 bursty terms is detected in our approach in window number 9. It is the triple of equal windows method. This is because window size is not increased for that high arrival rate of tweets in equal window method.

After processing window number 8 and 9, window size and threshold value are increased in adaptive method because ratio of skewness values of window 9 and 8 is greater than the ratio of skewness values of window 8 and 7.

After bursts are detected, window size is increased. But window size will shrink if arrival rate is declined. Numbers of detected events are depend on the number of correlated bursty terms contain in the same tweets. Sometimes bursty terms count is low but events count is high in window number 10 of (see Fig. 3). The only 9 bursty terms are detected but 6 events in window number 10. This is because there is high number of correlated burst terms in the tweets. The event list for Myanmar Earthquake 2016 are (1) Myanmar Earthquake, (2) Italy Earthquake, (3) Began Pagoda Damage, (4) Oklahoma shut fracking disposal water shuts, (5) Volunteer helping plan, and (6) Thai border also suffer earthquake.

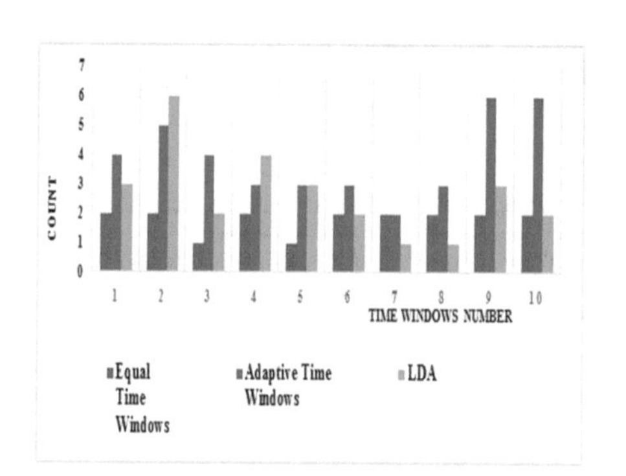

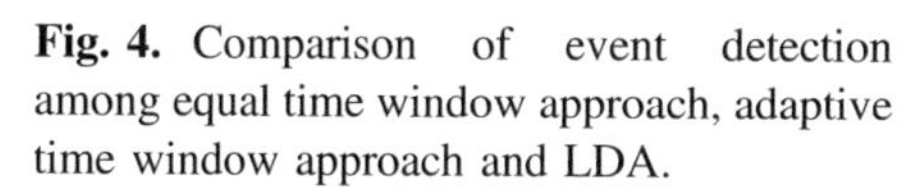

Fig. 4. Comparison of event detection among equal time window approach, adaptive time window approach and LDA.

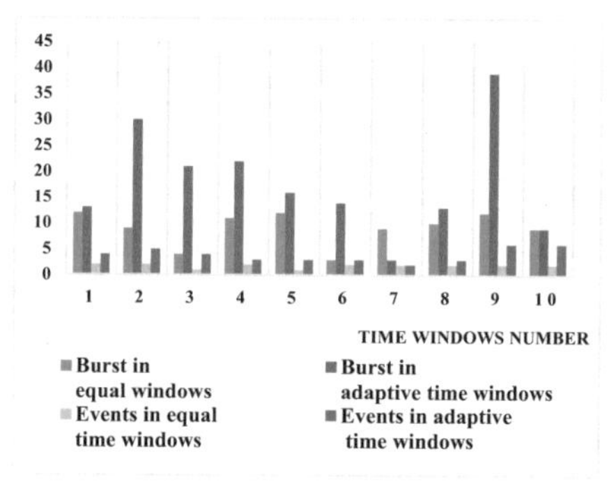

Fig. 5. Comparison of bursty terms detection, event detection in equal time windows and adaptive time windows in real time

For evaluation and comparison, LDA is used to compare event detections. LDA is used to define events and process over twitter tweets documents. Figure 4 shows the comparison results between our approach and LDA. Actually LDA detects topics and events but not all events and topics are bursty.

Moreover most of words in same group cannot give correlated words to form events. The proposed EventsDetection Algorithm grouped bursty terms from tweets. So the retrieved correlated bursty terms are more correlated in our approach than LDA method. Table 2 shows the performance score of handling keywords variants in two offline datasets using 3-fold cross validation. Hence compared to traditional preprocessing, our proposed preprocessing method (Algorithm 1) shows better result. Table 3 shows comparison results for tweet preprocessing with semantic keywords extraction and without semantic keywords extraction over two offline datasets. The precision score is higher because of proposed method contain possible words correction methods and also compare with dictionary. Besides, it has expanded abbreviations, solving slang words and typing error correction. Thus, result words are correct nearly 100%. But in our approach, emotional words are not solved.

Moreover our approach reduces the total words count of tweets in time window to 30%. It is because tweet signs and one third of these words are removed at cleaning stage. Figure 5 shows the comparison results of bursty terms detection between 10 min time windows and adaptive time windows in real time.

For our approach, 10 min time window is used for initial state of processing. In window number 9, maximum bursty term count is nearly 40 in our approach and about 10 in equal windows. Our approach is more burst detected. For window number 9, window sizes are different in 2.4 min but result is better than equal window. For adaptive window size is 12.4 min and other is 10 min. Beside Fig. 4 and Table 6 show the results of event detection over two methods. Events are defined from correlated

Table 2. Performance of semantic keywords extraction

Data set name	Precision	Recall	F1 Measure
Myanmar Democracy Government 2016	73.03	25.73	38.05
Myanmar Earthquake 2016	81.34	12.14	19.73

Table 3. Comparison of performance in semantic keywords extraction using precision score

Dataset name	Semantic keywords extraction	Without semantic keywords extraction
Myanmar Democracy Government 2016	73.03%	32.33%
Myanmar Earthquake 2016	81.34%	28.12%

bursty terms of tweets in the time window. Events are not only depending on bursty terms but also depend on the count of that terms which contain in tweets. The window number 2, 9 and 10 have more events than equal window approach because they might be difference in burst size and time window sizes. But adaptive window size is not more than double size of equal window size. The events detected are triple in our adaptive windows approach.

From the experiments, our approach is more effective than the equal time window size. So equal time window sampling is not applicable for real time system. The rate of tweet arrival is not stable in all time.

8 Conclusion

In this paper, we propose the approach to identify events using bursts via social media activities. We also proposed one optimization model to adjust sampling size of time window and threshold values in multiple activity streams to learn the bursty patterns. To better meaningful events, we proposed Semantic Keyword Extraction Algorithm for semantic keywords extraction. So semantically keywords extraction is described to get efficient bursty words. So in this work, we developed events analysis in real time windows sizes and threshold values in real time. Our experimental results showed both scalability and accurate events detection of our approach.

References

1. Aho, A.V., Ullman, J.D.: The Theory of Parsing, Translation, and Compiling, vol. 1, pp. 368–399. Prentice-Hall, Englewood Cliffs (1972)
2. Allan, J., Papka, R., Lavrenko, V.: On-line new event detection and tracking. In: Proceedings of the 21st Annual International ACM SIGIR Conference on Research and Development in Information Retrieval, pp. 37–45. ACM (1998)
3. Becker, H., Naaman, M., Gravano, L.: Beyond trending topics: real-world event identification on twitter. ICWSM **11**, 438–441 (2011)
4. Buntain, C., Lin, J., Golbeck, J.: Discovering key moments in social media streams. In: 2016 13th Annual IEEE Consumer Communications and Networking Conference (CCNC) (2016)
5. Diao, Q., Jiang, J., Zhu, F., Lim, E.-P.: Finding bursty topics from microblogs. In: Proceedings of the 50th Annual Meeting of the Association for Computational Linguistics: Long Papers, vol. 1, pp. 536–544. Association for Computational Linguistics (2012)
6. Fung, G.P.C., Yu, J.X., Yu, P.S., Lu, H.: Parameter free bursty events detection in text streams. In: Proceedings of the 31st international conference on Very large data bases, VLDB 2005, pp. 181–192. VLDB Endowment (2005)
7. Kleinberg, J.: Bursty and hierarchical structure in streams. In: Proceedings of the Eighth ACM SIGKDD International Conference on Knowledge Discovery and Data Mining, KDD 2002, pp. 91–101. ACM, New York (2002)
8. Lanagan, J., Smeaton, A.F.: Using twitter to detect and tag important events in live sports. In: Artificial Intelligence, pp. 542–545 (2011)
9. Lin, J., Efron, M., Wang, Y., Sherman, G., Voor-hees, E.: Overview of the TREC-2015 microblog track. In: Proceedings of the Twenty-Fourth Text Retrieval Conference, TREC Gaithersburg, MD (2015)
10. Khaing, P.P., New, N.: Adaptive methods for efficient burst and correlative burst detection. In: 2017 IEEE/ACIS 16th International Conference on Computer and Information Science (ICIS). IEEE (2017)

Community and Outliers Detection in Social Network

Htwe Nu Win[1(✉)] and Khin Thidar Lynn[2]

[1] University of Computer Studies, Mandalay, Mandalay, Myanmar
htwenuwin99@gmail.com
[2] Faculty of Information Science,
University of Computer Studies, Mandalay,
Mandalay, Myanmar
lynnthidar@gmail.com

Abstract. Challenges of detecting communities among users' interactions play the popular role for days of Social Network. The previous authors proposed for detecting communities in different point of view. However, similarity based on edge structure and nodes which cannot group into communities are still motivating. Considering the community detection is motivating from the similarity measurement to detect significant communities which are high tightly connected each other upon the edge structure and outliers which are unnecessary to group into the communities. This paper is proposed the approach of using similarity measure based on neighborhood overlapping of nodes to organize communities and to identify outliers which cannot be grouped into any of the communities based on Edge Structure. The result implies the best quality with modularity measurement which leads to more accurate communities as well as improved their density after removing outliers in the network structure.

Keywords: Social network · Community · Outlier

1 Introduction

Social networks are naturally modeled as graphs, which we sometimes represent it as a social graph. The entities are nodes, and an edge that connects two nodes if the nodes are interacted by the relationship that forms the network. For Facebook friendship graph which we will use in our work, social graphs are undirected and unweighted. The communities of complex networks are groups of nodes which are high tightly connected with nodes of the same group other than with less links connected with nodes of different groups [18]. Communities may be groups of friendship in social networks, sets of web pages concerning with the same topic and groups of cells with similar functions. While identifying the communities in graphs, nodes which cannot groups with any communities and need not be necessary group, will be identified as outliers. The early observer of outlier is Hawkins [2]. He said that "An outlier is an observation that deviates so much from other observations as to arouse suspicion that it was generated by a different mechanism". The groups of vertices which are similar to each other is naturally assumed as communities. The similarity between each pair of vertices can be

T. T. Zin and J. C.-W. Lin (Eds.): ICBDL 2018, AISC 744, pp. 58–67, 2019.
https://doi.org/10.1007/978-981-13-0869-7_7

computed with respect to some reference property, local or global, no matter whether they are connected by an edge or not. Vertices which have similarity value with the communities can be group into the corresponding communities. Our proposed approach is adopted from Hawkin's definition, nodes which have no any friendship or there are no common friends, so much deviate and saturated with individual lonely is defined outlier and community based on similarity measure on Edge structure.

This paper explores the use of neighborhood overlapping by using vertex similarity method for outlier and significant community detection. The heart of this approach is to represent the underlying dataset as an undirected graph, where a user refers to each node and friendship between two users represents each edge. Before we measure the similarity among neighborhood overlap, finding seed node by using the degree centrality is necessary which is designed to find nodes that are most "central" to the graph. We operate similarity from the most centrality node and its neighborhood nodes. The values of zero similarity are then used to identify as outliers.

To illustrate, consider graph of Fig. 1, which consists of 18 nodes and 25 edges. Upon applying approach method, three communities are represented by yellow, green and dark orange color. Continually, red color nodes are shown as outliers. It can be seen the significant communities and outliers in this toy example.

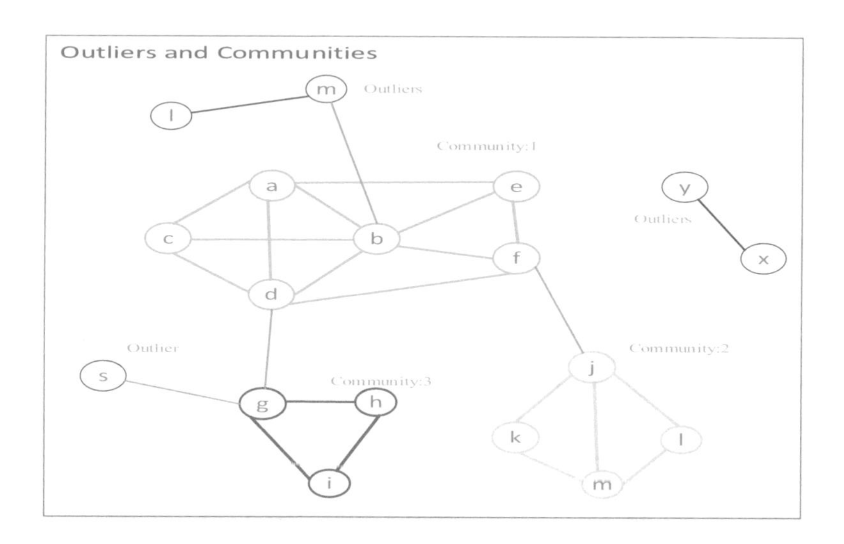

Fig. 1. Example of communities and outliers.

The rest of paper is organized as follows. Section 2 briefly surveys related work. In Sect. 3, we describe the background methodology of our work. And then, we briefly describe about our propose system in Sect. 4. In Sect. 5, we discuss about the experiment and evaluation of our work. Then, we concluded our work in Sect. 6 and talk about our future idea.

2 Related Work

Many approaches of outliers detection algorithm have been proved over years. Each trend has efficient and effective in their ways. After the time of Hawkins [2] definition, many different points of view in outliers were appeared. Graph based outlier detection were also appeared with the flow of researchers. SimRank method [3] identified the proximity measure of graph that quantified the closeness of two nodes in the graph, which based on their relationships with other objects and computed the similarity of the structural context in which the graph objects occur. Several variants of SimRank were also proposed by [5, 17]. On the other hand, [8] proposed the one could use to compute various measures associated with the nodes in the given graph structure, dyads, triads, egonets, communities, as well as the global graph structure. [9] proposed OddBall which is an algorithm to detect anomalous nodes in weighted graphs. Detecting anomalous sub-graphs using variants of the *Minimum Description Length* (MDL) principle was proposed by [1]. Outrank method [4] also used the MDL principle as well as other probabilistic measures to detect several types of anomalies (e.g. unexpected/missing nodes/edges). OutRank and LOADED [13] used similarity graphs of objects to detect outliers. MDL to spot anomalous edges was used by [6]. [7] used proximity and random walks, to assess the normality of nodes in bipartite graphs. [14] proposed the method of detecting outliers with community by using minimum valid size (mvs). If the minimum size is 2, it will be chosen the single node and marked it as outlier. In contrast to the above, we work with *undirected* and unweighted graph data with method of overlapping neighborhood. We explicitly focus on edge structure to detect outliers and significant communities with nodes similarity.

3 Background Theory

3.1 Graph

We consider basic definitions of a given network represented as a directed graph or undirected graph. Facebook friendship network which we will use in our work here, a graph may be undirected, meaning that there is no distinction between the two vertices associated with each edge.

The notation of a graph $G = (V,E)$ consists of two sets V and E. The elements of $V = \{v_1, v_2, \ldots, v_N\}$ are the nodes or vertices of the graph G where each vertex $v_i$is associated with the instance *xi* from the input data X and the cardinality of $|V|$ is N. The elements of $E = \{e_1, e_2, \ldots, e_M\}$ are links or edges between nodes and the cardinality of $|E|$ is M. An edge connecting the vertices v_i and v_j is denoted by e_{ij} [15]. The example of social network imagine as a graph is shown in Fig. 1.

3.2 Node Degree and Its Neighborhood

In network G, the degree of any node i is the number of nodes adjacent to i. The degree of node v is $d(v)$, that is, the number of edges associated with node v.Generally, the more degree that the node has, the more important it will be [10].

Two vertices v and u are called neighbors, if they are connected by an edge. Let Γ_i be the neighborhood of vertex i in a graph, i.e., the set of vertices that are directly connected to i via an edge. For a given node u, $N(u) = \{v|(u,v) \in E\}$ is a set of containing all neighbors of node u [10].

3.3 Communities

The communities of social network are groups of nodes, with more links connecting nodes of the same group and comparatively less links connecting nodes of different groups. Communities may be groups of related individuals in social networks. Identifying communities in a network can provide valuable information about the structural properties of the network, the interactions among nodes in the communities, and the role of the nodes in each community [18].

In an undirected graph $G(V,E)$, where the total number of node, $|V| = n$ and total number of edges,$|E| = m$ are defined. We can identify set of communities such that $Coms = \{V1', V2', \ldots\ldots, Vcn'\}$ where $\cup_{i=1}^{cn} V_i' \subseteq V_i$ and cn is the total number of communities Coms should satisfy, $Vi' \cap Vj' = \phi$[14].

3.4 Vertex Centrality

The importance of a node is determined by the number of nodes adjacent to it. The larger the degree of node, the more important the node is. Those high-degree nodes naturally have more impact are considered to be more important. The degree centrality is defined as

$$C_D(v_i) = d_i = \sum_j A_{ij} \tag{1}$$

When one needs to compare two nodes in different networks, a normalized degree centrality should be used,

$$C_D'(v_i) = d_i/(n-1). \tag{2}$$

Here, n is the number of nodes in a network. It is the proportion of nodes that are adjacent to node v_i[10].

3.5 Outliers

According to Hawkins [2], outliers can be defined as follows: "An outlier is an observation that deviates so much from other observations as to arouse suspicion that it was generated by a different mechanism". Most outlier detection schemes adopt Hawkin's definition of outliers and thus assume in our system is that outliers are nodes which have zero values of similarity measure in graph. Each node in a graph cannot be grouped into any of the communities is called outlier. As such, these outliers can be easily detected by existing distance or density based algorithms. However, in this paper we focus on outliers that might be concentrated in certain regions. We take edge structure based approach in graph to solve this problem. Outliers and communities can be defined by:

$$Outs = \{v | v \in V, \neg \exists Vi' \in Coms \wedge v \in Vi'\} = V - \bigcup_{i=1}^{cn} Vi' \tag{3}$$

Outliers directly identified by getting nodes from small communities [14, 19]:

$$Coms \bigcup Outs = v$$

$$Coms \bigcap Outs = 0$$

Where, *Coms* is the community and *Outs* is indicated as outlier.

3.6 Vertex Similarity

It can be assumed that communities are groups of vertices similar to each other. We can compute the similarity between each pair of vertices after searching seed nodes. Most existing similarity method are based on the measurement of distance called Euclidean, Manhattan and etc., Although, to considered the similarity between selected node and is neighborhood, Jaccard Similarity is more convenient in this work which we will measure the similarity based on the overlapped neighbor of seed nodes. Let N_i denote the neighbors of node v_i. Given a link (v_i, v_j), the neighborhood overlap is defined as

$$\begin{aligned} Overlap(vi, vj) &= \frac{number\ of\ shared\ friends\ of\ both\ v_i\ and\ v_j}{number\ of\ friends\ who\ are\ adjacent\ to\ atleast\ v_i\ and\ v_j} \\ &= \frac{|N_i \cap N_j|}{|N_i \cup N_j| - 2} \end{aligned} \tag{4}$$

We have -2 in the denominator just to exclude $v_i\, and\, v_j$ from the set $N_i \cup N_j$. If there are no overlap vertices in any two $N_i and N_j$ means $|N_i \cap N_j| = \emptyset$, it can be identified N_j as outliers of N_i. Assuming like that, our work identify outliers are appeared with among separated communities [10].

4 Proposed Approach

Network can be represented as graph for detecting outliers and significant communities. In this paper, we use concept of undirected graph in which two vertices have common node, if they share one or more of the same vertices. Let Γ_i be the neighborhood of vertex i in a network, i.e., the set of vertices that are directly connected to vi via an edge. Then the number of common friends of vi and vj is $|\Gamma_i \cap \Gamma_i|$. Contrary, nodes which have no any other node in common or sometime they can be isolated, we can identify them as outliers, can be given as $|\Gamma_i \cap \Gamma_i| = 0$. In our work, we propose the system to identify significant community with detecting outlier by using node similarity depend on this concept. Node similarity measurement can be considered in detecting communities, groups of vertices are similar to each other. The detected communities are defined by the computation of the similarity values of vertices and the corresponding seed node. This paper used the most common used method in considering of the similarity between selected node and its neighborhood is Jaccard Similarity. It is more convenient in this work which will be measured the similarity based on the neighborhood overlap of seed nodes. Let N_i denote the neighbors of node v_i.

Our work is based on the following intuitive properties:

- Outliers are defined by nodes which have no any intersect value to its related ones.
- Seed node is determined by degree centrality method which has the largest degree.
- Communities are calculated by Vertex Similarity among nodes that are linked to seed node.
- Nodes that belong to the same communities are likely to be more links connected to each other.
- Each community is likely to be fewer links outside the rest of the graph.
- Nodes in the same community are likely to share common neighborhood node.

The heart of this work has three main processes, firstly finding the intersect value of related node to detect outliers, then use the method of the degree centrality to determine seed nodes and finally use the method of neighborhood overlap based on vertex Similarity for detecting the communities.

5 Experiments

5.1 Description of Datasets

In this paper, a real undirected network, the popular studies in social network analysis, Zachary Karate Club Dataset [20] is used. In this Dataset statistics, "nodes" represents the number of friends; "Edges" represents the number of friendship in the network. There are 34 member nodes and 78 edges as shown in Fig. 2.

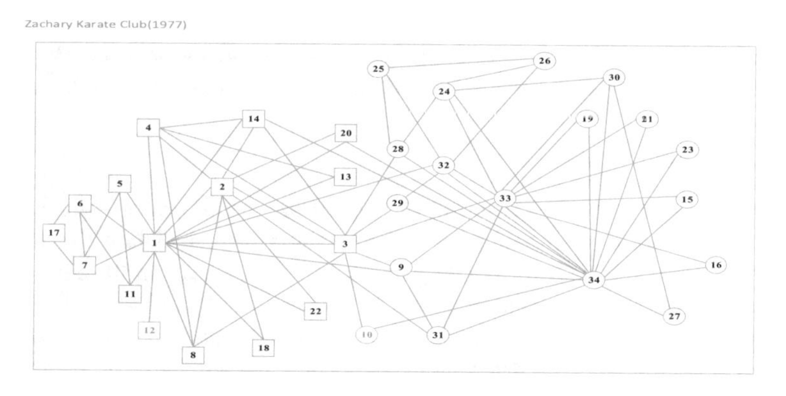

Fig. 2. Original network of Zachary Karate Club.

5.2 Evaluation Method

Generally, in thinking about the evaluation method of how good of a community, a set of nodes which have no ground truth community in undirected and unweighted graph, there are two criteria of interest. The first is the number of edges between the members of the cluster, and the second is the number of edges between the members of the cluster and the remainder of the network. The two groups are Multi-criterion scores and Single-criterion scores. Firstly, it is represented as Multi-criterion scores, combines both criteria (number of edges inside and the number of edges crossing) into a single objective function; the next criterion is the objective functions employs only a single of the two criteria (*e.g.*, volume of the cluster or the number of edges cut).

Multi-criterion Scores. Let $G(V, E)$ be an undirected graph with $n = |V|$ nodes and $m = |E|$ edges. Let S be the set of nodes in the cluster, where n_s is the number of nodes in S, $n_s = |S|$; m_s is the number of edges in S, $m_s = |\{(u, v) : u \in S, v \in S\}|$; and c_s, the number of edges on the boundary of, $c_s = |\{(u, v) : u \in S, v \notin S\}|$; and $d(u)$ is the degree of u.

It is considered the metrics $f(S)$ that capture the notion of a quality of the cluster. Lower value of score $f(S)$ signifies a more community-like set of nodes [16].

- Conductance: $f(S) = \frac{c_s}{2m_s + c_s}$ measures the fraction of total edge volume that points outside the community. The smaller the value of Conductance is, the better the community quality is [16].
- Expansion: $f(S) = \frac{c_s}{n_s}$ measures the number of edges per node that point outside the community. The smaller the value of Expansion, the better the community quality is [16].

Single-criterion Scores. Next it is also considered community scores that is a single criteria. Here it is considered the following two notions of a quality of the community that are based on using one or the other of the two criteria of the previous subsection:

- Volume: $\sum_{u \in S} d(u)$ is sum of degrees of nodes in S. The larger the value of this metric is, the better the community quality is [16].
- Edges cut: c_s is number of edges needed to be removed to disconnect nodes in S from the rest of the network. The smaller it is, the better the community quality is [16].

5.3 Results and Evaluation

The communities and outliers result of our evaluation are shown in Figs. 3, 4 and 5. Firstly, the outliers are defined with the zero value by calculating the method of intersection between their corresponding nodes which are based on the edge structure. In this paper, there are two outliers (node 10 and node 12) that had no friendship of common neighbors between different nodes. Then, seed nodes were determined by using vertex centrality. After determining seed node, the similarity measurement was processed the computation of overlapped value of common node between the corre-spondent seed node and its neighborhood. As shown in Fig. 3, there are two detected communities and two outliers by using the proposed approach.

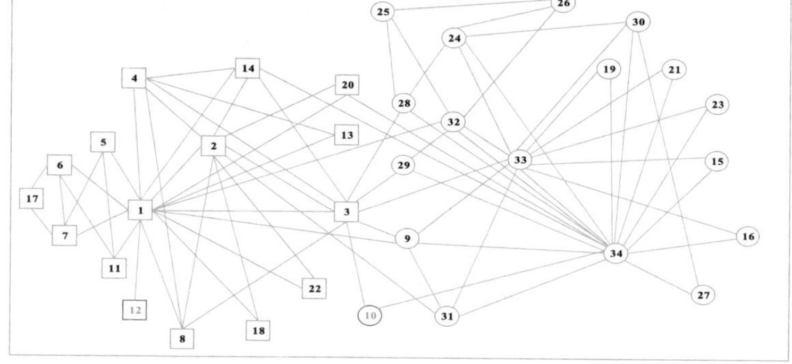

Fig. 3. Two Communities and Outliers by using proposed approach.

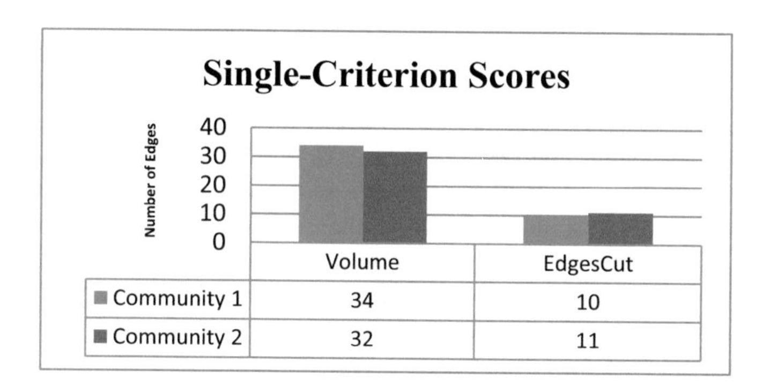

Fig. 4. Single-criterion scores

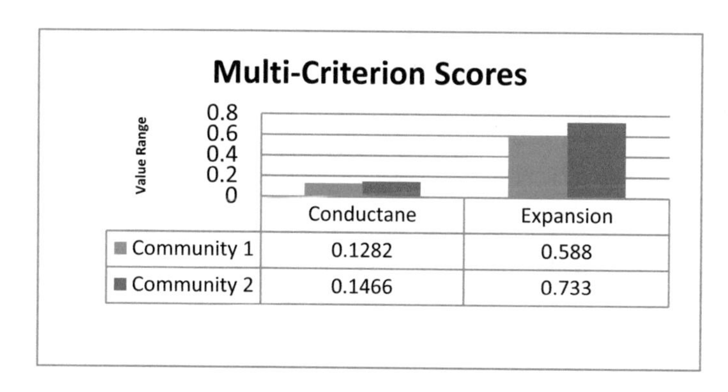

Fig. 5. Multi-criterion scores

This paper is measured by using two different quality of community as described in Sect. 5.2. Figure 4 showed the significant result based on the first two connections. In the last connectivity, the amount of the expansion value is very few in the expected number of links between the communities, therefore, it can be seen that the smaller conductance and expansion values are significant as shown in Fig. 5. Therefore, it can be decided that considering detecting outlier is effective to get the significant community in social network.

6 Conclusion

This paper proposed the approach to identify outlier and detect the community of social networks using the overlapping of neighborhoods with the method of vertex similarity. It described the detail computation to detect outliers with no common node and community with common node. The previous methods had been considered the structure of community without outlier. Therefore, this approach proposed outlier detecting method by using edge structure to get the effective community. In this paper, we based on edge structure of unweighted and undirected graph to detect the outlier and significant communities without thinking any information of the structure.

We used the standard real network dataset from Zachary Karate club [20]. It was used two quality measurement, single-criterion score and multi-criterion scores [16] to define how much our community is good and significant. By using the method of determining outlier with detecting community the density of our method's result is significant in measurement. The drawback of our method is only based on edge structure without considered any information such as profile of user. So, our future will be used feature of users to detect effective community.

References

1. Noble, C.C., Cook, D.J.: Graph-based anomaly detection. In: KDD, 2003, pp. 631–636 (2003)
2. Hawkins, D.: Identification of Outliers. Chapman and Hall, London (1980)
3. Jeh, G., Widom, J.: SimRank: a measure of structural-context similarity. In: Proceedings of the 8th ACM International Conference on Knowledge Discovery and Data Mining (SIGKDD), Edmonton, Alberta, pp. 538–543 (2002)
4. Moonesinghe, H.D.K., Tan, P.N.: Outrank: a graph-based outlier detection framework using random walk. Int. J. Artif. Intell. Tools **17**(1), 19–36 (2008)
5. Chen, H.-H., Giles, C.L.: ASCOS: an asymmetric network structure context similarity measure. In: IEEE/ACM International Conference on Advances in Social Networks Analysis and Mining (ASONAM), Niagara Falls, Canada (2013)
6. Chen, J.Y., Zaiane, O.R., Goebel, R.:. Detecting communities in large networks by iterative local expansion. In: International Conference on Computational Aspects of Social Networks (2009)
7. Sun, J., Qu, H., Chakrabarti, D., Faloutsos, C.: Neighborhood formation and anomaly detection in bipartite graph. In: ICDM, 27–30 November 2005

8. Henderson, K., Eliassi-Rad, T., Faloutsos, C., Akoglu, L., Li, L., Maruhashi, K., Prakash, B.A., Tong, H.: Metricforensics: a multi-level approach for mining volatile graphs. In: Proceedings of the 16th ACM International Conference on Knowledge Discovery and Data Mining (SIGKDD), Washington, DC, pp. 163–172 (2010)
9. Akoglu, L., McGlohon, M., Faloutsos, C.: Anomaly detection in large graphs. In: CMU-CS-09-173, November 2009
10. Tang, L., Liu, H.: Community Detection and Mining in Social Media. A Publication in the Morgan & Claypool Publishers series (2010). ISBN 9781608453559
11. Newman, M.E.J.: Modularity and community structure in networks. Proc. Natl. Acad. Sci. **103**(23), 8577–8582 (2006)
12. Newman, M.E.J., Girvan, M.: Finding and evaluating community structure in networks. Phys. Rev. E **69**, 026113 (2004)
13. Girvan, M., Newman, M.E.J.: Community structure in social and biological networks. Proc. Natl. Acad. Sci. U.S.A. **99**, 7821–7826 (2002)
14. Wang, M., Wang, C., Yu, J.X., Zhang, J.: Community detection in social networks: an indepth benchmarking study with a procedure oriented framework. In: 41st International Conference on Very Large Data Bases, Proceedings of the VLDB Endowment, 31 August–4 September 2015, Kohala Coast, Hawaii, vol. 8, no. 10 (2015). Copyright 2015 VLDB Endowment 21508097/15/06
15. Plantié, M., Crampes, M.: Survey on social community detection. In: Social Media Retrieval, Computer Communications and Networks, pp. 65–85. Springer (2013). ISBN 978-1-4471-4554-7
16. Chen, M., Nguyen, T., Szymanski, B.K.: On measuring the quality of a network community structure. In: Proceedings of the IEEE Social Computing Conference, Washington DC, 8–14 September, pp. 122–127 (2013)
17. Zhao, P., Han, J., Sun, Y.: P-rank: a comprehensive structural similarity measure over information networks. In: Proceedings of the 18th ACM Conference on Information and Knowledge Management (CIKM), Hong Kong, China, pp. 553–562. ACM (2009)
18. Fortunato, S.: Phys. Rep. **486**, 75 (2010)
19. Eberle, W., Holder, L.: Discovering structural anomalies in graph-based data. In: ICDM Workshops, pp. 393–398 (2007)
20. Zachary, W.W.: An information how model for conflict and fission in small groups. J. Anthropol. Res. **33**, 452–473 (1977)

Analyzing Sentiment Level of Social Media Data Based on SVM and Naïve Bayes Algorithms

Hsu Wai Naing(✉), Phyu Thwe(✉), Aye Chan Mon(✉), and Naw Naw(✉)

Department of Information Science, University of Technology (Yatanarpon Cyber City), Pyin Oo Lwin, Myanmar
hsuwainaing2054@gmail.com, pthwe19@gmail.com, polestar.mon20@gmail.com, nawnaw1986@gmail.com

Abstract. Social media is a popular network through which users can share their reviews about various topics, news, products etc. People use internet to access or update reviews so it is necessary to express opinion. Twitter is a hugely valuable resource from which insights can be extracted by using text mining tools like sentiment analysis. Sentiment analysis is the task of identifying opinion from reviews. The system performs classification by combining Naïve Bayes (NB) and Support Vector Machine (SVM). The system is intended to measure the impact of ASEAN citizens' social media based on their usage behavior. The system is developed for analyzing National Educational Rate, Business Rate and Crime Rate occurred in Malaysia, Singapore, Vietnam and our country, Myanmar. The system compares the performance of these two classifiers in accuracy, precision and recall.

Keywords: Opinion mining · Sentiment analysis · Twitter
Support vector machine (SVM) · Naïve bayes (NB) · Text classification

1 Introduction

Most of the people are using social media applications such as Twitter, Facebook to express their emotions, opinions and share views as the daily activity of their life. Sentiment Analysis area becomes the innovative research area for the analysis of public opinions behind certain topics, feelings and attitudes of people. Sentiment Analysis is the task of identifying whether the opinion expressed in a document is positive or negative or neutral about a given topic.

Sentiment analysis is a task to recognize writers' feelings as expressed in positive, negative and neutral comments. Machine learning sentiment analysis usually comes under supervised classification and under text classification techniques in specific.

Twitter is a social networking website which allows users to publish short messages that are visible to other users. These messages are known as tweets and can only be 140 characters or less in length [1]. The system is developed to analyze Educational Rate, Business Rate and Crime Rate occurred in Malaysia, Singapore, Vietnam and our

T. T. Zin and J. C.-W. Lin (Eds.): ICBDL 2018, AISC 744, pp. 68–76, 2019.
https://doi.org/10.1007/978-981-13-0869-7_8

country, Myanmar through tweets. In daily life, many users share their opinions and experiences on social media. In this paper, at first, the system crawls the real time social media data from twitter. And then, the system uses Naïve Bayes and Support Vector Machine for the classifying tasks in Sentiment. After that, the system displays the sentiment scores by using visualization techniques. The performance of classification is also analyzed using precision, recall and accuracy.

This paper is organized as follow; second section gives related work; third section that describes preprocessing stage. Fifth section gives about the proposed feature extraction and classification processes followed by the sixth section that shows the experimental results of sentiment analysis about Education, Business and Crime.

2 Related Work

Rambocas [2] performs the sentiment analysis by using the keyword-based approach. He identified the words based on the adjectives. These adjectives that indicate the sentiment polarity scores. Other researchers use Wordnet to get the collection of adjectives with polarity scores.

Weibe [3] performed the sentiment analysis by using the document and sentence level classification. He used the raw dataset from different product destinations such as automobiles, banks, movies and travel. He classified two categories such as positive and negative by calculating the positive and negative polarity scores of words. He identified positive class is that the number of positive words is more than negative otherwise negative class.

3 System Design Overview

Firstly, the system fetches the real time social media data about education, business and crime from Twitter. And then, the system implements sentiment analysis on these collected data. The input of the system includes Education, Business and Crime tweets. The twitter data cannot classify directly because it consists of noisy information. So, this noisy information is removed by pre-processing. After that, the system uses Supervised Machine Learning Algorithm (Naïve Bayes and Support Vector Machine) that can achieve competitive accuracy when it is trained using feature. The main task of this system is to perform social media sentiment analysis by applying machine learning approach of Artificial Intelligence (AI). And then, this system can compare the rate of change of Crime Sector, Business and Education Sector occurred in Malaysia, Singapore, Vietnam and our country, Myanmar. In the system design, there are three main components. They are pre-processing, feature extraction and classification. Figure 1 illustrates the overall system design.

At first, the system crawls tweets about Education, Business and Crime from Twitter. The language is as English using Twitter Streaming API. The extracted twitter data is needed to preprocess. In the pre-processing stage, transformation, tokenization, filtering and normalization are performed. And then, the system extracts meaningful features by using Term Frequency-Inverse Document Frequency (TF-IDF). Feature

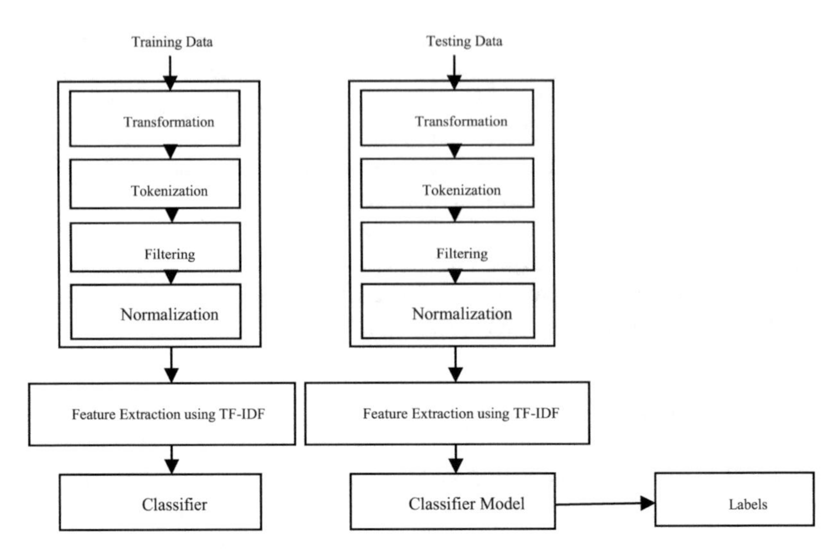

Fig. 1. The system design

extraction can make the classifier more effective by reducing the amount of data to be analyzed to identify the relevant features for further processing. After that, the system selects features as the input features of classifier (Naïve Bayes and Support Vector Machine). Finally, the system provides the percentage score of Education, Business and Crime sectors and displays according to their scores by using visualization techniques. The system also compared the performance of these two classifiers in accuracy, precision and recall.

4 Pre-preprocessing Stage

In preprocessing, the extracted data is cleaned and made ready for feeding it into the classifier. In this stage, there are four main processes:

4.1 Transformation

In Transformation stage, a clean tweet should not contain URLs, hashtags (i.e., #studying) or mentions (i.e. @Irene). The input tweets are transformed to lowercase. URLs are replaced with generic word URL. Then, @username is replaced with generic word URL. Then, @username is replaced with generic word AT_USER. Then, #hashtag is replaced with the exact same word without the hash. After that, all the punctuations are removed at the start and ending of the tweets. Additional whitespaces are replaced with a single whitespace. The next operation is to remove the vowels repeated in sequence at least three times [4]. Tweets about Education, Business and Crime from Myanmar, Singapore, Vietnam and Malaysia are preprocessed in the Transformation step.

4.2 Tokenization

Tokenization may be defined as the process of splitting the text into smaller parts called tokens, and is considered a crucial step in NLP [5]. The system tokenizes the uniformed sentence which got into smaller components (unigram).

4.3 Filtering

One of the major forms of pre-processing is to filter out useless data. In natural language processing, useless words (data), are referred to as stop words word (such as "the", "a", "an", "in"). Most search engines ignore these words because they are so common that including them would greatly increase the size of the index without improving precision or recall [6]. The system tokenizes the output tweets from Transformation step and then removes stopwords.

4.4 Normalization

In preprocessing step, Normalization process performs the important step. Normalization is a process that transforms a list of words to a more uniform sequence. By transforming the words to a standard format, the system leads to a more accurate classification. In the normalization step, lemmatization is applied. Lemmatization depends on correctly identifying the intended part of speech and meaning of a word in a sentence [7]. After the lemmatization, the root words are got and they are used for feature extraction step.

5 Feature Extraction

Transforming the input into the set of features is called Feature Extraction. If the extracted features are correctly chosen, it is expected that the features set will perform the desired task using the reduced representation instead of the full size input. In this system, Naïve Bayes Classifier and Support Vector Machine Classifier are trained on tf-idf weighted word frequency features. Term Frequency-Inverse Document Frequency (tf-idf) is a popular feature extraction method which reflects the relevance of a word in a particular document among the corpus. After Feature Extraction step with TF-IDF, the system selects features as the input features of classification. In this way, the system can get the essential features for the system and perform the best accuracy.

6 Classification

Sentiment Analysis is a current research area in text mining. It is the stem of natural language processing or machine learning methods. There are two sentiment analysis techniques such as unsupervised and supervised techniques. In unsupervised technique, classification is done by a function which compares the features of a given text against discriminatory-word lexicons whose polarity are determined prior to their use. In supervised technique, the main task is to build a classifier. The classifier needs training examples which can be labeled manually or obtained from a user-generated user-labeled online source. The system performs Sentiment Analysis by using supervised technique, Naïve Bayes Classifier and Support Vector Machine Classifier.

6.1 Naïve Bayes Classifier

Naïve Bays Classifier is one of Supervised Machine Learning Algorithm. Naïve Bayes Classifier can predict whether a new text message can be categorized as positive or negative or neutral. It is used to predict the probability for a given word to belong to a particular class. Pre-processed and Feature Extraction data is given as input to train input set using Naïve Bayes Classifier. That trained model is applied on test to generate positive or negative or neutral of Education, Business and Crime. First, Naïve Bayes Classifier computes the prior probability. Second, Naïve Bayes Classifier computes the conditional probability/Likelihood of each word attribute. Third, Naïve Bayes Classifier computes the posterior probability. Finally, Naïve Bayes Classifier determines the class.

6.2 Support Vector Machine Classifier

Support Vector Machine is supervised learning models with associated learning algorithm in machine learning. It analyzes data used for classification and regression analysis. Preprocessing and Feature Extraction is the input set to Support Vector Machine Classifier. SVM classifier generates positive, negative and neutral about Education, Business and Crime. Each feature about Education, Business and Crime is transformed into vector. SVM finds a linear hyperplane (decision boundary) that separates the features in such a way that the margin is maximized. And then, SVM classifier determines the (positive, negative or neutral). New testing data are mapped into the same space and predicted to belong to a category based on which side of the gap they fall.

6.3 Accuracy, Precision and Recall

Accuracy evaluates the performance of the system. The system performs accuracy, precision and recall of two classifiers. The system performs Precision as the exactness of two classifiers. The system performs Recall as the completeness of two classifiers [8].

The system computes the accuracy of Support Vector Machine Classifier and Naïve Bayes Classifier. It is calculated by number of correctly selected positive, negative and neutral words divided by total number of words present in the corpus. The system measures precision and recall of Support Vector Machine Classifier and Naïve Bayes Classifier by using NLTK metrics module.

7 Experimental Results

Sample tweet messages about education, business and crime are extracted from a particular Twitter account after getting prior permission. The extracted training dataset has 7540 tweets about education, 5414 tweets about crime and 8000 tweets about business. The extracted tweets are preprocessed such as transformation, tokenization, filtering, lemmatization. And then, the output words need to be meaningful features for the system (Figs. 2, 3 and 4).

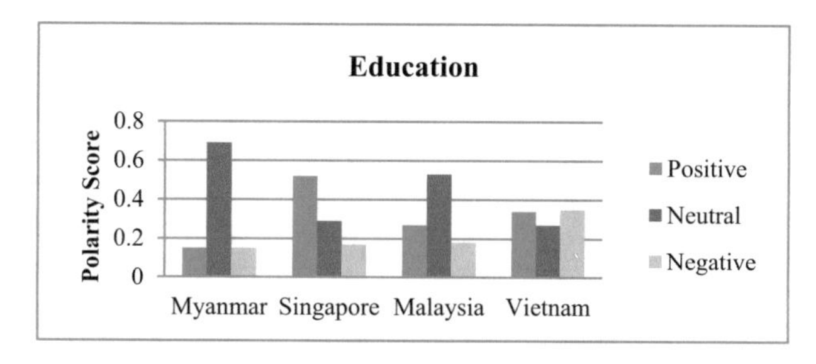

Fig. 2. Graphical analysis of education

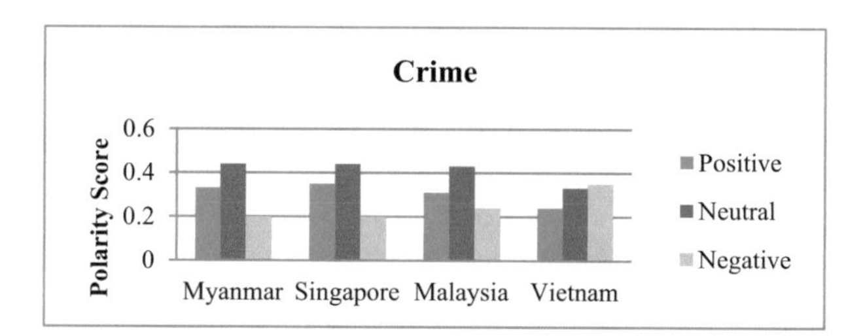

Fig. 3. Graphical analysis of crime

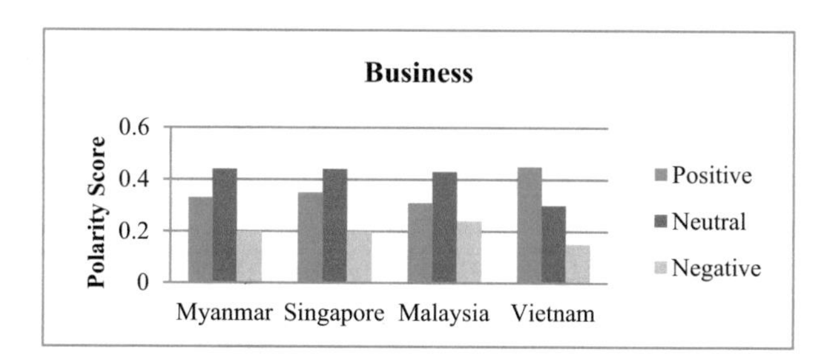

Fig. 4. Graphical analysis of business

The output feature words are input features of Naïve Bayes Classifier and Support Vector Machine Classifier. The system uses real time testing data on twitter. The testing datasets used in the application were retrieved from twitter using twitter4j API's (Table 1).

Table 1. Experimental results using support vector machine and Naïve Bayes

Location	Positive percentage	Negative percentage	Neutral percentage	Total percentage	Search key
Myanmar	15	15	69	99	Education
Malaysia	27	18	53	98	Education
Singapore	52	17	29	98	Education
Vietnam	34	35	27	96	Education
Myanmar	33	22	44	99	Crime
Malaysia	31	24	43	98	Crime
Singapore	35	20	44	99	Crime
Vietnam	24	35	33	92	Crime
Myanmar	45	35	10	90	Business
Malaysia	45	25	15	85	Business
Singapore	46	25	26	97	Business
Vietnam	45	15	30	90	Business

7.1 Performance Comparison

The system is intended to perform the accuracy comparison of Support Vector Machine Classifier and Naïve Bayes Classifier on the same training dataset. The system compares the precision and recall of these two classifiers on the same dataset.

Training Data 800 Testing Data 100	Support Vector Machine (SVM)	Naïve Bayes (NB)
Accuracy	0.91625	0.88125
Positive precision	0.967302452316	0.917098445596
Negative precision	0.97619047619	0.602564102564
Neutral precision	0.86189258312	0.904761904762
Positive recall	0.878712871287	0.876237623762
Negative recall	0.694915254237	0.796610169492
Neutral recall	1.0	0.902077151335

Training Data 700 Testing Data 100	Support Vector Machine (SVM)	Naïve Bayes (NB)
Accuracy	0.94	0.888571428571
Positive precision	0.993243243243	0.772093023256
Positive precision	0.993243243243	0.772093023256
Neutral precision	0.923232323232	0.978365384615
Positive recall	0.821229050279	0.927374301676
Negative recall	0.870967741935	0.790322580645
Neutral recall	0.995642701525	0.886710239651

Training Data 600 Testing Data 100	Support Vector Machine (SVM)	Naïve Bayes (NB)
Accuracy	0.951666666667	0.913333333333
Positive precision	0.983146067416	0.957575757576
Negative precision	0.991666666667	0.902985074627
Neutral precision	0.917218543946	0.893687707641
Positive recall	0.916230366492	0.82722513089
Negative recall	0.908396946565	0.923664122127
Neutral recall	0.996402877698	0.967625899281

The system performs the accuracy comparison of Support Vector Machine and Naïve Bayes Classifier on the same training dataset. Support Vector Machine classifier performs well in large dataset and gets the best accuracy. Naïve Bayes classifier performs well in small dataset and gets the best accuracy. Naïve Bayes classifiers is fast when decision making. But, Support Vector Machine classifier is slow when decision making.

8 Conclusion

This paper has presented Naïve Bayes and Support Vector Machine Classification on twitter to classify about Education, Business and Crime. The system is aimed to study machine learning model in the case of mining social media data for sentiment analysis. The system is developed for analyzing Educational Rate, Business Rate and Crime Rate occurred in Malaysia, Singapore and our country, Myanmar. The rate of change of these three sectors can be clearly compared by analyzing these conditions. The system is intended to contribute a lot of advantages for the Ministry of Education, Economists and Home Affairs in each country's government. The system can also analyze the accuracy, precision and recall of Support Vector Machine and Naïve Bayes Classifier. For Further Extension, the system can perform for Facebook and other social media application. The system can extend that input data is not only for text data but also for image data.

Acknowledgement. Firstly, I would like to appreciate Dr. SoeSoeKhaing, Pro-Rector, University of Technology (Yatanarpon Cyber City), for her vision, chosen, giving valuable advices and guidance for preparation of this article. And then, I wish to express my deepest gratitude to my teacher Dr. Hninn Aye Thant, Professor, Department of Information Science and Technology, University of Technology (Yatanarpon Cyber City), for her advice. I am also grateful to Dr. Yi Yi Myint, Assistant Lecturer, co-leader of our Research Development Team, University of Technology (Yatanarpon Cyber City), for giving me valuable advices. Last but not least, many thanks are extended to all persons who directly and indirectly contributed towards the success of this paper.

References

1. Twitter. http://www.businessdictionary.com/definition/Twitter.html
2. Rambocas, M., Gama, J.: Marketing Research: The Role of Sentiment Analysis, April 2013. ISSN 0870-8541
3. Weibe: Movie Review Dataset. http://www.cs.cornell.edu/people/pabo/movie-review-data, Accessed Oct 2013
4. Angiani, G., Ferrari, L., Paolo Fornacciari, T., Eleonoralotto, M.F., Manicard, S.: A Comparison between Preprocessing Techniques for Sentiment Analysis in Twitter (2017)
5. Tokenization. https://www.packpub.com/mapt/book/big_data_business_intelligence/01/20/ch01/vl1sec008/tokenization
6. Stop-Words-Collocation. https://streamhacker.com/2010/05/24/text-classification-sentiment-analysis-stopwords-collocation
7. Lemmatisation. https://en.m.wikipedia.org/wiki/Lemmatisation
8. Precision-Recall. https://streamhacker.com/2010/05/17/text-classification-sentiment-analysis-precision-recall/

Deep Learning and its Applications

Accuracy Improvement of Accelerometer-Based Location Estimation Using Neural Network

Noritaka Shigei[1(✉)], Hiroki Urakawa[1], Yoshihiro Nakamura[1], Masahiro Teramura[2], and Hiromi Miyajima[1]

[1] Kagoshima University, Kagoshima 890-0065, Japan
shigei@eee.kagoshima-u.ac.jp
[2] National Institute of Technology, Sasebo College, Sasebo 857-1193, Japan

Abstract. In this study, we propose a method for improving the accuracy of location estimation based on accelerometer and gyroscope sensors. The method utilizes a neural network for estimating the acceleration of the object. The effectiveness of the proposed method is demonstrated by experimental results.

Keywords: Location estimation · Accelerometer · Gyroscope Neural network

1 Introduction

Location estimation using accelerometer and gyroscope sensors has advantages of autonomous positioning and low cost [1]. On the other hand, its major drawback is of low accuracy of estimation resulting from disturbance and the instability of its installation state.

In this study, we propose a method for improving the accuracy of location estimation based on accelerometer and gyroscope sensors. The method utilizes a neural network for estimating the acceleration of the object. The effectiveness of the proposed method is demonstrated by experimental results.

2 Location Estimation Using Accelerometer and Gyroscope Sensors

2.1 Accelerometer and Gyroscope Sensors

The Adafruit 10-DOF IMU sensor module is equipped with an accelerometer sensor chip LSM303DLHC [2] and an gyroscope sensor chip L3GD20H [3]. The module detects the acceleration $(a_{\rm x}, a_{\rm y}, a_{\rm z})$ and the angular rate $(\omega_{\rm r}, \omega_{\rm p}, \omega_{\rm y})$ in

---

H. Miyajima—Former Kagoshima University.

T. T. Zin and J. C.-W. Lin (Eds.): ICBDL 2018, AISC 744, pp. 79–85, 2019.
https://doi.org/10.1007/978-981-13-0869-7_9

the directions of three axes as shown in Fig. 1. The sensing data contains high frequency noise components as shown in Fig. 2(a) High frequency noise can be removed by smoothing. Let $\boldsymbol{r}^{(p)} = (a_{\mathrm{x}}^{(p)}, a_{\mathrm{y}}^{(p)}, a_{\mathrm{z}}^{(p)}, \omega_{\mathrm{r}}^{(p)}, \omega_{\mathrm{p}}^{(p)}, \omega_{\mathrm{y}}^{(p)})$ be the p-th sensing data for $p \in \{1, 2, \cdots, P\}$. Then, the smoothed data $\tilde{\boldsymbol{r}}^{(p)}$ is calculated as follows:

$$\tilde{\boldsymbol{r}}^{(p)} = \frac{1}{W} \sum_{w=1}^{W} \boldsymbol{r}^{(p-W+w)}, \tag{1}$$

where W is the number of samples used for smoothing. Further, the sensing data is also biased. The bias can be removed by the calibration in which the mean value measured at the stationary state is subtracted from the sensing data.

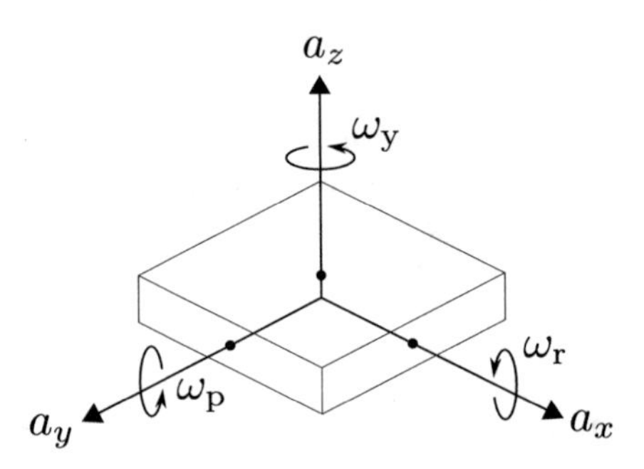

Fig. 1. Three axes of the accelerometer and gyroscope sensors.

2.2 Location Estimation Method

Assume that, (1) the sensor is attached to the mobile object whose location is estimated, (2) the acceleration and yaw angular velocity of the mobile object correspond to the sensing data $\boldsymbol{a}$ and ω_{y}. Let $\boldsymbol{a}_t$, $\boldsymbol{v}_t$ and $\boldsymbol{s}_t$ be the acceleration, the velocity and the displacement (from the initial position at time 0) at time t, respectively. Then, $\boldsymbol{v}_{t+1}$ and $\boldsymbol{s}_{t+1}$ are as follows:

$$\boldsymbol{v}_{t+1} = \boldsymbol{v}_t + \boldsymbol{a}_t \times \Delta t \tag{2}$$
$$\boldsymbol{s}_{t+1} = \boldsymbol{s}_t + \boldsymbol{v}_{t+1} \times \Delta t, \tag{3}$$

where $\boldsymbol{a}_t$, $\boldsymbol{v}_t$ and $\boldsymbol{s}_t$ are of the coordinates system on the mobile object and Δt is the sampling interval of sensing data.

The position of the mobile object at the time t, $\boldsymbol{S}_t$, is of the world (global) coordinates system as shown in Fig. 3. Let ϕ_t be the move direction of the mobile object against the X-axis of the world coordinates system. Then, the position of the mobile object at the time $t+1$, $\boldsymbol{S}_{t+1} = (X_{t+1}, Y_{t+1})$, is obtained as follows:

$$\phi_{t+1} = \phi_t + \omega_t \times \Delta t \tag{4}$$
$$X_{t+1} = X_t + |\boldsymbol{s}_{t+1}| \cos(\phi_{t+1} + \phi_0) \tag{5}$$
$$Y_{t+1} = Y_t + |\boldsymbol{s}_{t+1}| \sin(\phi_{t+1} + \phi_0) \tag{6}$$

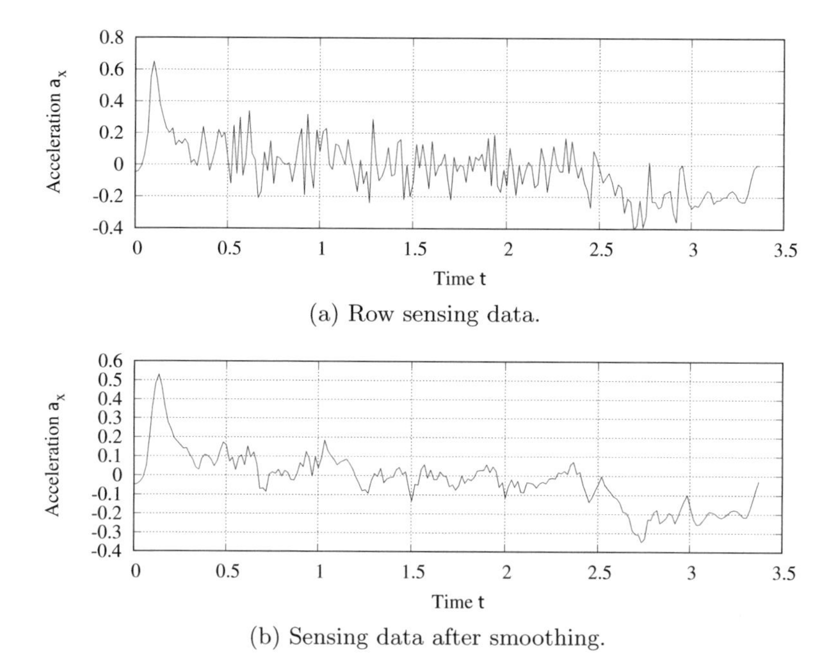

(a) Row sensing data.

(b) Sensing data after smoothing.

Fig. 2. Sensing data sample of a_{x}.

3 Location Estimation Method Using Neural Network

3.1 Neural Network Model

Only smoothing and calibration described in Subsect. 2.1 are insufficient for removing disturbances and biases introduced by circumstances and unsteady state that the mobile object is put in. In order to remove the factors degrading the estimation accuracy, we propose to use neural networks (NNs) to estimate the acceleration $\boldsymbol{a}_t$ in Eq. (2). Assume that the mobile object moves on the $X-Y$ plane. For this case, the acceleration $\boldsymbol{a}_t$ consists of x and y components. Each component of $\boldsymbol{a}_t$ is estimated by using a three-layered NN as shown in Fig. 4. The input of NN $\boldsymbol{r} = (r_1, r_2, \cdots, r_6)$ is the raw sensing data $\boldsymbol{r}^{(p)}$ or the smoothed data $\tilde{\boldsymbol{r}}^{(p)}$. The output of each inner unit h_i and the output of NN a are calculated as follows:

$$h_i = \frac{1 - \exp(-s_i)}{1 + \exp(-s_i)}, \quad s_i = \sum_{j=1}^{6} w_{1ij} r_j + \theta_i, \tag{7}$$

$$a = \frac{1 - \exp(-u)}{1 + \exp(-u)}, \quad u = \sum_{j=1}^{N} w_{2j} h_j + \theta. \tag{8}$$

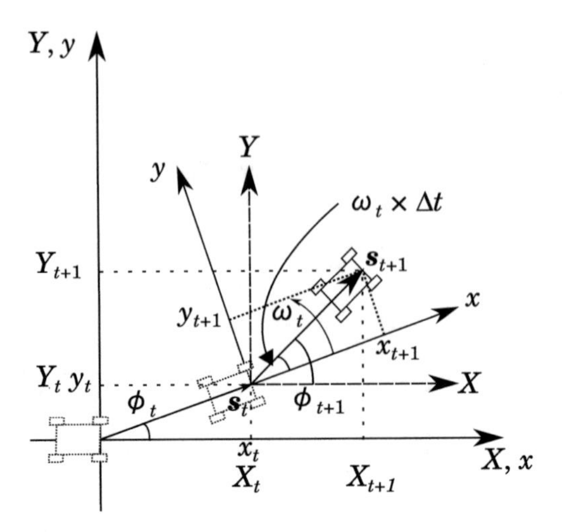

Fig. 3. The used coordinates system.

3.2 Training of NN

The NN described in the previous subsection is trained by using the training data. The training data consists of the sensing data of sensors and the acceleration data. The acceleration data is the reference signal (teacher) and is prepared by using other measurement system. In this study, the acceleration data is extracted from moving images as described later. Given the training data set $\{(\boldsymbol{r}^{(p)}, a^{(p)})|p \in \{1,2,\cdots,P\}\}$, then for every $p \in \{1,2,\cdots,P\}$, the training updates the parameters w_{1ij}, w_{2i}, θ_i and θ so as to minimize the following error E.

$$E = \frac{1}{2}\left(a(\boldsymbol{r}^{(p)}) - a^{(p)}\right)^2, \tag{9}$$

where $\boldsymbol{r}^{(p)}$ is the input data, $a^{(p)}$ is the reference signal and $a(\boldsymbol{r}^{(p)})$ is the output of NN for the input $\boldsymbol{r}^{(p)}$. Each parameter $b \in \{w_{1ij}, w_{2i}, \theta_i, \theta | i \in \{1,2,\cdots,N\}$, $j \in \{1,2,\cdots,6\}\}$ is updated as follows:

$$b \leftarrow b - \alpha\frac{\partial E}{\partial b}, \tag{10}$$

where α is the learning rate. The learning algorithm used in this study repeats the following procedure $T_{\max}$ times: (1) select randomly data number $p \in \{1,2,\cdots,P\}$ and (2) update the parameters w_{1ij}, w_{2i}, θ_i and θ by using Eq. (10).

3.3 Reference Signal Preparation

In this study, the reference signal $a^{(p)}$ is prepared from moving images. In the preparation, firstly, the position of the mobile object in k-th frame, (x', y'),

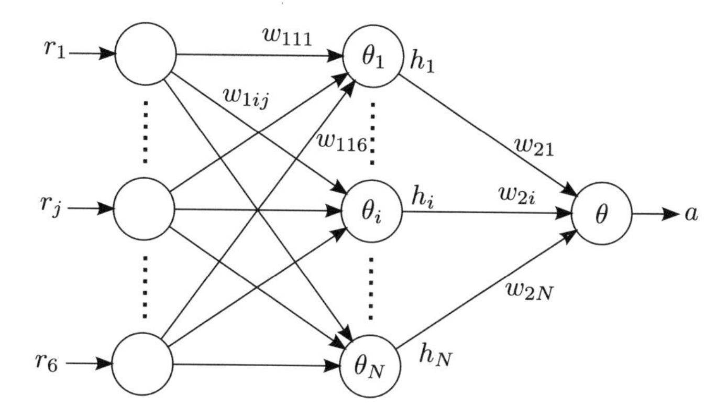

Fig. 4. Three layered NN for each component of $\boldsymbol{a}$.

is converted into the position on the $x - y$ plain, (x, y) by the Homography transformation as follows:

$$x = \frac{M_{11}x' + M_{12}y' + M_{13}}{M_{31}x' + M_{32}y' + M_{33}} \tag{11}$$

$$y = \frac{M_{21}x' + M_{22}y' + M_{23}}{M_{31}x' + M_{32}y' + M_{33}} \tag{12}$$

where the transformation matrix $M = (M_{ij})$ is calculated by solving the simultaneous equations obtained from the following equation with four coordinates pairs of (x', y') and (x, y).

$$\begin{bmatrix} x \\ y \\ 1 \end{bmatrix} = \begin{bmatrix} M_{11} & M_{12} & M_{13} \\ M_{21} & M_{22} & M_{23} \\ M_{31} & M_{32} & M_{33} \end{bmatrix} \begin{bmatrix} x' \\ y' \\ 1 \end{bmatrix}. \tag{13}$$

Then, from the obtained displacement $\boldsymbol{s}_t = (x_t, y_t)$, the reference signal $\boldsymbol{a}_t$ is obtained as follows:

$$\mathbf{v}_t = \frac{\boldsymbol{s}_{t+1} - \boldsymbol{s}_t}{\Delta t} \tag{14}$$

$$\mathbf{a}_t = \frac{\boldsymbol{v}_{t+1} - \boldsymbol{v}_t}{\Delta t}. \tag{15}$$

4 Experimental Evaluation

The evaluation is performed by leave-one-out cross-validation (LOOCV) with 18 data sets of sensing and reference data collected by running a mobile object on a straight course of 3 m 18 times. The experiment conditions are the number of hidden units of NN $N = 2$, the maximum learning iteration $T_{\max} = 500,000$, the learning rate $\alpha = 0.01$ and the sampling interval $\Delta t = 8$ ms or 9 ms.

Table 1 shows the error rate of estimated distance for the different number of samples used for smoothing W. "Conv." is the conventional method that uses the

Table 1. The error rate of estimated distance (%) for the different number of samples used for smoothing W.

W	1	2	3	4	5	6	7	8	9	10
Conv.	35.69	35.22	35.05	35.10	35.01	35.01	34.97	34.95	34.95	35.00
NN	25.38	19.96	14.70	15.20	12.57	12.13	13.26	14.12	14.07	15.10

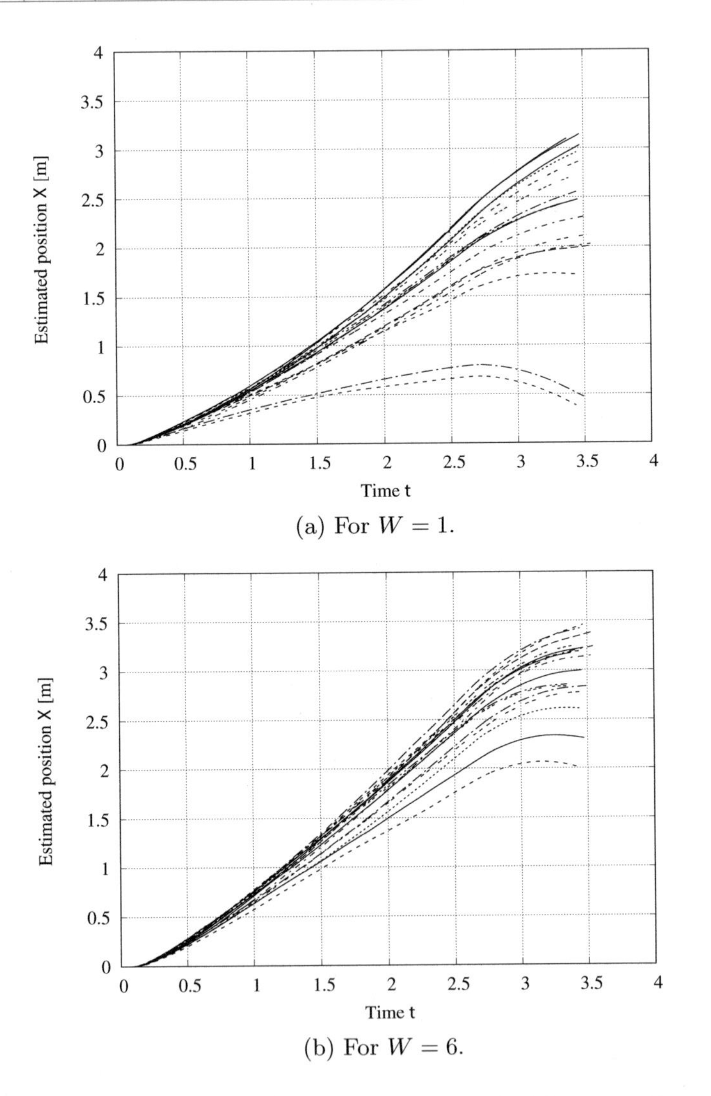

(a) For $W = 1$.

(b) For $W = 6$.

Fig. 5. The position X estimated by the proposed method for every 18 test cases.

smoothed and calibrated sensing data $\tilde{\boldsymbol{r}}^{(p)}$, and "NN" is the proposed method described in Sect. 3. According to the result, the conventional method and the proposed NN's one achieve the best accuracy for $W = 8{\sim}9$ and $W = 6$, respectively. For the case of no smoothing, i.e. $W = 1$, the proposed method achieves approximately 10.3% smaller error than the conventional one. Further, for the case of the best smoothing case, the proposed method achieves approximately 22.8% smaller error than the conventional one.

Figure 5 shows the position estimated by the proposed method for every 18 test cases. Ideally, the estimated position is 3 m at the end of time. From the figure, the cases of $W = 6$ are much closer to the ideal position than the ones of $W = 1$.

5 Conclusion

In this study, we proposed to use neural networks to improve the accuracy of location estimation based on accelerometer and gyroscope sensors. According to the experimental result, the proposed method reduced the estimated error of 34.97% to 12.13%.

Acknowledgment. This work was supported by JSPS KAKENHI Grant Number 26330108.

References

1. Hsu, C.H., Yu, C.H.: An accelerometer based approach for indoor localization, Symposia and workshops on ubiquitous, automatic and trusted computing, pp. 223–227 (2009)
2. STMicroelectronics, LSM303DLHC datasheet (2011)
3. STMicroelectronics, L3GD20H datasheet (2012)

Transparent Object Detection Using Convolutional Neural Network

May Phyo Khaing[1,2] and Mukunoki Masayuki[2(✉)]

[1] University of Technology (Yatanarpon Cyber City), Yatanarpon Cyber City, Myanmar
mayyphyokhaing@gmail.com

[2] Graduate School of Engineering, University of Miyazaki, Miyazaki, Japan
mukunoki@cs.miyazaki-u.ac.jp

Abstract. The detection of transparent object such as glass in the image is recently popular in computer vision researches. Among the various tasks of detecting objects in images, it is not an easy task to detect the presence of transparent objects in the image. The detection of transparent objects is very difficult to perform using classical computer vision algorithms since the appearance of transparent objects dramatically depends on its background and illumination conditions. In addition to the popularity of transparent object detection, deep learning is also giving high performance in object detection tasks. In this paper, we apply one of the Convolutional Neural Network called Single Shot MultiBox Detector (SSD) for transparent object detection task and evaluate the performance of the system. The results show that the application of deep learning method in detection of transparent objects can successfully perform the detection of transparent objects in images.

Keywords: Transparent object detection · Deep learning
Convolutional neural network

1 Introduction

Nowadays, the detection of different kinds of objects is increasingly challenging the computer vision researchers. Among these challenges, the detection of transparent objects has become a considerable problem in object detection task. Transparent objects are very widely used in our daily life and are existing in our domestic environment along with other objects. In contrast to the detection of other non-transparent (opaque) objects, transparent objects are hard to detect by regular image segmentation methods because these objects usually take the texture from their background and their appearances are similar to their surroundings.

Previously, the detection of transparent objects is performed by the classification between the glass-cover regions and other regions based on certain features of the transparent objects. Therefore, various features of the transparent object are needed to consider for each detection which leads to a slow detection process. In this research, transparent object detection is performed by taking the advantages of deep learning method, called Single Shot MultiBox Detector (SSD), which gives faster and accurate detection results.

T. T. Zin and J. C.-W. Lin (Eds.): ICBDL 2018, AISC 744, pp. 86–93, 2019.
https://doi.org/10.1007/978-981-13-0869-7_10

As the outline of the paper, Sect. 2 presents the previous works that are related to the research. Section 3 describes the method used in the detection of the transparent objects in images. In Sect. 4, the experimental results are included. Finally, Sect. 5 concludes the research of the detection of transparent objects using convolutional neural network and the future work of the system.

2 Related Works

The detection of transparent objects had been a difficult work because of their lack of own appearance. The research on detection of transparent objects has been increasingly focused along with the development of intelligent domestic service robotics and image search. In the domestic scenes, transparent objects are located among the other objects. For the detection of these transparent objects, Osadchy et al. [1] applied the specular highlights feature which makes glass objects different from the others. However, there was a requirement to have a light source. McHenry et al. [2] considered a number of features such as color similarity, blurring, overlay consistency and texture distortion in addition to highlights for transparent object detection purpose.

Fritz et al. [3] use an additive model of latent factors, method of a combination of SIFT and Latent Dirichlet Allocation (LDA) on a dataset of 4 transparent objects to generate transparent local patch appearance. The algorithm provides a useful result in the detection of transparent objects in different backgrounds.

Both of the detection and pose estimation of transparent objects have been proposed by Phillips et al. [4] and Lysenkov et al. [5] with the use of laser range finders and stereo and Kinect depth sensor, respectively. [4] use inverse perspective mapping with the assumption of two views of a test scene and that objects stay on a support plane. In [5], the fact that the Kinect sensor fails to estimate depth on specular or transparent surfaces is used to segment transparent objects from the images. And then, they perform 6 degree of freedom (6DOF) post estimation and recognition of transparent objects. However, both of these approaches cannot handle overlapping transparent objects. So, Lysenkov et al. propose [6] as an improvement to deal with the overlapped transparent objects.

As an interesting method of the segmentation of transparent object from a light-field image, Xu et al. [7] propose TransCut method using light field linearity, occlusion detector and graph-cut for pixel labeling. Unlike conventional methods which usually rely on the color similarity and highlight information, [7] use the overlay consistency and texture distortion properties for the segmentation of transparent object region in a light-field image.

In recent years, the traditional object recognition tasks have been shifted to the deep learning object recognition tasks. Along with the powerful and efficient results, deep neural network is also applied to recognize transparent objects. Lai et al. [8] use Region with Convolutional Neural Network (R-CNN) to recognize the transparent object in color image. R-CNN technique uses selective search [9] to extract the interested region proposals [10], and the efficiency of the selective search algorithm is improved in [8] by considering the highlight and color similarity features of the transparent objects in order to remove some region proposals that are not transparent. As an interesting application of later deep neural network, we use Single Shot Multibox Detector (SSD) [11] to detect transparent objects in images.

3 Methodology

For the detection of transparent object, the dataset is taken from the ImageNet ILSVRC dataset. The dataset provides some classes such as beaker, water glass, beer glass and wine glass. For each class, the images and annotation files are given. The annotation files are the files which are related to each image and describe the location of objects in the image along with their labels. For some class such as water glass, the annotation files are not readily given and we create them manually for using in training the neural network. An example of annotated image is shown in Fig. 1.

Fig. 1. Image with annotated bounding box

As a brief introduction of the SSD [11], it eliminates the proposal extraction stage and feature resampling stage from the architecture and makes predictions for detection of the objects based on features maps produced at different stages of the convolutional neural network. In other words, the features of the detect objects are extracted at different scales or resolutions of the feature maps and the predictions of the object label and bounding boxes are performed in a single pass, forming as a single shot multiscale or multibox detector. Since it also takes the feature maps from the layers which are closer to the original image, SSD can even perform the detection of objects in low resolution images. Therefore, SSD gives the accurate detection results on objects of different scales and different sizes, and the faster detection by single pass detection. The architecture of the SSD from [11] is given in the Fig. 2.

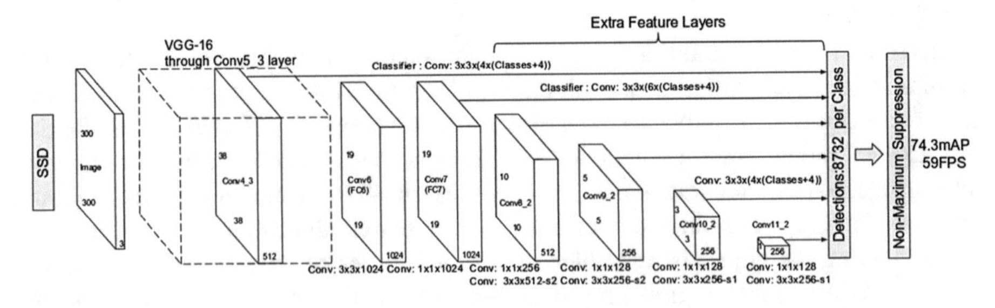

Fig. 2. The architecture of Single Shot Multibox Detector [11]

The convolutional neural network is trained with the images and bounding box annotation files of the transparent object classes. Using the trained network, transparent objects are detected from the images. In the detection output of SSD, the position of the detected objects is shown with the bounding boxes and, for each bounding box, the class label and the score for the class are described. In this system, we define the class label 1 for the transparent objects. Since the score of the detection is the probability of how much the detected region has the same features as the transparent objects, it is described by the value between 0 and 1. Some of the detection results of transparent objects in images are shown in Fig. 3.

Fig. 3. The detection results of transparent objects in images where class 1 represents the label of glass object

Basically, SSD can perform the detection of all transparent objects accurately. But, one problem found in SSD is that if the detection is performed on the non-transparent objects of the same shape as the transparent objects, SSD also classifies the non-transparent objects as the transparent objects. Although the detection of transparent objects using SSD gives very accurate results as shown in Fig. 3, some false detections are appeared in the results when non-transparent objects of similar shapes are detected as shown in Fig. 4.

Fig. 4. The false detection results of non-transparent objects which have the same shape of the transparent objects

The problem does not appear when the network is tested on other kinds of objects such as clock and this is because the features between the transparent objects and the clock are very distinct. In other words, some kinds of objects have their features which are very distinct from the others. In the case of transparent objects, their appearances are very simple and also they usually have no their own color. To solve the problem, the neural network needs to be trained to learn the distinct features between the transparent and non-transparent objects. Therefore, the neural network is trained again with a dataset of negative training images so that it can learn the distinct features between transparent objects and non-transparent objects of the same shape.

4 Experimental Results

We prepared a total of 2,700 images and their annotation files for training the convolutional neural network. The set contains 1,800 images of transparent objects and 900 images of non-transparent objects. Internally, 400 images for each class of beaker, beer glass and water glass and, 600 images of wine glass are contained as the subclasses of transparent objects. And then, 500 images of paper cup and, 200 images for each class of coffee cup and coffee mug are used as the subclasses of non-transparent objects. All of the training data are used from the ImageNet ILSVRC dataset.

After training the network with both of the positive and negative training images, the network can perform the correct detection between transparent objects and non-transparent objects. Here, the class label 1 is used for transparent objects and the class label 2 is used for non-transparent objects. The final detection results are shown in Fig. 5.

Fig. 5. The detection results of transparent and non-transparent objects where class 1 represents transparent object and class 2 represents non-transparent object

For performance evaluation of the detection results, a set of 300 testing images which contains 200 images of transparent object and 100 images of non-transparent object are used. For each of these test images, the ground-truth bounding boxes, in other words, the true locations of the objects are created to compared with the bounding boxes predicted by the trained network. An example of ground-truth bounding box and predicted bounding box is shown in Fig. 6.

Fig. 6. Ground-truth bounding box and predicted bounding box over the detected object

The number of ground-truth bounding boxes in each class of the testing images are described in the following Table 1.

Table 1. The number of ground-truth bounding boxes in each class of testing images.

		Num. of ground-truth bounding boxes
Transparent objects	Beaker	121
	Beer glass	92
	Water glass	104
	Wine glass	88
Non-transparent objects	Paper cup	92
	Coffee cup	64
	Coffee mug	82

Using the ground-truth bounding boxes and the predicted bounding boxes, we evaluate how precisely the network can detect the transparent objects in images. Intersection over Union (IoU), also called Jaccard Overlap index, is calculated for each detection.

$$\text{IoU} = \frac{A \cap B\,(\mathit{Area\ of\ Overlap})}{A \cup B\,(\mathit{Area\ of\ Union})} \tag{1}$$

In the above equation, A represents the area of the ground-truth box and B represents the area of the bounding box predicted by the network. Then, the IoU of each detection is calculated by dividing the overlap area of A and B by the union of the area of A and B. The higher IoU value means the more accurate the network can detect the object regions and the lower IoU value means the poor detection. Therefore, the IoU threshold value of 0.7 is used in this experiment for more precise detection of the transparent objects. And then, the overall precision and recall rate of the system are calculated on all of the test images.

Figure 7 compares the precision and recall of the detection results of the two training processes: the network trained only with the transparent object images and the network trained with both transparent and non-transparent object images. The precision and recall are calculated at different scores of the detection results. The precision rate of the system is the ratio of the number of true detections and the total number of detections. The recall rate of the system is the ratio of the number of true detections and the total number of ground-truth detections. In Fig. 7, Result 1 curve is the precision and recall of the network which is trained only with the transparent objects and Result 2 curve is the precision and recall of the network which is trained with both transparent and non-transparent objects. Changing the threshold for the output score of SSD from 0 to 1 by 0.1, we draw the curves. According to the experiment, the network trained with both transparent and non-transparent objects achieves higher performance result than the network trained with only transparent object classes. Considerably, the research achieves the mean average precision of 85.1% in the detection of transparent objects in Result 2 curve.

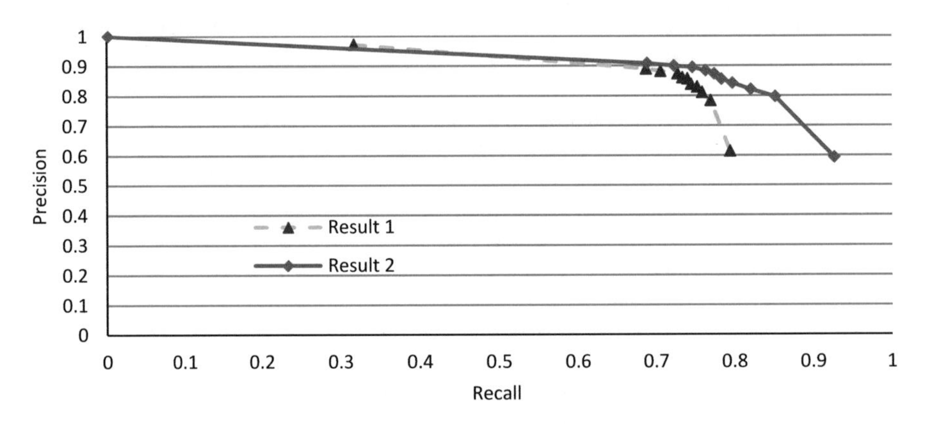

Fig. 7. Precision and recall curve of the detection of transparent objects evaluated over a set of 300 images containing both transparent and non-transparent objects

5 Conclusion

In the proposed system, the detection of transparent objects achieves accurate and considerable results. The problem of the network when testing on the non-transparent objects with the same shape of the transparent objects is introduced and the solution has been described. To sum up, the system has been shown that the detection of transparent objects is feasible without considering the special properties of the transparent objects.

Currently, the false detection has been solved by training the network with the negative examples of non-transparent objects of the same shapes. Instead of training the network for each same shape of the transparent and non-transparent objects, the future work will be to train the network to learn the distinguishable features between transparent objects and non-transparent objects of the same shape.

References

1. Osadchy, M., Jacobs, D., Ramamoorthi, R.: Using specularities for recognition. In: IEEE International Conference on Computer Vision, pp. 1512–1519 (2003)
2. McHenry, K., Ponce, J., Forsyth, D.: Finding glass. In: IEEE Computer Society Conference on Computer Vision and Pattern Recognition, pp. 973–979 (2005)
3. Fritz, M., Bardski, G., Karayev, S., Darrell, T., Black, M.: An additive latent feature model for transparent object recognition. In: Neural Information Processing Systems, pp. 558–566 (2009)
4. Phillips, C.J., Derpanis, K.G., Daniilidis, K.: A novel stereoscopic cue for figure-ground segregation of semi-transparent objects. In: IEEE International Conference on Computer Vision, pp. 1100–1107 (2011)
5. Lysenkov, I., Eruhimov, V., Bardski, G.: Recognition and pose estimation of rigid transparent objects with a kinect sensor. In: Robotics, Science and Systems, p. 273 (2013)
6. Lysenkov, I., Rabaud, V.: Pose estimation of rigid transparent objects in transparent clutter. In: IEEE International Conference on Robotics and Automation, pp. 162–169 (2013)
7. Xu, Y., Nagahara, H., Shimada, A., Taniguchi, R.: TransCut: transparent object segmentation from a light-field image. In: IEEE International Conference on Computer Vision, pp. 3442–3450 (2015)
8. Lai, P.J., Fuh, C.S.: Transparent object detection using regions with convolutional neural network. In: IPPR Conference on Computer Vision, Graphics, and Image Processing, pp. 1–8 (2015)
9. Uijlings, J.R., van de Sande, K.E., Gevers, T., Smeulders, A.W.: Selective search for object recognition. Int. J. Comput. Vis. **104**, 154–171 (2013)
10. Girshick, R., Donahue, J., Darrell, T., Malik, J.: Rich feature hierarchies for accurate object detection and semantic segmentation. In: IEEE Conference on Computer Vision and Pattern Recognition, pp. 580–587 (2014)
11. Liu, W., et al.: SSD: single shot multibox detector. In: Leibe, B., Sebe, N., Welling, M. (eds.) European Conference on Computer Vision 2016. LNCS, vol. 9905, pp. 21–37. Springer, Cham (2016)

Multi-label Land Cover Indices Classification of Satellite Images Using Deep Learning

Su Wit Yi Aung(✉), Soe Soe Khaing, and Shwe Thinzar Aung

Faculty of Information and Communication Technology,
University of Technology (Yatanarpon Cyber City), Pyin Oo Lwin, Myanmar
suwityiaung123@gmail.com, khaingss@gmail.com,
shwethinzaraung@gmail.com

Abstract. Accurate classification of land cover indices is important for diverse disciplines (e.g., ecology, geography, and climatology) because it serves as a basis for various real world applications. For detection and classification of land cover, remote sensing has long been used as an excellent source of data for finding different types of data attribute present in the land cover. A variety of feature extraction and classification methods in machine learning have been used to classify land cover using satellite images. In recent years, deep learning have recently emerged as a dominant paradigm for machine learning in a variety of domains. The objective of this paper presents the multi-labeled land cover indices classification using Google Earth Satellite images with deep convolutional neural network (DCNN). Since the lack of massive labeled land cover dataset, the own created labeled dataset for Ayeyarwaddy Delta is applied and tested with AlexNet. Then the results of land cover classification are compared with Multiclass-SVM using confusion matrices. According to the tested results, 76.6% of building index, 81.5% road index, 91.8% of vegetation index and 93.2% of water index can be correctly classified by using DCNN. The confusion matrix for Multiclass-SVM, 78.9% of building index, 72.7% road index, 94.2% of vegetation index and 98.1% of water index can be correctly classified.

Keywords: Multi-labeled land cover indices · DCNN · Multiclass-SVM

1 Introduction

Cyclone Nargis strike Myanmar during early May 2008. It caused the worst natural disaster in the recorded history of Myanmar. The cyclone made landfall in Myanmar on Friday, May 2, 2008, sending a storm surge 40 km up the densely populated Ayeyawaddy delta, causing catastrophic destruction and at least 138,000 fatalities. Thousands of buildings were destroyed in the Ayeyarwaddy Delta, state television reported that 75% of buildings had collapsed and 20% had their roofs ripped off. One report indicated that 95% of buildings in the Ayeyarwaddy delta area were destroyed. For this reason, it is needed to know land cover changes before and after Nargis Cyclone for regional planning, policy planning and understanding the impacts of disaster. Information about

T. T. Zin and J. C.-W. Lin (Eds.): ICBDL 2018, AISC 744, pp. 94–103, 2019.
https://doi.org/10.1007/978-981-13-0869-7_11

destruction during Nargis Cyclone and reconstruction after Nargis and land cover changes due to disaster will also be needed.

The main focus of this proposed work is to classify land cover and land use area of the Ayeyarwaddy delta using remote sensing satellite images. In the past few decades, the researchers applied remote sensing satellite images as the significant data types for classification of land use and land cover changes. The information and data obtained from land use and land cover classification have been applied in most of the applications and researches for finding the change detection of the study areas. Land use and land cover classification is carried out especially in urban areas and the obtained information can be applied in both qualitative and quantitative analysis of the findings in the study areas [1].

The physical material on the surface of the earth such as water, vegetation and building can be regarded as land cover. Therefore, for describing the data and information on the Earth surface, land cover is regarded as a fundamental parameter. Land use can be defined as the usage of land cover areas for various purposes such as urbanization, conservation or farming. Therefore, the exact land use and land cover information has been applied in urban planning, flood prediction and disaster management. Remotely sensed satellites captured the various images for the same scene with daily or weekly revisit and multiplicity of spectral bands. Such images have been used for urban land cover classification [2, 3], urban planning [4], soil test, and to study forest dynamics [5].

In this paper, four land cover indices: building, road/land, vegetation/forest and water are classified from Google earth images using deep convolutional neural network with AlexNet. The paper is organized as follow: Sects. 2 and 3 present related works and datasets and methodology. In Sect. 4 discussions is described. Finally, the conclusion is given in Sect. 5.

2 Related Works

In the past few years, there have been significant advances in remote sensing and high-resolution image processing, and a variety of Land Use and Land Cover (LULC) classification algorithms have been developed in the recent past. For land cover classification, various machine learning algorithms have been used. As an advanced machine learning approach, deep learning has been successfully applied in the field of image recognition and classification in recent years [6–10]. By mimicking the hicrarchical structure of the human brain, deep learning approaches, such as Deep Belief Networks (DBN), can exploit complex spatiotemporal statistical patterns implied in the studied data [11, 12]. For remotely sensed data, deep learning approaches can automatically extract more abstract, invariant features, thereby facilitating land cover mapping. In [13], Deep Belief Networks is applied for urban land cover classification using polarimetric synthetic aperture radar (PolSAR) data. Support vector machine (SVM), conventional neural networks (NN), and stochastic Expectation-Maximization (SEM) are then used to compare the results of DBN based classification approach. According to the experimental results, the DBN-based method outperforms three other approaches.

Two recently different deep learning architectures, CaffeNet and GoogLeNet are applied in [14] for land use classification in remote sensing images with two standard datasets. The well-known UC Merced Land Use dataset and Brazilian Coffee Scenes dataset with markedly different characteristics are used to classify land cover with deep learning. Besides conventional training from scratch, pre-trained networks that are only fine-tuned on the target data are used to avoid overfitting problems and reduce design time. Romero et al. [15] very recently proposed an unsupervised deep feature extraction for remote sensing image classification. The authors suggested the use of greedy layer-wise unsupervised pre-training coupled with an algorithm for unsupervised learning of sparse features. The proposed algorithm is tested on classification of aerial scenes, as well as land-use classification in very high resolution (VHR), or land-cover classification from multi- and hyper-spectral images.

The two combining techniques with DCNN: transfer learning (TL) with fine-tuning and data augmentation are employed in [16] for land cover classification of high-resolution imagery of UC Merced dataset. The CaffeNet, GoogLeNet, and ResNet architectures are used with these techniques to classify the well-known UC Merced data set to achieve the land–cover classification accuracies of $97.8 \pm 2.3\%$, $97.6 \pm 2.6\%$, and $98.5 \pm 1.4\%$.

3 Datasets and Methodology

The aforementioned related works applied different deep learning architecture with standard well-known labeled remote sensing datasets since the lack of massive labeled dataset for remote sensing. In this proposed system, the own dataset is created using the satellite images of Ayeyarwaddy Delta from Google Earth acquired between 2002 and 2017. Then, the acquired satellite images are labeled into four categories of land cover for classification with deep learning. The sample of input satellite images of Ayeyarwaddy Delta is shown in Fig. 1.

Fig. 1. Sample images of Ayeyarwaddy Delta dataset

The four land cover classes of the Ayeyarwaddy Delta dataset are as follow: building, road/land, vegetation/forest and water. Since the one scene of input contains multiple class of land cover, it is needed to consider the classification of multi-label classification framework. For multi-label classification framework, the two options are considered and compared as follow:

1. Feature extraction, training and classification using deep learning (AlexNet)
2. Feature extraction with deep learning and training and classification using Multi-class-SVM

The multi-label classification framework proposed by the system is shown in the following pseudo code:

```
1: SGRow ← Row/21
2: SGCol ← Col/21
3: Step ←SGRow/2
4: for i ← Step:Step:SGRow-Step
5:  for j ← Step:Step:SGCol-Step
6:         Tempfm = Img(i-Step:i+Step,j-Step:j+Step)
7:         Label = classifynet(Net,Tempfm)
8:  switch Label
9:  case Building
10:   Index(i-Step/2:i+Step/2,j-Step/2:(j+Step/2)←Blue
11: case Road
12:   Index(i-Step/2:(i+Step/2,j-Step/2:(j+Step/2)←Yellow
13: case Vegetation
14:   Index(i-Step/2:(i+Step/2,j-Step/2:(j+Step/2)← Green
15: case Water
16:   Index(i-Step/2:(i+Step/2,j-Step/2:(j+Step/2)← Black
17:   end
18: end
19: end
```

3.1 Multi-label Land Cover Classification with Deep Learning (AlexNet)

Firstly, the input satellite images are segmented into 25 × 25 pixel block for multi-label classification since single label category of image can be classified using AlexNet [17]. AlexNet Deep Convolutional Neural Network consists of two phases: feature extraction and classification each with various layers. Firstly, the segmented images are passed into the feature detection layer of deep convolutional neural network to detect different features of the input image. Feature Extraction Layer of AlexNet uses multiple layers of convolution, activation, max pooling, and normalization in various sequences. After features extraction, the images are classified in the fully connected layers of classification stage. The classification phase uses the extracted visual features and passes it into 4096 × 4096 × N multilayer perceptron (MLP) neural network. This phase utilizes rectified linear units [18] to provide a nonsaturating activation function defined as f (x) = max (0, x). The two 4096-D MLP layers are also configured to use 50% dropout to reduce overfitting [17]. The recently-introduced technique, called "dropout" [19], is used to set zero to the output of each hidden neuron with probability 0.5. The flow diagram of this approach is shown in Fig. 2.

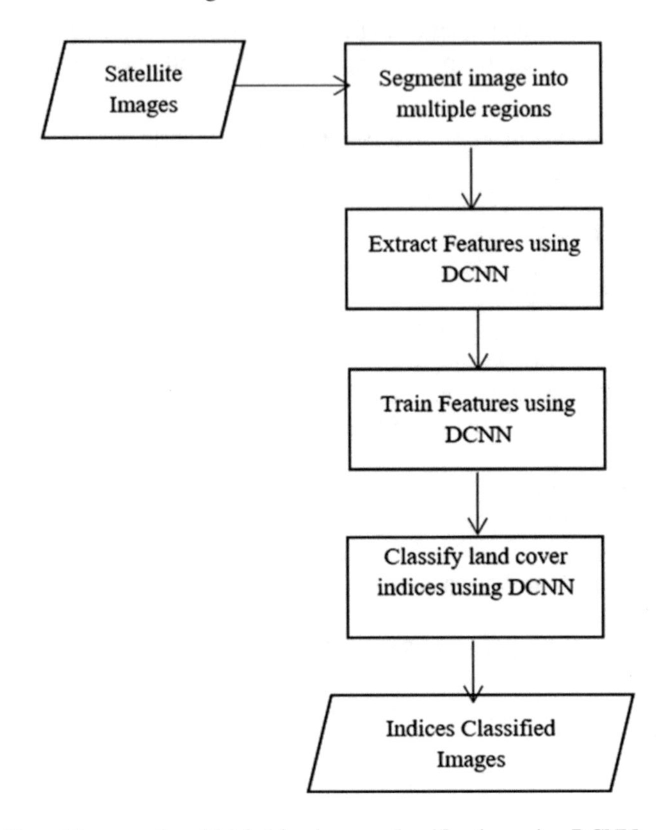

Fig. 2. Flow Diagram of multi-label land cover classification using DCNN as classifier

3.2 Multi-label Land Cover Classification with Multiclass-SVM

For land cover classification with multiclass-SVM, the input satellite images are segmented into 25 × 25 pixel block for multi-label classification. Then, the features are extracted from the segmented images using DCNN. After DCNN is used to extract features, then these features are trained and classified using Multiclass-SVM. The flow diagram of this approach is shown in Fig. 3.

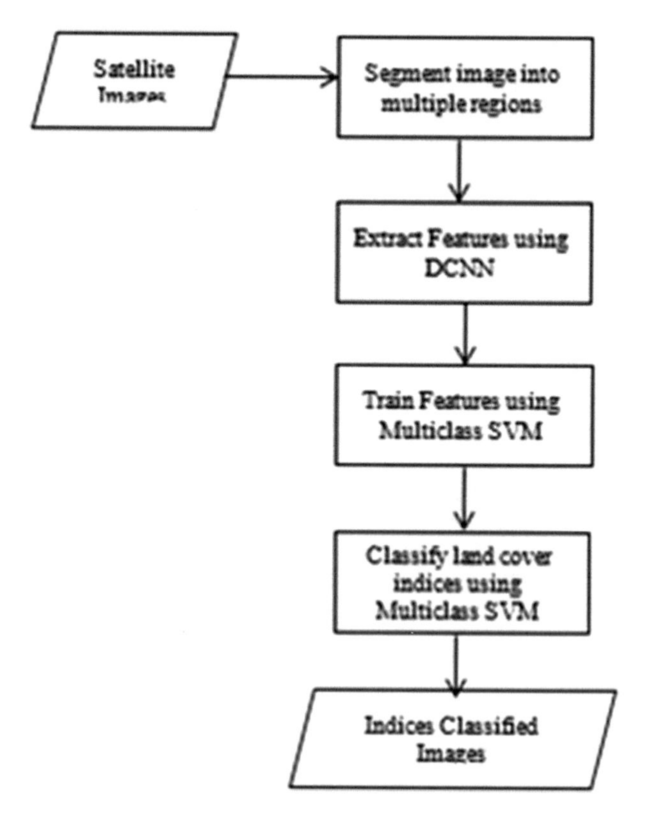

Fig. 3. Flow Diagram of multi-label land cover classification using Multi-class SVM as classifier

3.3 Experimental Results

The proposed system performance is evaluated on the Ayeyarwaddy Delta dataset using Deep Convolutional Neural Network and is in comparison with Multiclass-SVM for land cover classification. The number of images used for the experiment is shown in Table 1.

Table 1. Number of images used for experiment

Index	Total number of images
Building	582
Road	792
Vegetation	1923
Water	542
Total	**3839 -images**

The results of land cover classification for MyaungMya Region, Petye Region and Pathein region of Ayeyarwaddy Delta using DCNN and Multiclass-SVM is shown Figs. 4, 5 and 6.

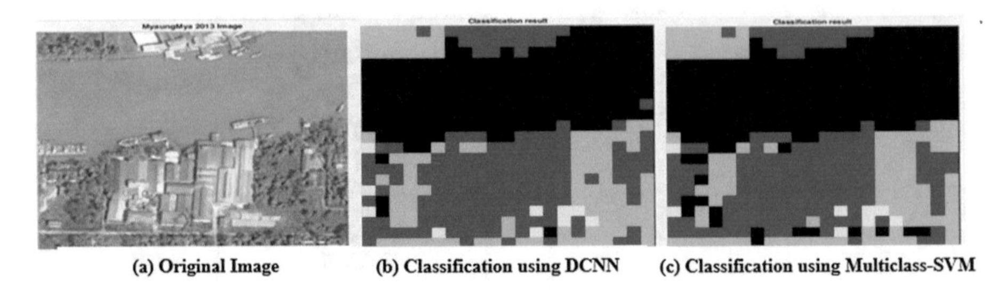

Fig. 4. Classification results of MyaungMya Region

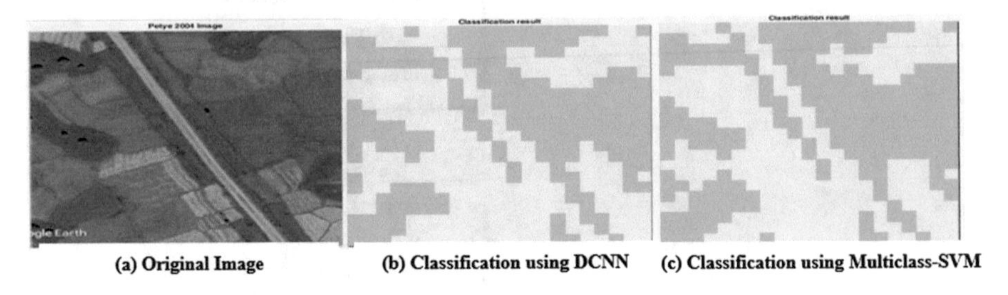

Fig. 5. Classification results of Petye Region

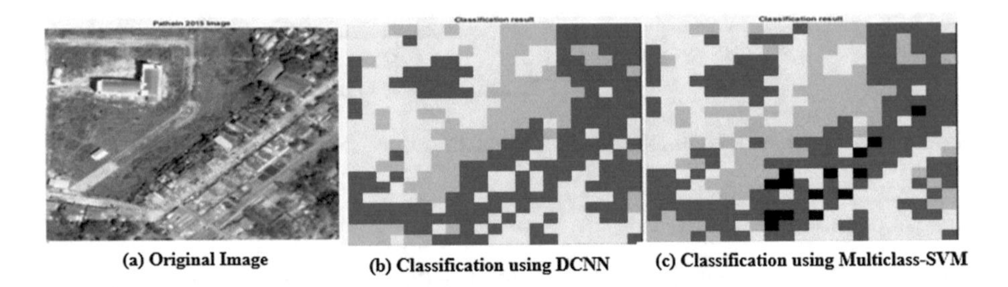

Fig. 6. Classification results of Petye Region

Figure 4(a) shows the original image of MyaunagMya region. Figure 4(b) and (c) show the multi-label classification results using DCNN and Multiclass-SVM. From the multi-label land cover classification image, the blue color is used to indicate the building index; yellow color is used for road/land index, green color for vegetation/forest index and black color for water index.

The original image of Petye region is shown in Fig. 5(a). The classification results using DCNN and Multiclass-SVM are shown in Fig. 5(b) and (c) and Pathein region for Fig. 6(a) and (b).

The classification results performance is calculated using confusion matrices for four land cover indices. The confusion matrix using DCNN as classifier is shown in Table 2.

Table 2. Confusion Matrix using DCNN as classifier

	Building	**Road**	**Vegetation**	**Water**
Building	0.7657	0.0286	0.1943	0.0114
Road	0.0210	0.8151	0.1555	0.0084
Vegetation	0.0243	0.0468	0.9185	0.0104
Water	0.0368	0	0.0307	0.9325

By using DCNN as classifier, 76.6% of building index, 81.5% road index, 91.8% of vegetation index and 93.2% of water index can be correctly classified. The confusion matrix using DCNN as classifier is shown in Table 3.

Table 3. Confusion Matrix using Multiclass-SVM as classifier

	Building	Road	Vegetation	Water
Building	0. 7886	0. 0229	0.1371	0.0514
Road	0.0672	0.7269	0.1975	0.0084
Vegetation	0.0295	0.0173	0.9428	0.0104
Water	0.0184	0	0	0.9816

By using Multiclass-SVM as classifier, 78.9% of building index, 72.7% road index, 94.2% of vegetation index and 98.1% of water index can be correctly classified.

4 Discussion

In this proposed system, the AlexNet framework is modified to classify the multi-label land cover classification since it consider for single label classification. According to the tested results, the remote sensing images can be used to classify multi-label land cover indices using Deep Convolutional Neural Network. The classification results of DCNN are then compared with Multiclass-SVM using confusion matrix which is used as classifier performance measurement.

5 Conclusion

This proposed system presents a technique for the multi-label land use and land cover classification for Ayeyarwaddy Delta using Google Earth satellite images. From the experimental results, it can be seen that AlexNet architecture can be used for multi-label land cover classification. Then, the classification results are compared with Multi-class SVM. The land cover classification information of this proposed system can be used to calculate land cover change detection for Ayeyarwaddy Delta before and after Nargis Cyclone as its future work. This land cover changes information aims to form valuable resources for urban planner and decision makers to decide the amount of land cover changes before and after the disaster and the impact of disaster.

References

1. Kaya, S., Pekin, F., Seker, D.Z., Tanik, A.: An algorithm approach for the analysis of urban land-use/cover: logic filters. Int. J. Environ. Geoinform. **1**(1), 12–20 (2014)
2. Zhang, J., Kerekes, J.: Unsupervised urban land cover classification using worldview-2 data and self-organizing maps. In: International Geoscience and Remote Sensing Symposium (IGARSS), pp. 201–204. IEEE (2011)
3. Cai, W., Liu, Y., Li, M., Zhang, Y., Li, Z.: A best-first multivariate decision tree method used for urban land cover classification. In: Proceedings of 8th IEEE conference on Geometrics, Beijing (2010)
4. Thunig, H., Wolf, N., Naumann, S., Siegmund, A., Jurgens, C., Uysal, C., Maktav, D.: Land use/land cover classification for applied urban planning – the challenge of automation. In: Joint Urban Remote Sensing Event JURSE, Munich, Germany, pp. 229–232 (2011)
5. Naydenova, V., Jelev, G.: Forest dynamics study using aerial photos and satellite images with very high spatial resolution. In: Proceedings of the 4th International Conference on Recent Advance on Space Technologies "Space in the Service of Society"-RAST 2009, Istanbul, Turkey, pp. 344–348. IEEE (2009). ISBN 978–1-4244-3624-6
6. Jones, N.: Computer science: the learning machines. Nature **505**(7482), 146–148 (2014)
7. Arel, I., Rose, D.C., Karnowski, T.P.: Deep machine learning—a new frontier in artificial intelligence research [research frontier]. IEEE Comput. Intell. Mag. **5**(4), 13–18 (2010)
8. Yu, D., Deng, L.: Deep learning and its applications to signal and information processing. IEEE Sig. Process. Mag. **28**(1), 145–154 (2011)
9. Bengio, Y.: Learning deep architectures for AI. Found. Trends Mach. Learn. **2**(1), 1–27 (2009)
10. Bengio, Y., Courville, A., Vincent, P.: Representation learning: a review and new perspectives. IEEE Trans. Pattern Anal. Mach. Intell. **35**(8), 1798–1828 (2013)
11. Hinton, G.E., Osindero, S., Teh, Y.-W.: A fast learning algorithm for deep belief nets. Neural Comput. **18**(7), 1527–1554 (2006)
12. Hinton, G.E., Salakhutdinov, R.R.: Reducing the dimensionality of data with neural networks. Science **313**(5786), 504–507 (2006)
13. Lv, Q., Dou, Y., Niu, X., Xu, J., Xu, J., Xia, F.: Urban land use and land cover classification using remotely sensed SAR data through deep belief networks, urban land use and land cover classification using remotely sensed SAR data through deep belief networks. J. Sens. **2015**, 10 pages (2015)
14. Castelluccio, M., Poggi, G., Sansone, C., Verdoliva, L.: Land Use Classification in Remote Sensing Images by Convolutional Neural Networks
15. Romero, A., Gatta, C., Camps-Valls, G.: Unsupervised deep feature extraction for remote sensing image classification. IEEE Trans. Geosci. Remote Sens. **54**, 1349–1362 (2016)
16. Scott, G.J., England, M.R., Starms, W.A., Marcum, R.A., Davis, C.H.: Training deep convolutional neural networks for land–cover classification of high-resolution imagery. IEEE Geosci. Rem. Sens. Lett. **14**, 469–470 (2017)
17. Krizhevsky, A., Sutskever, I., Hinton, G.E.: Imagenet classification with deep convolutional neural networks. In: Pereira, F., Burges, C., Bottou, L., Weinberger, K. (eds.) Advances in Neural Information Processing Systems, pp. 1097–1105. Curran & Associates Inc., Red Hook (2012)

18. Glorot, X., Bordes, A., Bengio, Y.: Deep sparse rectifier neural networks. In: Proceedings of International Conference on Artificial Intelligence and Statistics, pp. 315–323, April 2011
19. Hinton, G.E., Srivastava, N., Krizhevsky, A., Sutskever, I., Salakhutdinov, R.R.: Improving neural networks by preventing co-adaptation of feature detectors. arXiv preprint arXiv: 1207.0580 (2012)

Real-Time Hand Pose Recognition Using Faster Region-Based Convolutional Neural Network

Hsu Mon Soe(✉) and Tin Myint Naing

University of Technology (Yatanarpon Cyber City), Pyin Oo Lwin, Myanmar
hsumon.1740@gmail.com, utinmyintnaing08@gmail.com

Abstract. Hand gestures can be used as a modern technique to interact with machines. Hand gesture recognition has been an interesting research area in computer vision. In this paper, an approach using Faster R-CNN to recognize static hand gestures or postures from the real-time webcam input is presented. NUS Hand Pose Dataset II is used to train the Faster R-CNN. The proposed system is capable of recognizing 10 different hand postures shown to the webcam and control the VLC Media Player according to the recognized posture. The system is developed in Caffe deep learning framework. The system achieves acceptable accuracy on recognizing hand postures from real-time webcam input.

Keywords: Hand posture recognition · Faster R-CNN
NUS hand pose Dataset II · Deep learning · Caffe framework
VLC media player

1 Introduction

People use different kinds of gestures to communicate each other. It is a non-verbal communication between humans. Gestures are very useful for clarifying the meaning of speech, because gestures display consent, opposition or emotion, and gestures can express more in less time. Human uses head and hand gestures in daily life. Gestures are meaningful physical movements of fingers, hands, arms, or other parts of the body. Human hand has the capability of doing a large number of postures and movements. As it comprises of 27 bones, a large set of muscles and tendons, the hand model can easily make up 30 to 50 degrees of freedom [1]. Hand gestures can substitute the use of mouse and keyboard as input to computer. Gestural commands can be used to control computers and other intelligent machines.

Hand gesture recognition can be carried out in two approaches: static and dynamic. Static gestures, also called postures, are static forms of hand poses having a particular meaning. Dynamic gestures are organization of a sequence of static postures forming a single gesture and presented within a certain time period [6]. This system is developed to recognize 10 static hand postures shown to webcam and use the recognized postures as command to control personal computer. In this paper, the gestural commands are used to control the VLC Media Player.

T. T. Zin and J. C.-W. Lin (Eds.): ICBDL 2018, AISC 744, pp. 104–112, 2019.
https://doi.org/10.1007/978-981-13-0869-7_12

Humans are quite good at estimating the presence and posture of hands even between complex actions and strong constriction. But this task is relatively difficult for machines and computer vision systems [2]. Older approaches for hand gesture recognition include use of data gloves with sensors or wires attached, 3D model based approaches and appearance based approaches [3]. These approaches attained high recognition accuracy and low error rate when light conditions are stable. However, correct classification of gestures under various illuminations, and from different subjects is still challenging. After deep learning convolutional neural networks (CNNs) becoming popular with recognition tasks, CNNs have replaced the traditional flow for image recognition and computer vision tasks. Molchanov et al. [8] collected information of hand gestures from depth, color and radar sensors and mutually trained a convolutional neural network with it. Their system could precisely recognize 10 different gestures captured both indoors and outdoors in a car during the day and at night. Their approach needed short-range radar, a color camera and a depth camera. Neverova et al. [5] merged multi-scale and multi-modal deep learning to implement a method for gesture recognition and localization. Their method handled gesture classification and localization separately. In [4], Pavlo Molchanov et al. used 3D convolutional neural networks for recognizing drivers' dynamic hand gesture from intensity and depth data. Their method achieved a classification rate of 77.5% on the VIVA challenge dataset. Barros et al. [7] designed a Multichannel Convolutional Neural Network (MCNN) which can extract implicit features and recognize hand gestures. For real time recognition, their model needs images to be reduced in size.

Most of the hand gesture recognition methods (both static and dynamic), could detect only hands of large sizes (e.g. one-fourth of the whole image) except [7]. The capturing environment for input images usually has many constraints such as images must be captured with simple background and lighting condition needs to be stable. Some methods also constraint that the hand to be at the center of the image. In the proposed system, the input image has little constraints. The model can recognize the hand gestures correctly in complex background (skin-like objects and human faces in the scene are also allowed) and in various illumination conditions. Using Faster R-CNN [13], hand at various positions in the image and various sizes of hands (as small as one-tenth of the whole image) can also be detected.

The system is developed in Caffe deep learning framework and ZF Faster R-CNN is used. Faster R-CNN model is trained on NVIDIA Geforce GTX 1080 GPU with 8 GB GDDR5X video memory.

The overview of the system is described in Sect. 2. The background and workflow of Faster R-CNN is explained in Sect. 3. Section 4 presents the system set up, the training and testing results of the research. The conclusion and future work of this system is described in Sect. 5.

2 Overview of the Proposed System

This system employs Faster R-CNN for hand pose recognition. In the training phase, the raw images (RGB images without preprocessing) are used as input to the modified Faster R-CNN model. In Faster R-CNN, the CNN part extracts the features (from the lower-level features to higher-level features) and produces feature map. The RPN (Region Proposal Network) use the feature map to generate region proposals. The ROI pooling uses the feature map and region proposals to make classifications. The output of training phase is a new Caffe model. This model is used to make detections on the real-time webcam input. The output of the model is the class of the hand pose along with a bounding box and the class score (the probability of being the predefined hand pose). The recognized hand pose is used to as command to control the VLC Media Player. The following figure shows how the system works. The input video is taken from regular webcam and images from the input frame are fed to Faster R-CNN network. When a hand posture is detected, the system generates a keyboard command to control the VLC Media Player according to the detected posture class (Fig. 1).

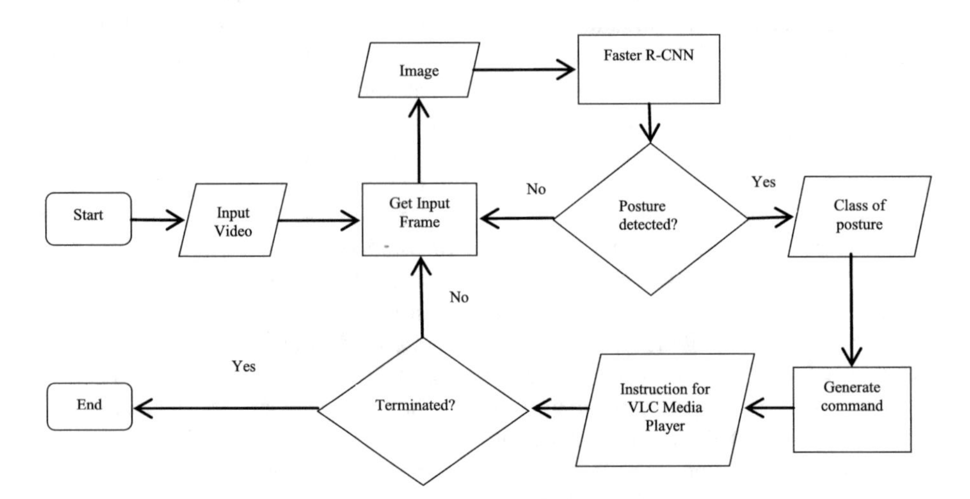

Fig. 1. System Design of the proposed method

3 Methodology

Convolutional Neural Network (CNN) became popular because of their effectiveness in image recognition and classification. With image classification, convolutional neural networks are suitable because they take images as input and output the class of the objects in the image. For CNNs to be able to perform detection tasks, three generations of region-based CNNs (R-CNN), Fast R-CNN and Faster R-CNN were developed.

3.1 Region-Based CNNs

Region based CNN (R-CNN) [11] is introduced by Girshick et al. for object detection. In R-CNN, region proposals are produced by Selective Search. The proposed region is a region which has high probability to contain an object. Every part of image in each region proposal is warped and fed to a CNN (for example AlexNet). CNN extracts a feature vector for each region. This feature vector is used as input to SVM for classification of the object in the proposed region. The feature vector is also fed to a bounding box regressor to detect the object location. R-CNN performs well on object detection with great accuracy but it takes a long time because it runs a separate CNN for each object proposal.

3.2 Fast R-CNN

Fast R-CNN [12] runs CNN first on the entire input image and shares the features between object proposals. It uses ROI pooling to extract features for each region proposals from the feature map produced by the CNN. Fast R-CNN trains the CNN, the classifier and the bounding box regressor in a single model. Instead of using separate SVM classifier, Fast R-CNN adds a softmax layer for classification. It also adds a linear regression layer to output the bounding boxes. But it still needs a region proposal method to generate object proposals.

3.3 Faster R-CNN

Both R-CNN and Fast R-CNN use external object proposal techniques such as Selective Search or EdgeBoxes to generate region proposals. Faster R-CNN is, actually, the modification of Fast R-CNN. It adds a Region Proposal Network (RPN) for generating object proposals to Fast R-CNN. The RPN reuses the feature maps which results from the last convolutional layer of CNN so it is cost-free for region proposals generating. In Faster R-CNN, region proposals and classification are done in a single network.

Faster R-CNN is composed of two modules: the Fast R-CNN and the Region Proposal Network (RPN). The RPN works as a region proposer and Fast R-CNN uses these region proposals to perform object detection. An RPN is a fully convolutional network constructed on top of the CNN.

3.3.1 Region Proposal Network (RPN)

The task of RPN is to guess the object bounds and objectiveness scores at each location of the image. The input to RPN is an image (of any size and with no preprocessing) and the output is a set of rectangular object proposals with their objectiveness scores.

For generating region proposals, a sliding network is run over the last convolutional feature map output of CNN. Several region proposals are predicted at each sliding window location. The sliding network takes an $n \times n$ spatial window from the convolutional feature map and output a 256d (for ZF here) intermediate layer. This intermediate layer is followed by two 1×1 convolution layers: classification layer and regression layer. Classification layer outputs the probability of being an object and probability of not being an object. Regression layer outputs the coordinates of the bounding box. Faster R-CNN employs anchors with different scales and ratios to achieve translation invariance. Thus, Faster R-CNN can take input of various sizes and detect objects at various positions and of different aspect ratios.

4 Experimental Results

This system uses Caffe (Convolution Architecture for Feature Extraction) deep learning framework to train the model. So, system environment for Caffe framework is first set up. A pre-trained Faster R-CNN model is adopted and it is fine-tuned to train the NUS Hand Pose Dataset II.

4.1 Setting up

This work is carried out on Ubuntu 16.04 OS and used CUDA 8.0 and cuDNN 5.1. The CPU and GPU specifications are: CPU: Intel Core i7-7700 and GPU: GeForce GTX 1080 8 GB. Setting up Caffe framework includes installing BLAS (Basic Linear Algebra Subprograms), OpenCV and other necessary packages such as pip and Numpy, etc. OpenCV version 3.0.0 is installed. OpenCV supports image and data pre-processing, high-level machine learning tasks, generic image I/O and display interfaces into a high-level programming language.

Caffe is grabbed from the BVLC GitHub repository and original Faster R-CNN model by R. B. Girshick is grabbed from GitHub.

4.2 Dataset

This research uses NUS Hand Pose Dataset II [10]. This dataset is employed because it contains images with hands of various sizes captured in varying lighting conditions. The dataset consists of: 2000 hand images and 750 hand images with human noise and 2000 background images. The hand posture detection and recognition results using this dataset are reported in [9]. This dataset is available for academic research purposes free of cost.

4.3 Training Faster R-CNN

Training the NUS Hand Pose Dataset II on Faster R-CNN using Caffe framework includes 3 main stages. First, the dataset is prepared in the form of Pascal VOC dataset. Secondly, the network and pre-trained model are modified in order to classify the hand postures according to the NUS Hand Pose Dataset II. Finally, the Faster R-CNN is trained using the NUS Hand Pose Dataset II and results a new Caffe model that recognizes the different hand postures.

For training NUS Hand Pose Dataset II, all 2000 hand images are put in the Images folder and their annotations in the Annotations folder. Among these 2000 images, 1800 images are used for training and 200 are used for testing. The dataset does not contain the annotations of images so an application named LabelImg is used to annotate the images. For network preparation, a pre-trained Faster R-CNN (here ZF Faster R-CNN) is adopted and modified to be able to classify 10 different hand postures of [10]. In this research, the 10 hand postures are assumed as 10 different objects. The network is trained to recognize these 10 postures correctly. Actually, it is trained to classify 11 classes (the background and 10 hand postures).

After the dataset and the network model are prepared, the network is fine-tuned into one which could recognize the different postures in the input image according to NUS Hand Pose Dataset II. The result of fine-tuning is a new Caffe model. Here, a total of 1800 images with hand postures are used for training. The images are of .jpg types and have the size of 160 × 120 pixels. The training takes around 30 min for 10000 iterations on Geforce GTX 1080 GPU.

4.4 Controlling VLC Media Player

The system uses 10 hand postures to control VLC Media Player. The postures are named with alphabets 'a' to 'j'. The corresponding actions on VLC to the postures includes: (a) swift to previous track, (b) play and pause the video, (c) volume up, (d) start and stop recording, (e) close the media player, (f) volume down, (g) mute/unmute, (h) switch to full-screen mode, (i) stop the video and (j) swift to next track. Controlling VLC Media Player is done by generating keyboard events as VLC allows the keyboard shortcuts to control it. Figure 2 shows the demonstration of controlling VLC to pause by showing posture 'b' from the webcam. The person in Figs. 2 and 4 is the author herself performing the gestures as predefined by the NUS Hand Pose Dataset II.

Fig. 2. Controlling VLC Media Player to pause by posture 'b'

4.5 Testing Results

First, the modified network is tested on images from the dataset. Secondly, the network is tested on webcam input. Figure 3 shows the detections of the testing images from the dataset. The class of detected posture is shown with bounding boxes in red color. The detection results on webcam input are shown in Fig. 4. The class of detected posture is shown with bounding boxes in green color. The class score and class of posture of each bounding box are described with white-color text.

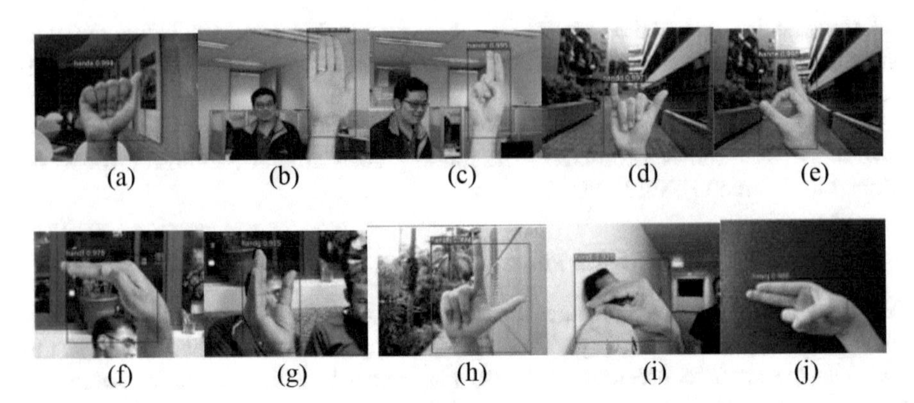

Fig. 3. Detections on images from dataset

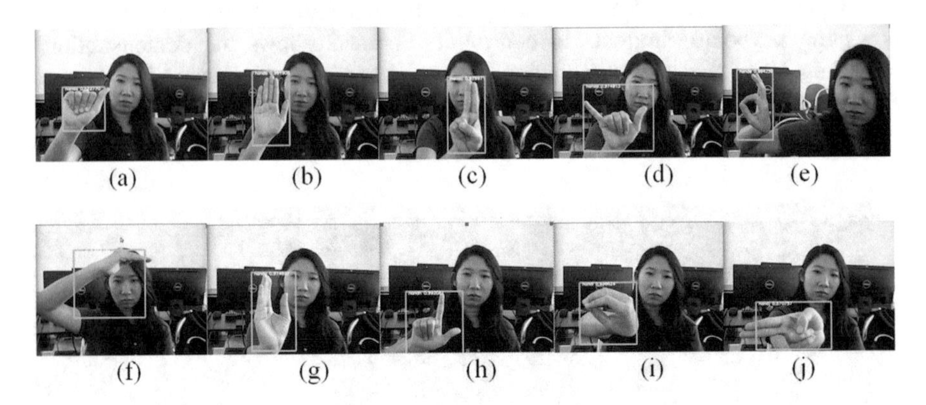

Fig. 4. Detections on images from webcam

The accuracies of the model on the testing images from dataset and on real-time webcam input are presented in Table 1. The postures are named with alphabet 'a' to 'j'. For calculating accuracy, 20 testing images (without human noise) and 75 images (with human noise) from the dataset are used for each class. For real-time testing, each gesture is shown to webcam 120 times. The performance evaluation of the model is summarized in Table 2. Table 2 shows the average time the model takes for detection per image and average accuracy of the model. The accuracy for each class of posture is calculated as (4.1). Average accuracy in Table 2 is calculated by (4.2).

$$\text{Accuracy} = (\text{true positives} + \text{true negatives})/\text{total number of samples}. \quad (4.1)$$

$$\text{Average Accuracy} = \text{Sum of accuracies of each class}/\text{number of classes}. \quad (4.2)$$

The accuracy decreases either because the images contain hands of very small sizes or because of deformation due to motion. It is not because of complex background or lighting conditions. Images taken from webcam in real-time testing also have complex background (and skin like objects in the scene) and of varying lighting conditions.

Table 1. Accuracy for each class of posture

Posture class	Hand images from dataset	Images from webcam input
Posture 'a'	80.36%	86.66%
Posture 'b'	89.2%	99.4%
Posture 'c'	65.82%	77.77%
Posture 'd'	100%	94.26%
Posture 'e'	85.22%	87.87%
Posture 'f'	100%	97.46%
Posture 'g'	80.16%	83.16%
Posture 'h'	100%	95.77%
Posture 'i'	100%	75.93%
Posture 'j'	100%	65.38%

Table 2. Performance evaluation of the model

Type of input	Testing time	Average accuracy
Hand images from dataset	0.030 s	89.95%
Images from webcam	0.035 s	86.12%

5 Conclusion and Future Work

In this paper, the use of Faster R-CNN for real-time hand pose recognition is presented. Use of Faster R-CNN solves the problems of complex background and varying lighting conditions in input images. No preprocessing is needed to train the data (RGB images). The only drawback is that it cannot detect very small objects. The model can detect and recognizes hand poses correctly with high accuracy but cannot detect small hands (i.e. hands smaller than one-tenth of the whole image). The performance can be improved by using more training images. This research attempts to use gestural commands to control the VLC Media Player. The system can be extended to be able to control other applications on personal computer.

References

1. Thippur, A., Ek, C.H., Kjellström, H.: Inferring hand pose: a comparative study of visual shape features. In: 10th IEEE International Conference and Workshops on Automatic Face and Gesture Recognition (FG), pp. 1–8. IEEE (2013)
2. Fanelli, G., Gall, J., Van Gool, L.: Real time head pose estimation with random regression forests. In: 2011 IEEE Conference on Computer Vision and Pattern Recognition (CVPR), pp. 617–624. IEEE (2011)
3. Simion, G., Gui, V., Oteşteanu, M.: A brief review of vision based hand gesture recognition. In: Recent Researches in Circuits, Systems, Mechanics and Transportation Systems (2009). ISBN 978-1-61804-062-6

4. Molchanov, P., Gupta, S., Kim, K., Kautz, J.: Hand gesture recognition with 3D convolutional neural networks. In: Proceedings of the IEEE Conference on Computer Vision and Pattern Recognition Workshops (2015)
5. Neverova, N., Wolf, C., Taylor, G.W., Nebout, F.: Multi-scale deep learning for gesture detection and localization. In: Computer Vision - ECCV 2014 Workshops on ECCV 2014, pp. 474–490
6. Ibraheem, N.A., Khan, R.Z.: Vision based gesture recognition using neural networks approaches: a review. Int. J. Hum. Comput. Interact. (IJHCI) **3**(1), 1–14 (2012)
7. Barros, P., Magg, S., Weber, C., Wermter, S.: A multichannel convolutional neural network for hand gesture recognition. In: Proceedings of the 24th International Conference on Artificial Neural Networks (ICANN 2014), 15–19 September 2014
8. Molchanov, P., Gupta, S., Kim, K., Pulli, K.: Multi-sensor system for driver's hand-gesture recognition. In: 2015 11th IEEE International Conference and Workshops on Automatic Face and Gesture Recognition
9. Pisharady, P.K., Vadakkepat, P., Loh, A.P.: Attention based detection and recognition of hand postures against complex backgrounds. Int. J. Comput. Vis. **101**(3), 403–419 (2013)
10. Pramod Kumar, P., Vadakkepat, P., Poh, L.A.: The NUS hand posture datasets II. ScholarBank@NUS Repository, 11 June 2017
11. Girshick, R., Donahue, J., Darrell, T., Malik, J.: Rich feature hierarchies for accurate object detection and semantic segmentation. In: CVPR (2014)
12. Girshick, R.: Fast R-CNN. In: Proceedings of the IEEE International Conference on Computer Vision, pp. 1440–1448 (2015)
13. Ren, S., He, K., Girshick, R., Sun, J.: Faster R-CNN: towards real-time object detection with region proposal networks. In: IEEE International Conference on Computer Vision (ICCV) (2016)

Data Mining and its Applications

School Mapping for Schools of Basic Education in Myanmar

Myint Myint Sein(✉), Saw Zay Maung Maung, Myat Thiri Khine, K-zin Phyo, Thida Aung, and Phyo Pa Pa Tun

University of Computer Studies, Yangon, Myanmar
{myint, sawzaymaungmaung, myatthirikhine, kzinphyo, tdathida, phyopapatun}@ucsy.edu.mm

Abstract. The school mapping is essentially required to detect and accomplish the educational facilities of the schools. The android application of school mapping is developed to check easily and perform the investment decisions by the educational council or government. Especially, this school mapping is generated for basic educational schools in whole Myanmar. The information of school such as location, building, laboratory, electricity, drinking water and water supply, boys and girls, teachers, students, staffs and number of vacant positions are collected for each school and store by SQLite database. A new index structures is constructed by combining B-tree and inverted file to quickly retrieve the school data. The developed system will provide to understand easily the situation of the interested school and support decision making on the planning of future school development. The generated School mapping application can be performed everywhere by mobile phone with android version 4 and above.

Keywords: School mapping · Basic education · Rural development Spatial query · SQLite database · Index structure · Android application

1 Introduction

The education in Myanmar is principally managed and supported by the Ministry of Education (MOE). Basic education sector contains the Elementary school education, Lower Secondary school education (Middle school) and Upper Secondary school education (High school). In general, all children in the age group of 6 to 11 have admission to elementary school. They access to secondary schools after completing the five years elementary education. Each three years courses of middle school and high school education are included in secondary education. Each school should be obtained sufficient provisions such as adequate number of teachers, qualitative learning, assist, teaching materials, equipments, class rooms, laboratories and so on. These are very important to facilitate successful completion of basic high school education (secondary education). More than 45,900 basic education schools are generated by the government in the whole Myanmar. It is difficult to know the information and situation of a desired school from a large numbers of schools at once. School mapping is very important tool for planning to expansion of educational capabilities of those schools. The generated

T. T. Zin and J. C.-W. Lin (Eds.): ICBDL 2018, AISC 744, pp. 115–122, 2019.
https://doi.org/10.1007/978-981-13-0869-7_13

school mapping system is provided to user for searching the facilities and requirements of desired school by using their mobile phone or tablet.

Many existing school mapping systems [1–4] are generated for small specific region of their nation and only consider for limited number of schools which less than 100 schools. Some school mapping systems are developed the based on the web application and some are based on the GIS application. Mendelsohn [1] proposed the school mapping program for Somalia country. Khobragade and Kale presented the school mapping system using GIS for Aurangabad City [2]. Many researches are carried by using B-Tree and various databases for mapping applications [5–9]. Graefe [5] reviews the fundamentals of B-trees and query processing techniques related to B-trees. Wang [6] defined an approach using dual index structure and introduced a Shortest-Distance-based Tree (SD-Tree) to preserve and reuse the network connectivity and distance information. This approach is supported to reduce the continuous query update cost when the query point location is updated.

Bender *et al.* [7] used the B-tree query with different sized atomic keys to build the atomic key dictionaries. Mate *et al.* [8] modified the android based continuous query processing of road network for location based services of Mumbai City Guide application. It provides to search easily user requirements such as Hotel, Hospital, School and so on.

Android application of school mapping system is developed for all basic education schools in nine states and nine regions of Myanmar (Fig. 1). A new index structure that combines b-tree and inverted file is developed based on the index structure for Myanmar keyword query [10, 11] to search the basic education schools data effectively and efficiently. Google map is applied for confirming and describing the school's location.

	STATES	REGIONS
1	Kachin	Yangon
2	Kayah	Naypyi Taw
3	Kayin	Magway
4	Chin	Sagaing
5	Mon	Ayeyarwaddy
6	Rakhine	Mandalay
7	Shan (East)	Pegu(East)
8	Shan (North)	Pegu(West)
9	Shan(South)	Taninthayi

Fig. 1. The map of Myanmar nation with states and regions

2 System Design

The overview of the system is described in Fig. 2. B-tree with inverted file is applied to retrieve the school data effectively and efficiently. The searching process take places depend on the states and regions which are chosen by the user and the results are display to the users on the mobile devices. The developed android application system stored the data in SQLite database inside android phone. SQLite is a relational database management system, similar to Oracle, MySQL, PostgreSQL and SQL Server. It is intuitive to both Android and iOS, and every application can create and use SQLite database as needed. The main advantage of SQLite is not required the distinct server process to install, setup, configure, initialize, manage, and troubleshoot.

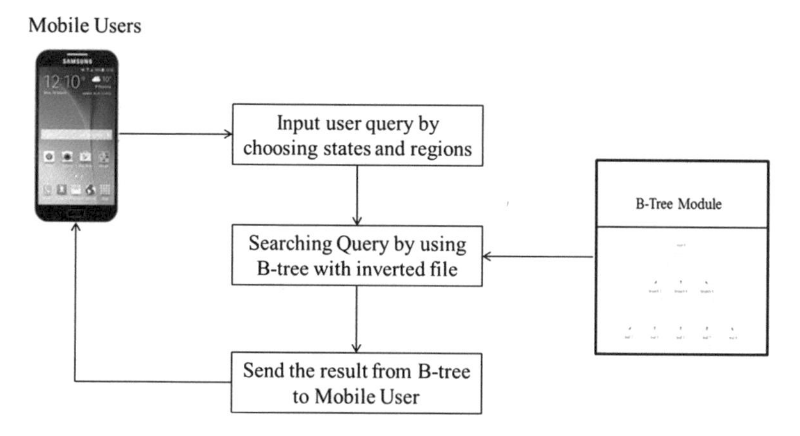

Fig. 2. System overview

The detail school information is divided into eight groups such as teacher, students, building, water & electricity, need, location and staff, according to their related items. Some contents in each group are shown in Fig. 3.

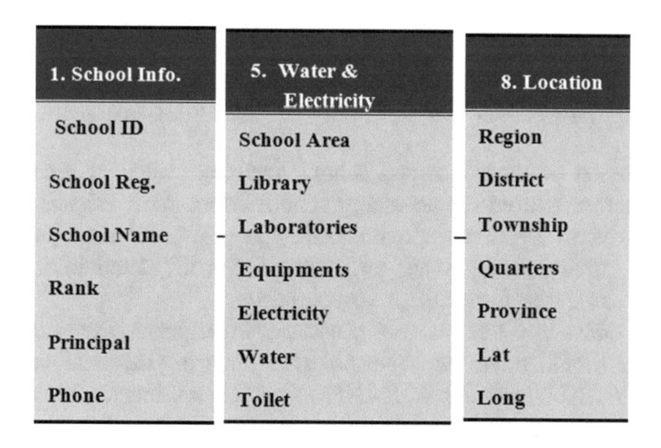

Fig. 3. Contents of some groups

3 Index Structure for Searching School Data

The new index structure for school mapping is shown in Fig. 4. It combined the B-tree and inverted file. B-tree is tree data structure and it enables the specific ordering of node collection. The advantages of B-tree format is that can be improved data access efficiency for large data set. Before constructing the b-tree, the two arrays are built. The first array is created with the name of Regions and States in Myanmar. In another one, it constructed with the name of townships according to their Regions and States. Then, b-tree is built with the school's name based on their township name. After that, the inverted file is combined with the tree to easily retrieve the geo-location of the school, principals, phone number and the other detail information. In the inverted file, it contains ID and their respective keyword to retrieve the information of the school data. To know the detail information of schools, the information of students, staffs, teachers and any others factors is retrieved from their respective document according to the ID.

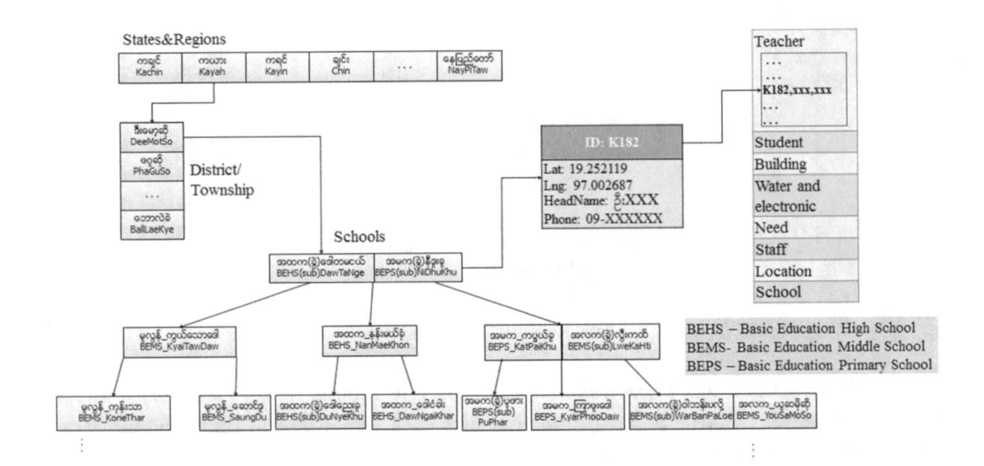

Fig. 4. Index structure for school mapping

4 System Implementation

For implementation of the developed school mapping system, B-tree with inverted index data structure is used for accessing school information efficiently. Firstly, this system is showed in Fig. 5 as a main prototype pages. In main page, the school information is categorized by states and regions. So, the detail information for the desired school can query from this prototype page.

The basic education schools are generated in eighteen states and regions of Myanmar. User should select the township after choosing the state or region. As an example, the YANKIN township of YANGON region is selected (See Fig. 6 (a)). The number of High schools, number of Middle schools, number of Primary schools and Total number of schools in the selected township are illustrated as Fig. 6 (b),

Fig. 5. Main page of mobile apps.

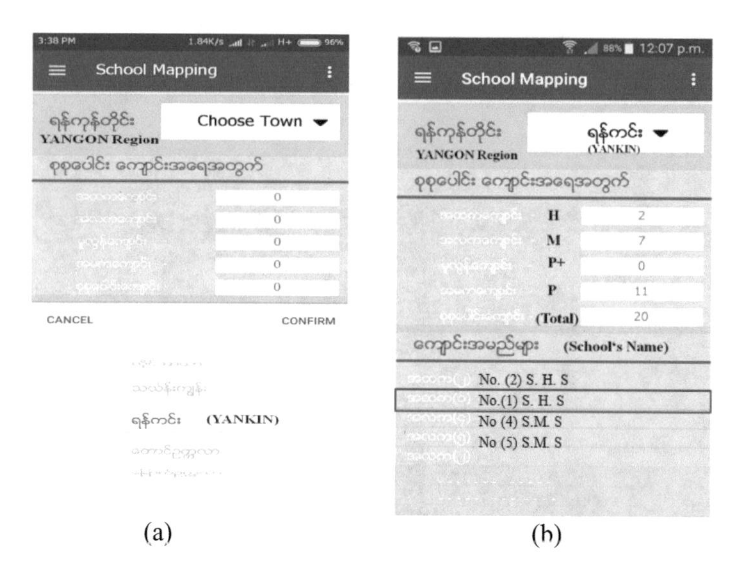

Fig. 6. Illustration the selection of region and school.

respectively. The desired school name can be selected from these 20 schools. The desired information for a school is obtained by using B-tree with inverted index data structure.

If No.(1) High school is selected, the information of the principle name, contact phone number and location map will come into view as Fig. 7 (a). The detail information can get from the segmented eight groups (Fig. 7 (b)).

For each school, the teacher information is created by using rank information and teacher's position. The number of allowed positions, current and vacant positions are

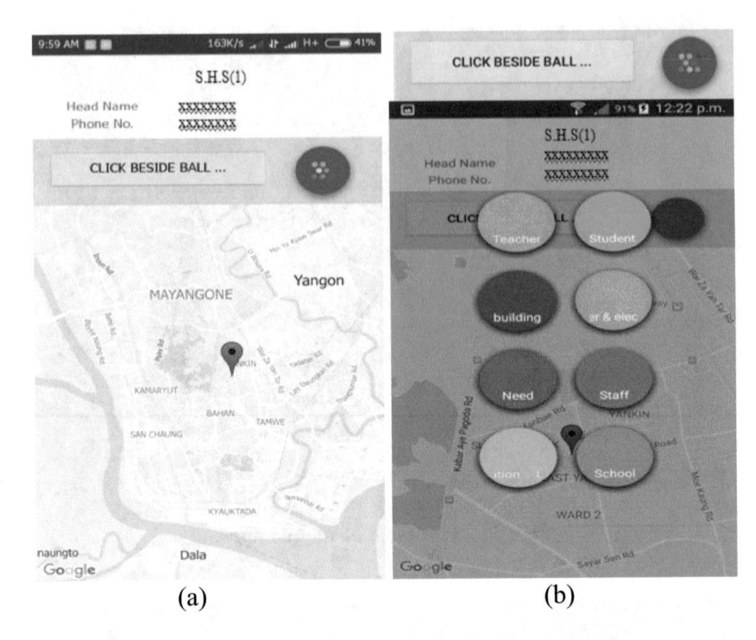

Fig. 7. Detail information from the segmented eight group

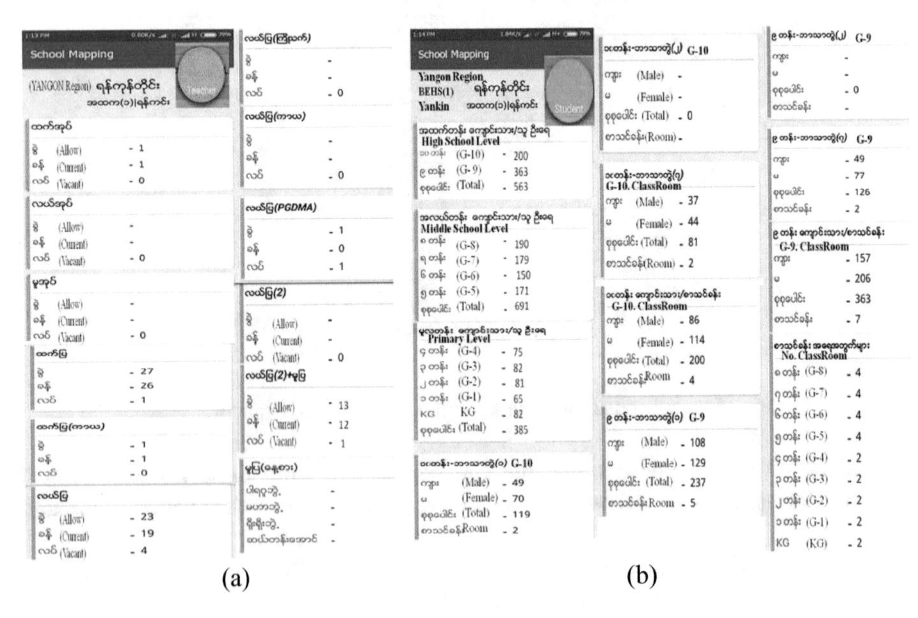

Fig. 8. Teacher and student information

listed by the acending order of rank as shown in Fig. 8 (a). As the information of teacher in each school, there are many students and their information is also by B-tree index structure. So the information of students can query and the results are shown in Fig. 8 (b).

The number of students and number of class rooms described by the three levels of education such as elementary level, lower secondary school level (Middle school) and upper secondary school education level (High school). Kindergarten (KG), Grade 1 (G − 1) to Grade 10 (G − 10) are included in these eleven years education system. The total number of students in each education level can investigate easily from this description. And also the general staff information in each school is also included in this system by dividing with rank positions and this information. The building information can be detected by choosing the building groups. Moreover, this system is given the other related information of each school such as the total number of class rooms, library information, land distance, and the latitude and logitude of school location, etc.,.

Eighteen states and regions of Myanmar are used in this testing. Each states and regions of Myanmar has approximately between 400 (minimum) and 4000 (maximum) schools. The total numbers of basic education schools has over 46 thousand. The searching time is shown in Fig. 9.

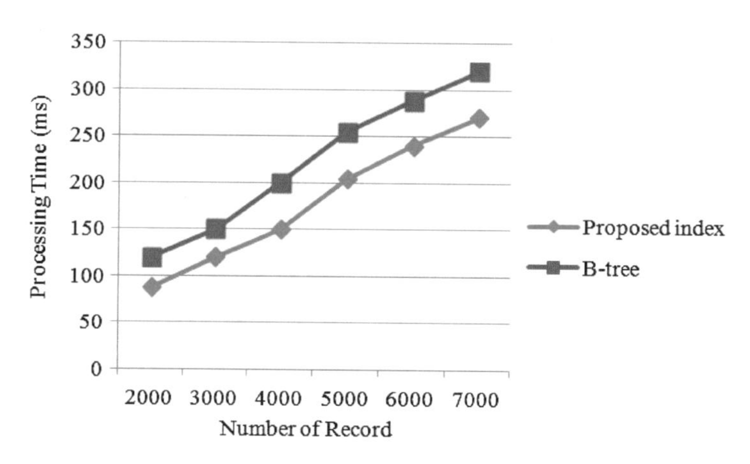

Fig. 9. Searching time

5 Conclusions

In this proposed work, the database is created based on the information of Ministry of Education in Myanmar. A new index structure using B-tree with inverted file is used to retrieve the school information. The location of each school is confirmed by using Google Earth. The user can easily find the information of all basic education schools in Myanmar by their mobile phone with android version 4 and above. So, the information of schools can manage and update within short time. This application can extend for Universities and Colleges in Myanmar by changing the related database.

Acknowledgements. We'd like to acknowledge to the Ministry of Education (MOE), Myanmar, for supporting the collected data of basic education schools and permission to accomplish this project. Also, the authors would like to appreciate to Dr. Mie Mie Thet Thwin, Rector of UCSY for allowing and supporting this project. And we thank to Ms. Wai Mar Lwin, Ms. Thazin Moe and members of GIS Lab., UCSY for their helps of data arrangements.

References

1. Mendelsohn, J.: An assessment of school mapping in Somalia. UNICEF (Somalia) School Mapping Report, pp. 1–28, August–September 2007
2. Khobragade, S.P., Kale, K.V.: School mapping system using GIS for Aurangabad city. J. IJIRCCE **4**(10), 17110–17119 (2016)
3. Makino, Y., Watanabe, S.: The application of GIS to the school mapping in Bangkok. In: Proceeding of E-Leader Conference, Bangkok (2018)
4. Olubadewo, O., Abdulkarim, I.A., Ahmed, M.: The use of GIS as educational decision support system (EDSS) for primary schools in local government area of Kano State NIGERIA. SAVAP Int. J. **4**(6), 614–624 (2013)
5. Graefe, G.: Modern B-tree techniques. Foundation and Trends in Database, vol. 3, no. 4 (2010)
6. Wang, H., Zimmermann, R.: Processing of continuous location-based range queries on moving objects in road networks. IEEE Trans. Knowl. Data Eng. **23**(7), 1065–1078 (2011). https://doi.org/10.1109/TKDE.2010.171
7. Bender, M.A., Hu, H., Kuszmaul, B.C.: Performance guarantees for B-trees with different-sized atomic keys. In: ACM 978-1-4503-0033-9/10/06 (2010)
8. Mate, P., Chavan, H., Gaikwad, V.: Android based continuous query processing in location based services. Int. J. Adv. Res. Comput. Sci. Softw. Eng. **4**(5), 134–138 (2014)
9. Aung, S.N., Sein, M.M.: K-nearest neighbors approximate keyword search for spatial database. In: Proceedings of 9th International Conference on Technological Advances in Electrical, Electronics and Computer Engineering (ICTAEECE), Bangkok, Thailand, 7th February 2015, pp. 65−68 (2015)
10. Khine, M.T., Sein, M.M.: Geo-spatial index structure for Myanmar keyword query. In: Proceedings of the 15th International Conference on Computer Applications (ICCA2017), Yangon, Myanmar, February 2017
11. Khine, M.T., Sein, M.M.: Poster: index structure for spatial keyword query with myanmar language on the mobile devices. In: Processing MobiSys 2016 Companion Proceedings of the 14th Annual International Conference on Mobile System, Application, and Services Companion, Singapore, p. 43 (2016)

GBSO-RSS: GPU-Based BSO for Rules Space Summarization

Youcef Djenouri[1], Jerry Chun-Wei Lin[2(✉)], Djamel Djenouri[3], Asma Belhadi[4], and Philippe Fournier-Viger[5]

[1] IMADA, Computer Science Department, Southern Denmark University, Odense, Denmark
djenouri@imada.sdu.dk
[2] School of Computer Science and Technology, Harbin Institute of Technology Shenzhen Graduate School, Shenzhen, China
jerrylin@ieee.org
[3] CERIST Research Center, Algiers, Algeria
ddjenouri@acm.org
[4] RIMA, USTHB, Algiers, Algeria
abelhadi@usthb.dz
[5] School of Humanities and Social Sciences, Harbin Institute of Technology Shenzhen Graduate School, Shenzhen, China
philfv8@yahoo.com

Abstract. In this paper, we present a novel GBSO-RSS algorithm to deal with exploration and mining of association rules in big data, with the big challenge of increasing computation time. The GBSO-RSS algorithm is based on meta-rules discovery that gives to the user the summary of the rules' space through a meta-rules representation. This allows the user to decide about the rules to take and prune. We also adapt a pruning strategy of our previous work to keep only the representatives rules. As the meta-rules space is much larger than the rules space, a new GPU-based approach called GBSO-RSS approach is proposed for efficient exploitation. The proposed approach has been compared on big database instances, and the results illustrate the acceleration on the summarization process. Further experimentation reveals the superiority of GBSO-RSS compared to Berrado approach in terms of number of satisfied association rules.

Keywords: Big data · Association rules · summarization
GPU architecture

1 Introduction

Association-rule mining (ARM) is a fundamental data mining task, which consists of discovering hidden patterns in transaction databases. It is applied in many real world problems such as: Constraint Programming [1,2], Information

T. T. Zin and J. C.-W. Lin (Eds.): ICBDL 2018, AISC 744, pp. 123–129, 2019.
https://doi.org/10.1007/978-981-13-0869-7_14

Retrieval [3] and Business Intelligence [4,6]. The ARM problem can thus be simply defined as follows. Let $I = \{i_1, i_2, \ldots, i_n\}$, $T = \{t_1, t_2, \ldots, t_m\}$ be a set of n different items and m transactions, respectively. Each transaction $t_i \subseteq I$ for $i \in \{1, 2, \ldots m\}$. An association rule is represented by $X \rightarrow Y$, where $X \subset I$, $Y \subset I$, and $X \cap Y = \emptyset$. Two fundamental thresholds are commonly used to verify the interestingness of the association rules: (i) The support of a rule $X \rightarrow Y$ is the number of transactions containing $X \cup Y$, and (ii) The confidence of $X \rightarrow Y$ is defined as: $\frac{support(X \cup Y)}{support(X)}$.

ARM became a real challenge when mining big databases, and it becomes challenging for the users to exploit high number of selected rules. In [26], the summary of relevant rules are obtained by applying the association rules mining from the set of rules. A set of meta-rules is then consequently generated to derive relationships between the rules. Using the meta-rules, the user can easily decide which rules will be kept and which ones will be withdraw. Alternative approaches proposed in [6–9] aim to explore the bio-inspired methods to reduce the runtime of Berrado‘s work. These algorithms need high computational time when mining big databases. To deal with this challenging topic, a GPU-based version that benefit from the massively GPU threaded is suggested in this paper. The suggested method has been compared on big databases, and the results illustrate the acceleration on the summarization process. Further experimentation reveals the superiority of GBSO-RSS compared to Berrado approach in terms of number of satisfied association rules.

The outline of the paper is presented as follows: Sect. 2 stated the GPU-based association rules mining algorithms. Section 3 illustrates the major contributions of this work. Section 4 shows the experiments of the proposed approach and the obtained results. We conclude the paper by some remarks and future perspectives.

2 Literature Review

In the two last decade, many sequential approaches have been explored for ARM problem solving [5,10–14]. Nevertheless, These algorithms are high time intensive and generate high number of useless rules when applied to big data instances. The high time consuming problem for ARM approaches have been deeply explored in previous works using GPU architecture. Indeed, extensions of GPU-based algorithms for ARM are developed in progress [15,16].

Wenbin et al. in [21] developed a Apriori parallel named Pure Bitmap Implementation (PBI). For the PBI approach, the itemsets and the transactions are stored in bitmap data structures. The itemsets are then represented as the bitmap structure, where the number of rows is set to the number of k itemsets and the number of columns is set to the number of its items. In addition, bit (i, j) is set as 1 if the item j belongs to the itemset i, 0 otherwise. A transaction structure is a bitmap, where the number of rows is the number of itemsets and the number of columns is the number of transactions, bit (i,j) is set to 1 if the itemset i belongs to the transaction j, 0 otherwise.

Zhou et al. [22] proposed an Apriori-Like algorithm GPU-FPM (GPU for Frequent Pattern Mining), that applies the vertical representation to store the database. Since the limitation of the memory usage in the GPU, the speedup attend ×15 when mining big transactional databases. Syed et al. [23] develop a GPU-based Apriori implementation which performs in two steps: (i) The generation of itemsets is first performed on GPU host, where each block computes the support of a set of itemsets, and (ii) The derived itemsets are sent back to the CPU for generating and evaluating the corresponding rules to each itemset. The limitation of this approach is the communication costs between CPU and GPU. Cui et al. [24] suggested the Cuda-Apriori algorithm. The set of transactions is first divided among the GPU threads. After that, the k-candidates are then generated and explored, which will be stored in the global memory. Each thread only handles one candidate by its assigned transactions. A synchronization between blocks is established during each iteration for determining the global support of each candidate. Zhang et al. [25] suggested a GPApriori algorithm using two data structures for itemsets counting acceleration. This algorithm reaches (×100) of speedup for small databases, but, the speedup is decreased when mining large databases. These results are explained by the fact that GPArpriori requires high number of threads divergence.

Several GPU-based works for bio-inspired approaches used in ARM are respectively presented [17–20]. For [17], it applies the evolutionary approach to solve the problem of ARM based on the GPUs model. The discovered rules are then sorted in the memory, and the antecedent and consequent rules are then concurrently evaluated. The jth thread of the SEGPU [18], each single itemset is then evaluated on GPU, and for each block, it checks whether the current itemset is included in the one part of the transaction in the database. For the MEGPU [19], the multiple itemsets are then evaluated by the GPU. Each block is occupied by a single itemset, and each thread checks the current itemset to the set of transactions assigned to it. Moreover, [20] propose a comparative analytics of three bio-inspired approaches for handling the massive transactional database based on the GPU architecture. All these works deal with the intensive-time consuming for ARM problem. However, the process still generates high number of useless rules. In this paper, we propose an intelligent mechanism and the GBSO-RSS algorithm to summarize the obtained rules.

3 GBSO-RSS Algorithm

Summarizing rules aims to discover relationships between the set of derived rules. In the big data context, the number of relationships between rules is too high, consequently, only the most frequent relationships are extracted. The first work that addresses this issue is presented in [26]. Meta rules are derived from the set of association rules saved in a transactional representation, its limit is the high computational time meta rule mining process. This is due by applying the Apriori algorithm. Thus, the number of generated rules is much more greater than the number of transactions. In this case, GPU computing and swarm intelligence

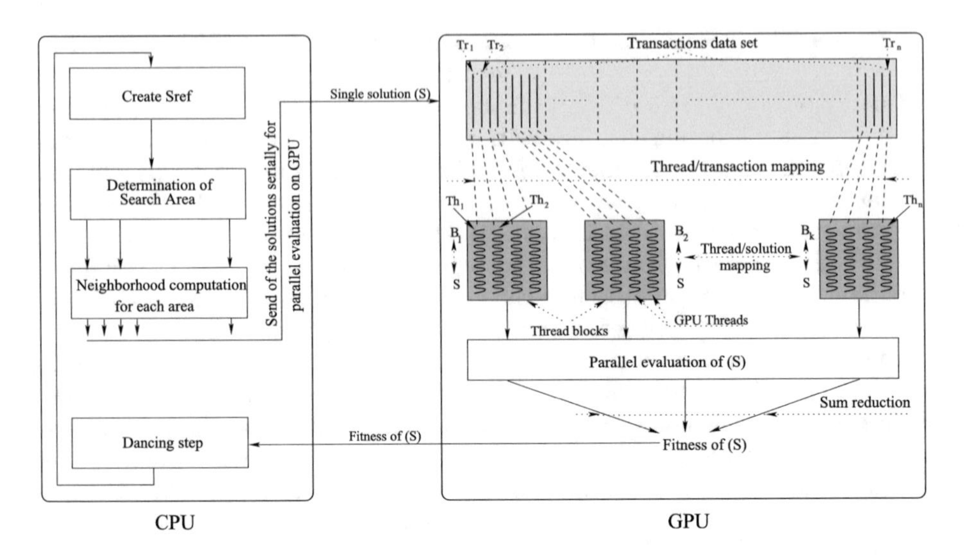

Fig. 1. GBSO-RSS Framework.

are investigated rather than exact and sequential methods. In this section, we propose a new approach of meta rules discovery allowing the summarization of a great amount of rules in a reasonable time. The approach is an adapted version of BSO-ARM algorithm [10], which employs the massive threaded provided in GPU architecture. This yields new algorithm called GBSO-RSS (GPU-based Bees Swarm Optimization for Rules Space Summarization). Figure 1 shows the framework of GBSO-RSS algorithm, the determination of search area and the neighborhood search are performed on CPU, whereas, the evaluation of the solutions is parallelized using the massive threaded of GPU architecture. GBSO-RSS algorithm follows the following steps, first the solution reference is initialized randomly on CPU, then the determination of search area is applied sequentially from the solution reference to found K different regions, each one is assigned to one bee. After that, each bee explores sequentially its regions by applying the neighborhood search which is obtained by changing from a current bee one bit in random way. Based on this simple operation, n neighbors are generated, all neighbors will be evaluated on GPU. The best neighbor of each bee is put on the dance table. Then, the communication among bees is done, and the best solution obtained by all bees is the returned. This best solution became the solution reference for the next iteration. The above processes are then repeated until the number of maximum iterations is achieved.

4 Experimentations

The experiments have been carried out using a CPU host coupled with a GPU device. In the following, several tests have been carried out to validate the proposed approach. GBSO-RSS approach is first evaluated to the sequential version in terms of execution time. The quality of the meta-rules generated by the GBSO-RSS vs. the Berrado approach is then compared.

Table 1. Speed up of the parallel approach compared to the sequential version in terms of execution time.

Number of rules (in million)	Sequential version/GBSO-RSS
1	35
2	65
5	102
10	343
100	834
1000	1067

Table 2. (%) of the satisfied rules for GBSO-RSS and Berrado approach.

Number of transactions (in million)	GBSO-RSS	Berrado
1	76	72
2	72	67
5	70	62
10	68	60
100	60	53

Table 1 shows the speed up of GBSO-RSS compared to the sequential version with different number of rules. While increasing the number of rules from 1 million to 1000 million rules, the speed up of the parallel approach is enhanced. For example, when the number of rules is 1000 million, the execution time of GSum-BSO is 1067 times better than sequential version. The proposed approach improves considerably the execution time. The previous experiment reveals that GBSO-RSS outperforms the sequential version in terms of CPU Time. Nonetheless, optimization methods give a near solution and not an optimal solution. In fact, the question in this issue is how the parallelization influences the quality of the returned meta-rules. Table 2, presents the (%) of satisfied rules (that exceeds the minimum support and the minimum confidence constraints), of GBSO-RSS and Berrado approach [26] with different number of transactions as input. While increasing the number of transactions from 1 million to 100 million transactions, GBSO-RSS outperforms Berrado approach in all cases. The previous experiment reveals that GBSO-RSS better summarizes the rules space compared to the baseline Berrado approach.

5 Conclusion

In this paper, a new approach for summarizing the rules space in the big data context has been proposed. The summary of rules space has been presented by new meta-rules. The user uses the information provided by the meta-rules to extract only the relevant rules. Moreover, GPU-based parallel bees swarm optimization algorithm has been developed for the meta-rules discovery process to benefit from the power offered by these devices. In this solution, the generation of the solutions and search process are performed on CPU. However, the massive threads are used to evaluate a single solution in GPU. Several experiments have been performed. The results show that the GPU-based approach accelerates the process of summarizing rules without reducing the quality of resulted meta-rules. Further experiments reveal that the proposed method better summarizes the rules space compared to the baseline Berrado approach. As a future work, we plan to other HPC computing for summarizing big association rules space in real time.

Acknowledgment. This research was partially supported by the National Natural Science Foundation of China (NSFC) under grant No. 61503092, by the Shenzhen Technical Project under JCYJ20170307151733005 and KQJSCX20170726103424709.

References

1. Djenouri, Y., Habbas, Z., Djenouri, D.: Data mining-based decomposition for solving the MAXSAT problem: toward a new approach. IEEE Int. Syst. **32**(4), 48–58 (2017)
2. Djenouri, Y., Habbas, Z., Djenouri, D., Fournier-Viger, P.: Bee swarm optimization for solving the MAXSAT problem using prior knowledge. Soft Comput., 1-18 (2017)
3. Djenouri, Y., Belhadi, A., Belkebir, R.: Bees swarm optimization guided by data mining techniques for document information retrieval. Expert Syst. Appl. **94**(15), 126–136 (2018)
4. Djenouri, Y., Belhadi, A., Fournier-Viger, P.: Extracting useful knowledge from event logs: a frequent itemset mining approach. Knowl. Based Syst. **139**, 132–148 (2018)
5. Agrawal, R., Imielinski, T., Swami, A.: Mining association rules between sets of items in large databases. In: ACM SIGMOD Record, vol. 22, no. 2. ACM (1993)
6. Djenouri, Y., Drias, H., Bendjoudi, A.: Pruning irrelevant association rules using knowledge mining. Int. J. Bus. Intell. Data Min. **9**(2), 112–144 (2014)
7. Djenouri, Y., Comuzzi, M.: Combining apriori heuristic and bio-inspired algorithms for solving the frequent itemsets mining problem. Inf. Sci. **420**, 1–15 (2017)
8. Djenouri, Y.,Comuzzi, M.: GA-Apriori: combining Apriori heuristic and genetic algorithms for solving the frequent itemsets mining problem. In: Pacific Asia Conference on Knowledge Discovery and Data Mining, pp. 138-148. Springer, Cham, May 2017
9. Djenouri, Y., Habbas, Z., Djenouri, D., Comuzzi, M.: Diversification heuristics in bees swarm optimization for association rules mining. In: Pacific-Asia Conference on Knowledge Discovery and Data Mining, pp. 68-78. Springer, Cham, May 2017

10. Djenouri, Y., Drias, H., Habbas, Z.: Bees swarm optimisation using multiple strategies for association rule mining. Int. J. Bio-Inspir. Comput. **6**(4), 239–249 (2014)
11. Gheraibia, Y., Moussaoui, A., Djenouri, Y., Kabir, S., Yin, P.Y.: Penguins search optimisation algorithm for association rules mining. CIT J. Comput. Inf. Technol. **24**(2), 165–179 (2016)
12. Djenouri, Y., Drias, H., Habbas, Z.: Hybrid intelligent method for association rules mining using multiple strategies. Int. J. Appl. Metaheuristic Comput. (IJAMC) **5**(1), 46–64 (2014)
13. Djenouri, Y., Bendjoudi, A., Nouali-Taboudjemat, N., Habbas, Z.: an improved evolutionary approach for association rules mining. In: Bio-Inspired Computing-Theories and Applications, pp. 93–97. Springer, Heidelberg (2014)
14. Djenouri, Y., Comuzzi, M., Djenouri, D.: SS-FIM: single scan for frequent itemsets mining in transactional databases. In: Pacific-Asia Conference on Knowledge Discovery and Data Mining, pp. 644-654. Springer, Cham, May 2017
15. Djenouri, Y., Belhadi, A., Fournier-Viger, P., Lin, J.C.W.: An hybrid multi-core/GPU-based mimetic algorithm for big association rule mining. In: International Conference on Genetic and Evolutionary Computing, pp. 59-65. Springer, Singapore, November 2017
16. Djenouri, Y., Djenouri, D., Habbas, Z., Belhadi, A.: How to exploit high performance computing in population-based metaheuristics for solving association rule mining problem. In: Distributed and Parallel Databases, pp. 1-29 (2018)
17. Cano, A., Luna, J.M., Ventura, S.: High performance evaluation of evolutionary-mined association rules on GPUs. J. Supercomput. **66**(3), 1438–1461 (2013)
18. Djenouri, Y., Bendjoudi, A., Mehdi, M., Nouali-Taboudjemat, N., Habbas, Z.: Parallel association rules mining using GPUS and bees behaviors. In: 2014 6th International Conference of Soft Computing and Pattern Recognition (SoCPaR), pp. 401-405. IEEE, August 2014
19. Djenouri, Y., Bendjoudi, A., Mehdi, M., Nouali-Taboudjemat, N., Habbas, Z.: GPU-based bees swarm optimization for association rules mining. J. Supercomput. **71**(4), 1318–1344 (2015)
20. Djenouri, Y., Bendjoudi, A., Djenouri, D., Comuzzi, M.: GPU-based bio-inspired model for solving association rules mining problem. In: 2017 25th Euromicro International Conference on Parallel, Distributed and Network-based Processing (PDP), pp. 262-269. IEEE, March 2017
21. Fang, W., Lu, M., Xiao, X., He, B., Luo, Q.: Frequent itemset mining on graphics processors. In: Proceedings of the Fifth International Workshop on Data Management on New Hardware, pp. 34-42. ACM (2009)
22. Zhou, J., Yu, K.M., Wu, B.C.: Parallel frequent patterns mining algorithm on GPU. In: 2010 IEEE International Conference on Systems Man and Cybernetics (SMC), pp. 435-440. IEEE (2010)
23. Adil, S.H., Qamar, S.: Implementation of association rule mining using CUDA. In: 2009 International Conference on Emerging Technologies, ICET, pp. 332-336. IEEE (2009)
24. Cui, Q., Guo, X.: Research on parallel association rules mining on GPU. In: Proceedings of the 2nd International Conference on Green Communications and Networks 2012 (GCN 2012), vol. 2, pp. 215-222. Springer, Heidelberg (2013)
25. Zhang, F., Zhang, Y., Bakos, J.: Gpapriori: GPU-accelerated frequent itemset mining. In: 2011 IEEE International Conference on Cluster Computing, pp. 590-594. IEEE (2011)
26. Berrado, A., Runger, G.C.: Using metarules to organize and group discovered association rules. Data Min. Knowl. Disc. **14**(3), 409–431 (2007)

Machine Learning Based Live VM Migration for Efficient Cloud Data Center

Ei Phyu Zaw(✉)

University of Computer Studies, Pathein, Pathein, Myanmar
zaw.eiphyu@gmail.com

Abstract. Increasing users' service demands, large-scale IaaS cloud data centers are used in everywhere. IaaS cloud data centers run with thousands of heterogeneous servers and so virtualization is state-of-arts in today IT trend to reduce the energy consumption. Virtualization is a methodology of logically dividing computer resources. It allows multiple virtual machines, with heterogeneous operating systems to run side by side on the same physical machine. Migration operation system instance across distinct physical hosts is a useful tool for administrators of data centers. Live migration is done by performing most of the migration while the operating system is still running, achieving very little downtime. By carrying out the majority of migration while OSes continue to run, we achieve impressive performance with minimal service downtime and total migration time. In this paper, machine learning based working set prediction is proposed to reduce the total migration time. It uses the prediction model with historical data during the live VM migration process. At first, it trains experimental dataset which includes the performance parameters collected from various workloads by machine learning techniques to build the best prediction model and then predict the working set which can affect the total migration time. We evaluated the effectiveness of the working set prediction algorithm with various workloads with simulation model and the experimental result shows that this method can more reduce the total migration time in live VM migration than XEN's default pre-copy based live migration.

Keywords: Virtual machine · Live VM migration · Machine learning
Total migration time

1 Introduction

Infrastructure–as–a–Service (IaaS) cloud data centers including AmazonEC2, offer several types of virtual machines (VMs) that differ in their amount of resources based on the pay–as–you–go model. IaaS cloud data centers currently consist of many thousands or even millions of heterogeneous servers and each server may host a set of heterogeneous VMs. Therefore, the ever increasing heterogeneity for both the physical servers and the VMs needs to be managed efficiently in order to achieve the following key goals: maximize resource utilization and reduce the energy costs.

In virtualized clusters and data centers, the powerful capability is live VM migration. It means that although running the operating system, VM migration can be performed. The main process of VM migration is transferring its memory pages from

T. T. Zin and J. C.-W. Lin (Eds.): ICBDL 2018, AISC 744, pp. 130–138, 2019.
https://doi.org/10.1007/978-981-13-0869-7_15

one virtual machine to another. In previous work of live VM migration, firstly, the whole memory pages are transferred and the memory pages which are updated in migration process are transferred again in next time. During the whole migration process, the updated memory pages are transferred again and again and as a result, the total migration time is prolonged. To improve the live VM migration performance, it is important to reduce the transferred memory pages.

In this algorithm, the memory pages which will need in the future are predicted as the working set. After that, the memory pages except the working set is transferred in push phases and in the stop phase, the working set and system state are transferred. The proposed system is named as WSPML.

The rest of this paper is organized as follows: Sect. 2 related background theory is discussed and the system design of WSPML is described in Sect. 3. Section 4 concerns with the simulation model of this system. In Sect. 5, experimental results are described. Finally, Sect. 6 concludes the paper.

2 Background Theory

2.1 Live Virtual Machine Migration

Virtualization promises to dramatically change how data centers operate by breaking the bond between physical servers and the resource shares granted to customers. Live migration, a key benefit brought by virtualization technologies, refers to the process of transferring the run time data of a VM from one physical host (the source) to another machine (the destination). The execution of the VM is not interrupted during most of the transfer. This allows administrators to manage system resources on-the-fly and simplifies maintenance.

The approach best suited for live migration of virtual machines is *pre-copy [1]*. These include hypervisor-based approaches from VMware [5], Xen [1], KVM [3], OS-level approaches that do not use hypervisors from OpenVZ [6]. Pre-copy technique incorporates iterative push phases and a stop-and-copy phase which lasts for a very short duration. In short, the pages to be transferred during round *'n'* are only the ones dirtied during round *'n − 1'*.

Another novel strategy post-copy is also introduced into live migration of virtual machines. In this approach, all memory pages are transferred only once during the whole migration process and the baseline total migration time is achieved. But the downtime is much higher than that of pre-copy due to the latency of fetching pages from the source node before VM can be resumed on the target.

2.2 Machine Learning Techniques

Machine Learning (ML) Techniques is the ability to learn from previous executions what are the most important variables to take into account to build the model. The techniques used in WSPML could predict the working set includes the memory pages which will be used in near future by using the collected metrics from resources related with the live VM migration process. ML techniques are:

i. **Linear Regression** is a statistical modelling tool that assumes a linear relationship between the inputs and outputs. It may be used for the prediction of unknown output values where all input values are known. It is widely used today to help for capacity planning analysis or detecting anomalies on the system.
ii. **Reduced Error Pruning REPTree** is a model based tree. Starting at the leaves, each node is replaced with its most popular class. If the prediction accuracy is not affected then the change is kept. Reduced error pruning has the advantage of simplicity and speed.
iii. **M5P** implementation uses decision trees to model for nonlinearity. Decision trees are tree-based data structures that classify data based on conditionals related to independent inputs. Checks are made at each node, owing down the tree until a classification is made at a leaf. Where a traditional decision tree has simple classifications at the leaves, M5P instead builds linear regression models at the leaves. This allows us to use linear regression on data sets that have some nonlinearity.

3 System Design of Machine Learning Based Live VM Migration

This system design consists of two main processes as shown in Fig. 1. In machine learning evaluation process, ML is used to build the prediction model and the experimental dataset which includes the performance parameters collected from various workloads is divided into training set and testing set. An initial prediction model is built by performing only parameters on training set. Then, prediction accuracy is compared with a predefined threshold value of prediction rate.

After a prediction model is constructed, it is evaluated once more by 5-fold cross validation for an unbiased prediction evaluation. At last, the prediction model is evaluated using testing set in terms of correlation coefficient to take into account the accuracy of the prediction model. This process is done by various ML techniques and the best model is selected by the accuracy of the model.

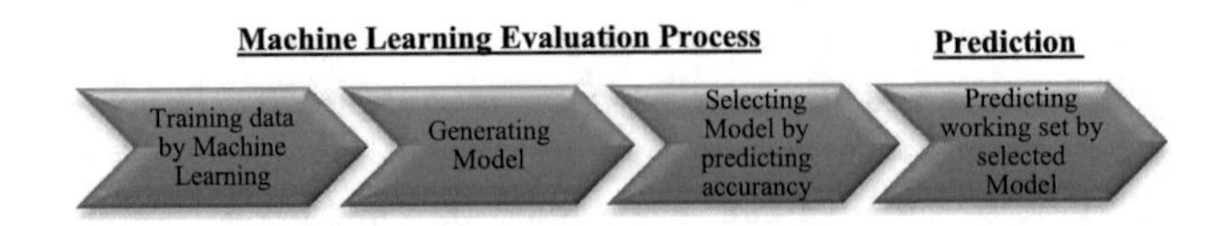

Fig. 1. Processes machine learning based live VM migration

In the prediction working set process, the system use the prediction model with historical data during the live VM migration process and trains the model parameters. The model reads the statistical data of memory updating and other performance parameters such as memory dirtying rate and transmission rate and simulates the migration process to reduce the total migration time.

3.1 Machine Learning Evaluation Process

Machine Learning is used to predict the working set in live VM migration process. Due to the complexity of modeling the complex environments and with low knowledge a priori about them, ML is used to automatically build the model from a set of metrics easily available in VM like memory dirtying rate and transmission rate. The techniques used in this paper could predict the working set if the metrics related with live VM migration process is collected. The idea is that ML algorithms evaluated in prediction of the working set if the state of the system (including workload) does not vary in the future. However, if the situation changes (the consumption speed changes) the model has to be able to recalculate the working set under the new circumstances.

For this reason, a set of derived metrics as variables is added to achieve a more accurate prediction. All the metrics are collected from the full experiment running live VM migration process with two physical servers. After that, the working set occurs in the live VM migration is identified. The new dataset is used as input for the process, which has the main task to add variables derived from the metrics monitored. The resulting data set is used to build the model, which would be validated using different workloads and transmission rates. Besides, it is prepared to develop a set of models and choose the best one according to different parameters selecting the more accurate result. Machine Learning evaluation process includes in two steps as follows:

- Model Selection: Comparing different Machine Learning algorithms in order to choose the (approximately) best one.
- Model Assessment: Having chosen a final algorithm, estimate its generalization error on new data.

The main goal of this sub-session is to conduct an evaluation of different ML algorithms to choose the best one. Model selection process requires that ML algorithms given are trained with the training data set, and later compared according to the estimated test error using a completely different data set, called validation data set or development data test set.

According to Table 1, M5P offers a good trade-off between accuracy and training. WSPML needs quick predictions with low computational cost and M5P accomplishes this constraint. Finally, other important point for choosing M5P is that M5P models are human interpretable.

Table 1. Accuracy of prediction model using different ML algorithms

	LinuxIdle	Dbench	TPC-C	SPECWeb	LinPack
M5P	0.9977	0.9635	0.9975	0.9937	0.9941
LinearRegression	0.9939	0.9535	0.9636	0.9757	0.9729
REPTree	0.9415	0.9182	0.9347	0.9576	0.9563

3.2 Prediction Process

In traditional pre-copy approach, the memory pages are copied over multiple iterations followed by a final transfer of the processor state. WSPML can provide a “win-win” by reducing total migration time closer to its equivalent time achieved by non-live VM migration. WSPML construct an simulation model to predict the working set using historical data during the live VM migration process. Due to the complexity of modeling the real environments and with low knowledge a priori about them, WSPML decided to use ML to automatically build the model from a set of metrics which affect the total migration time of live VM migration such as memory dirtying rate and network transmission rate.

Working set includes the memory pages which updated frequently during the migration process. It should be skipped during the iterative pre-copying and transferred at the stop-and-copy phase. The sized of working set is approximately proportional to the pages dirtied in each iteration. That is:

$$\mathrm{Ws} = \mathrm{y} * \mathrm{Ti} * \mathrm{DR}$$

where y is the ratio of correlation with the memory dirtying rate and the duration of each iteration. The relationship in a linear equation model is as follow:

$$\mathrm{y} = \mathrm{w}_0 + (\mathrm{w}_1 * \mathrm{Ti}) + (\mathrm{w}_2 * \mathrm{DR})$$

where w_0, w_1 and w_2 are the model weights to be learned.

WSPML estimate the model weights by running the various workloads and train the model by executing the migration of each workload one by one using WEKA tools.

4 Machine Learning Based Live VM Migration Simulation Model

In this sub-section, the simulation model for WSPML and pre-copy based live VM migration is described. Assuming that the memory could be processed in the propose framework as pages with a fixed size equal to $\boldsymbol{P}$ bytes, $\boldsymbol{M}$ is the memory size, the bandwidth available for the transfer is constant and equal to $\boldsymbol{B}$ bytes per second, $\boldsymbol{W_S}$ is the working set list and $\boldsymbol{W_M}$ is the modified page per second. Therefore, the time cost to transfer the memory pages in the propose framework is

Execution time for *push phase* of the proposed framework:

$$T_i = \frac{(M_i - W_s)P}{B}$$

where i = 1,2,3,…n and $M_n \leq W_M$ and $M_1 = M$ for first iteration and then modified memory page size is updated by:

$$M_{i+1} = T_i W_M$$

Execution time for stop phase of the proposed framework:

$$\mathrm{T_{stop}} = \frac{\mathrm{W_s P}}{\mathrm{B}}$$

Total Migration Time = $T_{pre} + \sum_{i}^{n} T_i + T_{stop}$
And the time cost to transfer the memory pages in XEN is
Execution time for push phase of XEN :

$$T_i = \frac{M_i P}{B}$$

where i = 1,2,3,…n and $M_n \leq W_M$ and $M_1 = M$ for first iteration and then modified memory page size is updated by:

$$M_{i+1} = T_i W_M$$

Execution time for stop phase of XEN,

$$T_{stop} = \frac{W_s P}{B}$$

Total Migration Time = $\sum_{i}^{n} T_i + T_{stop}$

Using the simulation models for both approaches, WSPML reduces the transfer memory page using the working set during the pre-copying process of live VM migration. But the total migration of WSPML includes not only the migration time for pre-copying process and last iteration but also the overhead for prediction process. The simulation algorithm for WSPML is as shown in Fig. 2. During each migration process, the amount of dirtied pages is tracked and skipped paged in each iteration of pre-copying to calculate the value of y.

```
1.assign the memory size of VM
2.  for i=0 to no. of iteration do
3.      calculate Migration Time for current round
4.      calculate coefficient from historical data
5.      predict the working set size using Machine Learning
6       calculate the transferred memory size for next round
7.      if working set size is greater than dirtying rate then
8.          calculate the transferred memory size for last round
9.          calculate the migration time for last round
10          calculate the time for stop phase
11.         Break
12.     end if
13.  end for
14. calculate the total migration volume
15. calculate the total migration time
```

Fig. 2. WSPML's procedure

With the migration time and memory dirtying rate of each pre-copying iteration, WSPML build a set of independent equations based on the experimental results and derive the value of each parameter (w_0, w_1, w_2). In regression, the R2 coefficient of determination is a statistical measure of how well the regression line approximates the real data points. An R2 of 1.0 indicates that the regression line perfectly fits the data. R2 of the simulation model of WSPML is 0.92813.

5 Evaluation

In this section, the proposed system evaluate the framework for working set prediction algorithm used by the following workloads and then analyse how to improve the performance of live VM migration with traditional pre-copy based live VM migration algorithm in XEN.

5.1 Experimental Environment

For our experiment platform, the proposed system used six identical computer servers as a cluster. The two servers act as the source and destination for live migration with gigabit Ethernet. The third one is for storage server which used iSCSI protocol for shared storage. The other three servers run as the client with various workloads. Citrix Xen-5.6.0 for VMM and modified Linux 2.6.18.8 for guest OS is used in the proposed system. All of the servers and clients shared the same network and connected by a Gigabit LAN.

For evaluating the performance of WSPML, we select several representative applications in virtualization environment as the workloads of migrated VM: **Linux idle** is an idle Linux OS for daily use. This workload is used as a frame of reference for comparison. **TPC-C** is an on-line transaction processing (OLTP) benchmark. TPC-C simulates a complete environment where a population of terminal operators executes transactions against a database. We configure 1000 terminals threads and 500 database connections in our experiments. **Dbench** is an open source benchmark emulating the file system load. This benchmark can simulate a variety of real file servers by executing create/write/read/delete operations on a large number of directories and files with different sizes. We configure 10 simultaneous connections to generate reasonable disk load. We use the Intel optimized MP **LINPACK** benchmark to perform massive vector and matrix operations, which produces severe CPU and memory pressure.

5.2 Evaluation Results

Total migration time is a key performance metric that would be the most concern in some scenarios such as performance degradation. The system conducted a set of experiments to validate the effectiveness of the WSML. For each workload, the system compared the total migration time with the estimates using the simulation model and traditional pre-copy approach.

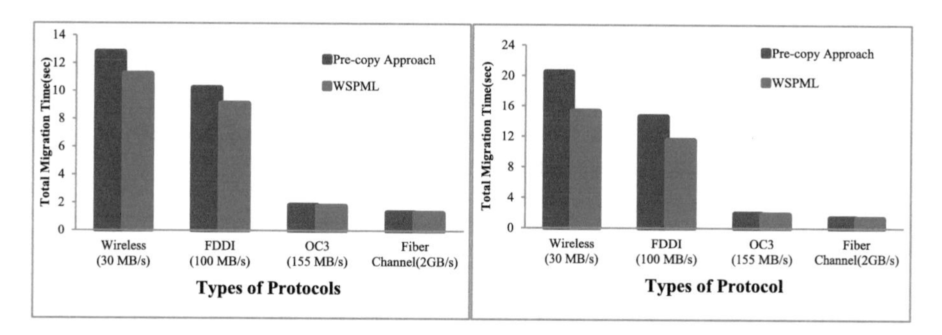

Fig. 3. Comparison of total migration time (a. LinuxIdle) and (b.TPC-C)

Figure 3(a) shows the variation of total migration time with the different types of transmission rate using LinuxIdle workload compared with pre-copy approach and WSPML. Using this workload, WSML can more reduce in the migration process using wireless and FDDI than traditional pre-copy approach. Figure 3(b) compares the metric of the total migration time as the various types of protocols for traditional pre-copy approach and WSPL in TPC-C workload. This figure shows that WSPML can more reduce in slow network transmission rate such as wireless protocol but can little reduce in fast network transmission rate such as fiber channel.

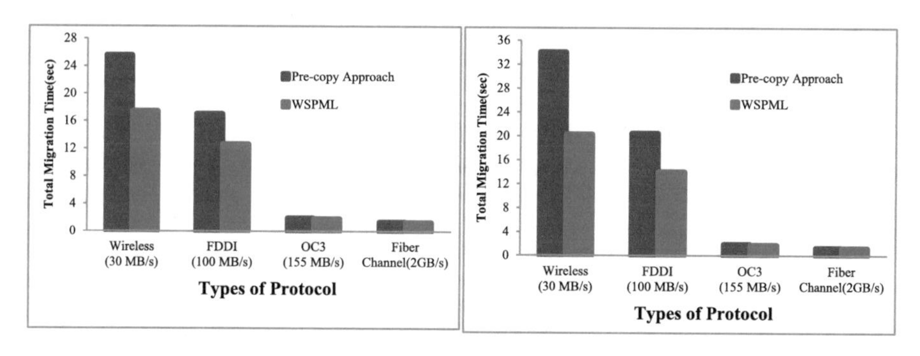

Fig. 4. Comparison of total migration time (a.Dbench) and (b.SPECWeb)

The result shows the total migration time for WSPML and XEN in TPC-C workload using different types of protocol in Fig. 4(a). Using wireless connection, the total migration time of WSPML is more significantly reduced than the total migration time of traditional pre-copy approach. Figure 4(b) illustrates the total migration time for the proposed approach and XEN's default pre-copy migration for SPECWeb workload using different transmission protocols. In this workload, WSML can reduce nearly 14 s than the traditional pre-copy approach using slow network connection such as wireless protocol.

Total migration time is mostly correlated with the amount of dirty memory pages that should be transferred during the migration process. As shown in Figures, SPEC-Web exhibit very short total migration time in Wireless protocols than the traditional pre-copy approach because of their high memory dirtying rate and slow transmission rate. LinuxIdle is not significantly reduce the total migration time than the pre-copy approach in every types of transmission rate. However, guided by WSPML, the system always more reduce the total migration time than XEN's default pre-copy approach using various workloads and transmission rate.

6 Conclusion

To address the problem of high energy use in IaaS cloud data centers, it is necessary to eliminate inefficiencies while using computing resources. Computing resources consist of multiple types including CPU, memory, disk, and network bandwidth and all need to be considered while designing energy–efficient mechanisms for resource management. WSPML has conducted an extensive evaluation of Machine Learning algorithms to predict the working set which affect on the performance degradation of VM with the various types of workload. The system evaluated the proposed approach to reduce the total migration time during the migration process and have conducted a set of experiments to evaluate the M5P in front of Linear Regression in different workloads and transmission rate to predict the working set. From our experiments we can conclude that M5P seems a good accurate predictor and showed that WSPML more reduces the total migration time than the traditional pre-copy approach during the migration process.

References

1. Clark, C., Fraser, K., Hand, S., Hansen, J.G., July, E., Limpach, C., Pratt, I., Warfield, A.: Live migration of virtual machines. In: Proceedings of the 2nd USENIX Symposium on Networked Systems Design and Implementation (2005). Tavel, P.: Modeling and simulation design. AK Peters Ltd. (2007)
2. Hines, M.R., Gopalan, K.: Post-copy based live virtual machine migration using adaptive pre-paging and dynamic self-ballooning. In: Proceedings of the ACM/Usenix International Conference on Virtual Execution Environments (VEE 2009), pp. 51–60 (2009)
3. Kivity, A., Kamay, Y., Laor, D.: KVM: the linux VM monitor. In: Proceedings of Ottawa Linux Symposium (2007)
4. Li, L., Vaidyanathan, K., Trivedi, K.S.: An approach for estimation of software aging in a web server. In: Proceedings of the 2002 International Symposium on Empirical Software Engineering, Washington, DC, USA, p. 91. IEEE Computer Society (2002)
5. Nelson, M., Lim, B., Hutchines, G.: Fast transparent migration for virtual machines. In: Proceedings of the USENIX Annual Technical Conference (USENIX 2005), pp. 391–394 (2005)
6. OpenVZ. (Virtuozzo). http://www.openvz.com/
7. Zhang, Q., Cherkasova, L., Mi, N., Smirni, E.: A regression-based analytic model for capacity planning of multi-tier applications. Cluster Comput. **11**(3), 197–211 (2008)
8. Vaidyanathan, K., Trivedi, K.S.: A measurement-based model for estimation of resource exhaustion in operational software systems. In: Proceedings of the 10th International Symposium on Software Reliability Engineering, Washington, DC, USA, p. 84. IEEE Computer Society (1999)

Dynamic Replication Management Scheme for Distributed File System

May Phyo Thu, Khine Moe Nwe(✉), and Kyar Nyo Aye(✉)

Faculty of Computer Science, University of Computer Studies (Yangon), Yangon, Myanmar
mayphyothu.mptl@gmail.com, khinemoenwe@ucsy.edu.mm, kyarnyoaye@gmail.com

Abstract. Nowadays, replication technique is widely used in data center storage systems to prevent data loss. Data popularity is a key factor in data replication as popular files are accessed most frequently and then they become unstable and unpredictable. Moreover, replicas placement is one of key issues that affect the performance of the system such as load balancing, data locality etc. Data locality is a fundamental problem to data-parallel applications that often happens and this problem leads to the decrease in performance. To address these challenges, this paper proposes a dynamic replication management scheme based on data popularity and data locality; it includes replica allocation and replica placement algorithms. Data locality, disk bandwidth, CPU processing speed and storage utilization are considered in the proposed data placement algorithm in order to achieve better data locality and load balancing effectively. Our proposed scheme will be effective for large-scale cloud storage.

Keywords: Replication · Data popularity · Data locality · Storage utilization Disk bandwidth

1 Introduction

Cloud storage is a technology that allows us to save files in storage and then access those files via Cloud. The cloud storage system convergences data storage among multiple servers into a single storage pool and provides users with immediate access to a broad range of resources and applications hosted in the infrastructure of another organization via a web service interface [6].

At present, the existing Cloud storage products are Google (Google File System GFS), Amazon (Simple Storage Service S3), IBM (Blue Cloud), Yahoo (Hadoop Distributed File System HDFS) etc. HDFS provides reliable storage and high throughput access to application data. In HDFS, data is split in a fixed size (e.g., 32 MB, 64 MB, and 128 MB) and the split data blocks (chunks) are distributed and stored in multiple data nodes with replication.

In HDFS, to provide data locality, Hadoop tries to automatically collocate the data with the computing node. Data locality is a principal factor of Hadoop's performance. Data locality means the degree of distance between data and the processing node for the data. There are three types of data locality in Hadoop: node locality, rack locality

T. T. Zin and J. C.-W. Lin (Eds.): ICBDL 2018, AISC 744, pp. 139–148, 2019.
https://doi.org/10.1007/978-981-13-0869-7_16

and rack-off locality. Uniform data replication is used in current implementations of MapReduce systems (e.g., Hadoop). The concept of popularity of files is introduced to replication strategies for selecting a popular file in reality. In this paper, therefore, data popularity based replication method is proposed to overcome the problems of static replication in HDFS and to support better efficiency in cloud storage. Firstly, the rate of change of file popularity is calculated by analyzing the access histories with first order differential equation. Secondly, the replication degree for each file is calculated according to the rate of change of file popularity. Finally, the replicas will be placed based on proposed data placement algorithm.

The rest of this paper is organized as follows. Section 2 describes related works and proposed system architecture is presented in Sect. 3. Section 4 presents performance evaluation and finally, Sect. 5 describes the conclusion and future work.

2 Related Works

In cloud storage environment, data can be stored with some geographical or logical distance and this data is accessible to cloud based applications. A cost effective replication management scheme for cloud storage cluster was proposed by Wei [2]. That paper aimed to improve file availability by centrally determining the ideal number of replicas for a file, and an adequate placement strategy based on the blocking probability. One approach, Latest Access Largest Weight (LALW) algorithm [7], that was proposed by Chang and Chang for data grids. LALW found out the most popular file in the end of each time interval and calculated a suitable number of copies for that popular file and decides which grid sites were suitable to locate the replicas.

Hunger and Myint compared two data popularity-based replication algorithms: PopStore and Latest Access Largest Weight (LALW) [1]. In that paper, both algorithms found more popular files according to the time intervals through the concept of Half-life. However, this paper did not consider for load balance in replica placement. Scarlett [5] adopted a proactive replication scheme that periodically replicates files based on predicted data popularity. It focused on data that receives at least three concurrent accesses. However, it did not consider node popularity caused by co-location of moderately popular data.

In DARE [3], the authors proposed a dynamic data replication scheme based on access patterns of data blocks during runtime to improve data locality. However, removing the replicated data was performed when only the available data storage was insufficient. Thus, it had a limit to provide the optimized replication factor with data access patterns.

In [8], the authors proposed a delay scheduling method that focused on the conflict between data locality and fairness among jobs. However, delay scheduling made assumptions that might not hold universally and these assumptions made it difficult for delay scheduling to adapt to changes in workload, network conditions, or node popularity. In [4], the authors proposed an efficient data replication scheme based on access count prediction in a Hadoop framework. Although this scheme was designed to improve data locality, it considered file level replication did not consider block level replication.

3 Proposed System Architecture

The goal of proposed system is to design an adaptive replication scheme that seeks to increase data locality by replicating "popular" data while keeping a minimum number of replicas for unpopular data. Because the nature of data access pattern is random, a method that predicts the rate of change of file popularity for the next time slot is required. The proposed system flow diagram is shown in Fig. 1.

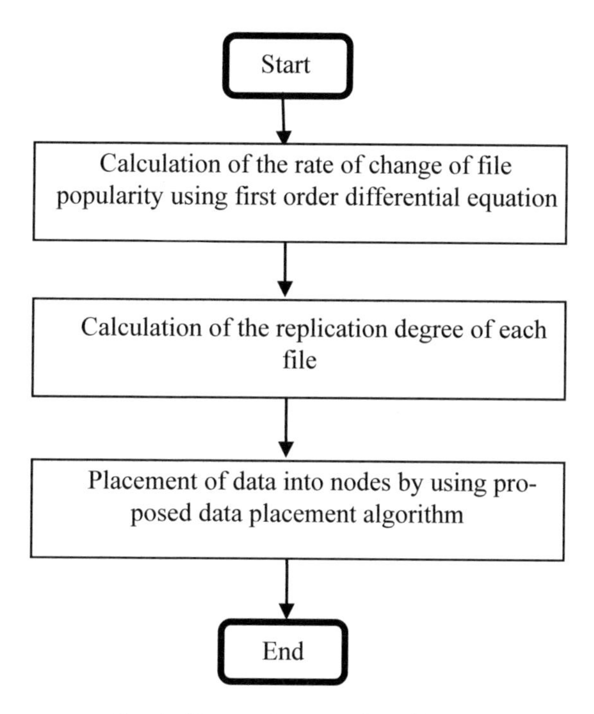

Fig. 1. Proposed system flow diagram

The proposed scheme includes three-step processes, the rate of change of file popularity will be calculated using first order differential equation in the first step and the number of replicas of each file will be calculated in the second step and then the replicas will be placed into nodes based on proposed data placement algorithm in the third step.

3.1 Proposed Popularity Growth Rate Algorithm

In this step, the rate of change of file popularity will be calculated using first order differential equation. LALW and Pop-Store algorithms applied half-life strategy which means the weight of the records in an interval decays to half of its previous weight. The idea of popularity is the assumption that the rate at which a popularity of an item grows at a certain time is proportional to the total popularity of the item at that time. In mathematical terms, if P(t) denotes the total population at time t, then

$$\frac{dp}{dt} = kP(t) \tag{1}$$

where P(t) denotes population at time t and k is the growth constant or the decay constant, as appropriate. If $k > 0$, we have growth, and if $k < 0$, we have decay. It is a linear differential equation which solve into

$$P(t) = P_0 e^{kt} \tag{2}$$

Then,

$$k = \frac{\ln\left(\frac{P(t)}{P_0}\right)}{t} \tag{3}$$

Where P_0 is the initial population, i.e. p (0) = P0, and k is called the growth or the decay constant. In this step, the rate of change of file popularity will be calculated by using Yahoo Hadoop audit log file [9] as data source. Users can enable audit logging from the NameNode. The Yahoo HDFS User Audit log format is shown in Fig. 2.

2016-12-1011:11:59,693INFO
org.apache.hadoop.hdfs.server.namenode.FSNamesystem.audit: **ugi**=hduser
ip=/134.91.100.59 **cmd**=delete **src**=/app/hadoop/tmp/test.txt **dst**=null **perm**=null

Fig. 2. HDFS user audit log format

To get the frequency count of each file, user audit log is split into small files based on timeslot duration and number of records. Then the required fields are extracted such as Date, Time, IP and src. After that, the user access frequency is counted from src source link from Fig. 2. In each time slot, the access frequency is counted and stored for individual files. Then the rate of change of file popularity of each file is calculated on each time slot according to Table 1 and Fig. 3.

Table 1. Notations used in popularity growth rate algorithm

Notation	Description
$P(t_f)$	Popularity values of file f
$AF(t_f)$	Total access frequency of file f at each time slot
inLog	The input log file
k	The rate of change of file popularity

Algorithm 1. Popularity Growth Rate Algorithm

Input: inLog
Output: *k*
1. Read inLog
2. Calculate access frequency of each file by using
$$P(t_f) = AF(t_f),\ \forall\, f \in F$$
3. Calculate the rate of change of file popularity k of each file by substituting $P(t)$ as $P(t_f)$ in (3)
4. return k.

Fig. 3. Popularity growth rate algorithm

According to the rate of change of file popularity, replica degree for each file is considered as follows. If the rate of change of file popularity is greater than 0, then existing replica degree is increased by one. If the rate of change of file popularity is less than 0, then existing replica degree is decreased by one. If the rate of change of file popularity is equal to 0 then existing replica degree is remained unchanged. Otherwise, if the accessed file is new and there is no access record history, the replica degree for this file will be assigned 3 as like HDFS default replica number.

3.2 Proposed Data Placement Algorithm

After determining the number of replicas, we will consider how to place these replicas efficiently in order to improve data locality and load balancing. In this step, let me assume that the jobs will have to access this replica in the next time slot. The incoming job is broken into tasks and each map task is assigned into nodes within the cluster. There is one map task per input block.

In this system, the input data file is divided into 64 MB blocks and place them into blocks within the cluster. For instance, if the replica for this file is 3 and this file has 4 blocks, then the total replica block number of this file is 12. Let the maximum number of replicas be the number of nodes in the cluster and the minimum number of replicas be 1.

Suppose that at the assigned node, there is no replica block for the incoming map task. In this case, this system considers for improvement of node locality. In this case, the remote data retrieval is performed by loading the replica data block into this node. While loading this data block, if the load factor of this node is less than the predefined threshold, this replica data block is loaded into this node. Otherwise, the replacement is performed by replacing this replica data block with existing block into this node.

The proposed data replacement algorithm is based on Least Recently Used (LRU). It will be more reliable than the LRU and will have the more efficient results than the LRU algorithm because it considers access frequency for replacement. The proposed enhanced LRU replacement algorithm is shown in Fig. 4.

The existing Hadoop block placement strategy does not take into account DataNodes' utilization, which leads to in an imbalanced load. This policy assumes that all nodes in the cluster are homogeneous, and randomly place blocks without considering

Algorithm 2. Enhanced LRU Algorithm

Step 1. When loading the replica data block into the assigned node, it will calculate total number of access frequencies (TAF) for all blocks in that node.

Step 2. If only one block with minimum TAF is found, that block will be selected to evict from that node.

Step 3. If one or more minimum TAF blocks are found, least recently accessed block (outgoing block) will be selected to evict from that node as LRU.

Fig. 4. Enhanced LRU algorithm

any nodes' resource characteristics, which decreases self-adaptability of the system. Therefore, this system considers the heterogeneous environment for nodes in the cluster. We need to consider the load factor such as storage utilization, disk bandwidth and CPU processing speed. During the process of placement, the storage utilization, disk bandwidth and CPU processing speed of DataNode are important factors to affect the load balancing in HDFS. Therefore, the capacity of DataNode stored should be proportional to its total disk capacity, in the condition of effective load balancing. We can carry out the storage utilization model as

$$U(D_i) = \frac{D_i(use)}{D_i(total)} \tag{4}$$

Where, $U(D_i)$ is the storage utilization of the ith DataNode. $D_i(use)$ is the used disk capacity of the ith DataNode, and its unit is GB. $D_i(total)$ is the total disk capacity of the ith DataNode, it is a fixed value of each DataNode, and its unit is GB.

Then, we can carry out the disk bandwidth model as

$$BW(D_i) = \frac{T_b}{T_s} \tag{5}$$

Where, $BW(D_i)$ is the disk bandwidth of the ith DataNode. T_b is the total number of bytes transferred, and T_s is the total time between the first request for service and the completion of the last transfer.

Then, the CPU processing speed is used as one of the important factors and each node has different CPU processing speed due to the heterogeneous environment.

Among these three factors, storage utilization is set as first priority, disk bandwidth as second priority and CPU processing speed as last priority.

So, we put the coefficients of storage utilization, disk bandwidth and CPU processing speed are set as $\propto$, β and γ. Therefore, we can carry out the load factor model as

$$LF(D_i) = \alpha U(D_i) + \beta BW(D_i) + \gamma SP(D_i) \tag{6}$$

The predefined threshold Ti of the ith cluster is assumed as the sum of maximum storage utilization, maximum disk bandwidth and maximum CPU processing speed in cluster is divided by the number of nodes in the cluster. Therefore, we can carry out the predefined threshold of cluster Ci as

$$T_i = \frac{Max_i(U) + Max_i(BW) + Max_i(SP)}{N} \tag{7}$$

Where, T_i is the predefined threshold of the ith cluster and N is the number of nodes in the ith cluster. If the load factor of this node is less than the predefined threshold, this replica data block is loaded into this node. The proposed data placement algorithm is shown in Table 2 and Fig. 5.

Table 2. Notations used in data placement algorithm

Notation	Description
DN	DataNodes list
BW	Bandwidth
U	Storage utilization
RP	Replica List
MT	Map task list
SP	CPU processing speed
C	Cluster list
LF	Load factor list

4 Performance Evaluation

In this section, the probabilistic model is applied to evaluate the analysis of availability. In this model, data file F is stripped into n blocks denoted as B = {b1,b2,b3,…,bn} and stored into different DataNodes. There are rn replicas of each block in data file F, and all blocks at the same site will have the same available probability as all blocks are stored in DataNodes with the same configuration. Assume that NameNode do not fail any condition. To analyze the data availability of this system, assume that node failures are independent with failure probability. If a node fails, all of the blocks will lose on the nodes. The block availability of data block Bn is denoted as BA_n. $P(BA_n)$ is the probability of data block Bn in an available state. $P(\overline{BA_n})$ is the probability of data block Bn in an available state. Let the number of replicas of data block bn be r_n and the available and unavailable probability of each replica of data block bn are $P(b_n)$ and $P(\overline{b_n}) = 1 - P(b_n)$. So, the availability and unavailability of data block Bn are calculated as

$$P(BA_n) = 1 - (1 - p(b_n))^{r_n} \tag{8}$$

$$P(\overline{BA_n}) = (1 - p(b_n))^{r_n} \tag{9}$$

Algorithm 3. Data Placement Algorithm

Input: DataNodes List DN= {DN_1, DN_2,.., DN_n }, Replica List RP ={ RP_1, RP_2, RP_3,…., RP_n }, Map Task List MT = {MT_1,MT_2,MT_3,…,MT_n}, Load Factor List LF = {LF_1,LF_2,LF_3,…., LF_n}, Predefined Threshold T_i, Cluster List C = {C_1, C_2, C_3,…, C_n}
Output: DataNodes List DN
for each incoming map task **MT do**
  for each DataNode **DN do**
    Check node locality of task **MT_i**
    if there is node locality **then** assign task **MT_i** to that DataNode **DN_i**
    else
      Perform remote data replica retrieval for task **MT_i**
        Calculate storage utilization **U** of this assigned DataNode **DN_i** using (4)
        Calculate disk bandwidth BW of this assigned DataNode **DN_i** using (5)
        Check CPU processing speed **SP** of this assigned DataNode **DN_i**
        Calculate load factor **LF_i** for this assigned DataNode **DN_i** using (6)
      Calculate predefined threshold **T_i** for the cluster **C_i**
      if LF_i > predefined threshold **T_i then**
        Perform replacement using algorithm 2
        Place replica **RP_i** for this task on that DataNode **DN_i**
      break
      else
      Place replica **RP_i** for this task on that DataNode **DN_i**
      break
      end if
    end if
  end for
end for

Fig. 5. Data placement algorithm

Consider the probability of each replica block bn in the system of n DataNodes that containing b data blocks replicated with the replication factor of br. So, the probability of bn available is

$$p(b_n) = \sum\nolimits_{f=0}^{n} p(failure) \times p(\text{no - loss}) \quad (10)$$

where *p*(*failure*) is the probability that there are exactly failures in the system with n DataNodes. So, *p*(*failure*) is

$$p(failure) = \binom{n}{f} \times p^f \times (1-p)^{(n-f)} \quad (11)$$

$p(\text{no\_loss})$ is the probability of not losing data in the system with f failures. To calculate $p(\text{no\_loss})$, firstly we compute the probability of losing a block of data when there is f failure that is defined as $p\left(\overline{single_{block}}\right)$. The two are related as

$$p(no\_loss) = 1 - p\left(\overline{single_{block}}\right)^{b} \tag{12}$$

The probability of losing a block of data when there is f failure in n DataNodes is

$$p\left(\overline{single_{block}}\right) = \binom{f}{br} / \binom{n}{br} \tag{13}$$

Therefore, we get

$$p(b_n) = \sum_{f=0}^{n} \binom{n}{f} \times p^f \times (1-p)^{(n-f)} \times (1 - \binom{f}{r} / \binom{n}{r})^b \tag{14}$$

To retrieve the data file F, we need to get all block $\{b_1, b_2, b_3, \ldots, b_n\}$. Any block unavailable will cause file unavailable. The file availability of file f_n is denoted as FA_n. $P(FA_n)$ is the probability of file f_n in an available state. $P\left(\overline{FA_n}\right)$ is the probability of file f_n in unavailable state. The availability and unavailability of file f_n are $P(FA_n)$ and $P\left(\overline{FA_n}\right) = (1 - P(FA_n))$. Then, the availability and unavailability of file f_n are calculated as

$$P(FA_n) = (1 - (1 - p(b_n)^{r_n})^{b_n} \tag{15}$$

$$P\left(\overline{FA_n}\right) = 1 - (1 - (1 - p(b_n)^{r_n})^{b_n} \tag{16}$$

To evaluate the availability of data block, the 1 GB of data is stored in Cluster based storage system. The average failure rate of each DataNode in the system is 0.01, 0.02, 0.03, 0.04 and 0.05 (Fig. 6).

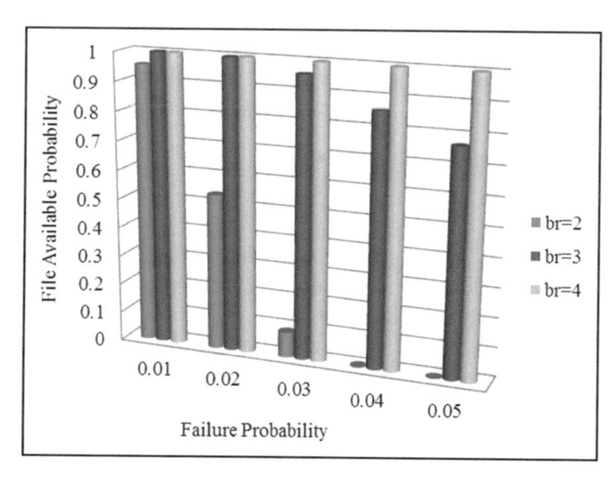

Fig. 6. File availability with replication factor (r) 2, 3 and 4 in the eighty DataNodes

It can be observed from the evaluation results, and file availability is a rapidly fall on replication factor 2 in the eighty number of data nodes in the system. According to the evaluation results, the file availability depends on the replication factors and failure probability. From the experiment, it is obvious that the expected availability values can be satisfied by increasing replication degree in the existence of failure on cluster.

5 Conclusion

In cloud storage environment, data can be stored with some geographical or logical distance and this data is accessible for cloud based applications. In this paper, a dynamic replication management scheme is proposed for cloud storage. At each time intervals, the proposed system collects the data access history in cloud storage. According to access frequencies for all files, the rate of change of file popularity rate can be calculated and replicated them to suitable DataNodes in order to achieve load balance and node locality of system. As a future work, many experimental evaluations have to be carried out in order to get the efficiency of proposed data placement algorithm. In addition, many experimental evaluations have to be performed in order to get better threshold value and load factor value. And as well, replica deallocation will be considered for overall system improvement.

References

1. Hunger, A., Myint, J.: Comparative analysis of adaptive file replication algorithms for cloud data storage. In: 2014 International Conference on Future Internet of Things and Cloud (2014)
2. Gong, B., Veeravalli, B., Feng, D., Zeng, L., Wei, Q.: CDRM: a cost-effective dynamic replication management scheme for cloud storage cluster. In: 2010 IEEE International Conference on Cluster Computing, September 2010, pp. 188–196 (2010)
3. Abad, C.L., Lu, Y., Campbell, R.H.: DARE: adaptive data replication for efficient cluster scheduling. In: IEEE International Conference on Cluster Computing (CLUSTER 2011), pp. 159–168 (2011)
4. Lee, D., Lee, J., Chung, J.: Efficient data replication scheme based on hadoop distributed file system. Int. J. Softw. Eng. Appl. **9**(12), 177–186 (2015)
5. Ananthanarayanan, G., et al.: Scarlett: coping with skewed content popularity in mapreduce clusters. In: Proceedings of Conference on Computer. Systems (EuroSys), pp. 287–300 (2011)
6. Gobioff, H., Ghemawat, S., Leung, S.-T.: The Google file system. In: Proceedings of 19th ACM Symposium on Operating Systems Principles (SOSP 2003), New York, USA, October 2003
7. Chang, H.-P., Chang, R.-S., Wang, Y.-T.: A dynamic weighted data replication strategy in data grids. In: 2008 IEEE/ACS International Conference on Computer Systems and Applications, March 2008, pp. 414–421 (2008)
8. Zaharia, M., Borthakur, D., Sen Sarma, J., Elmeleegy, K., Shenker, S., Stoica, I.: Delay scheduling: a simple technique for achieving locality and fairness in cluster scheduling. In: Proceeding of European Conference Computer System (EuroSys) (2010)
9. https://webscope.sandbox.yahoo.com

Frequent Pattern Mining for Dynamic Database by Using Hadoop GM-Tree and GTree

Than Htike Aung(✉) and Nang Saing Moon Kham

Information Science Department,
Computer University Yangon, Yangon, Myanmar
{thanhtikeaung, moonkhamucsy}@ucsy.edu.mm

Abstract. Since origination of mining, frequent pattern mining has become a mandatory issue in data mining. Transaction process for mining pattern needs efficient data structures and algorithms. This system proposed tree structure, called GMTree (Generate and Merge Tree)-GTree (Group Tree), which is a hybrid of prefix based incremental mining using canonical order tree and batch incrementing techniques. Proposed system make the tree structure more compact, canonically ordered of nodes and avoids sequential incrementing of transactions. It gives a scalable algorithm with minimum overheads of modifying the tree structure during update operations. It operates on extremely large transaction database in dynamic environment which is especially expected to give better results in this case. The proposed system used Apache Hadoop and hybrid GMTree-GTree. The results shows Hadoop implementation of algorithm performs more times better than in Java.

Keywords: Canonical · Frequent pattern · GMTree · GTree · Hadoop
Hybrid

1 Introduction

Over the past few decades, many researchers have proposed many algorithms for discovering frequent item sets from a given data set [1, 4, 7, 14, 16]. Generally, the data set can be of two types static or dynamic. Most algorithms focus on a static data set. However, static data mining algorithms which can't be applied to a dynamic data set. Specially, a real time method is preferred to a batch processing method. Mining on streaming data can be categorized into the following types on the basis of the window concept: Landmark, damped and sliding-window. Landmark model, mainly focus on data set that is observed from a fix time in the past to present time. In the damped approach, frequent item set are extracted in data stream where every transaction of the data is allotted a weight that reduce with age. The sliding-window model, the item sets are collected in a certain interval of time from the present time.

The proposed system's algorithm are sliding-window technique [6, 12] which moves per unit batch. The proposed system efficiently representing the transaction makes use of base-tree that is constructed from GMTree, which is almost the same as CanTree. The

T. T. Zin and J. C.-W. Lin (Eds.): ICBDL 2018, AISC 744, pp. 149–159, 2019.
https://doi.org/10.1007/978-981-13-0869-7_17

change between the two is that when a new arrives, similar items generally form a single node in the new tree for comparison (not need each new item compared as a CanTree.) and which is not need restructure the entire tree same as FPTree.

GMTree node represents a set of nodes that have the same data item in the base tree (GMTree or parent GTree). That is the proposed algorithm called Hadoop GMTree-GTree. It is very sample and efficient. The algorithm has the following properties.

- Single database scan.
- Usage of sliding-window
- Similar items generally form a single node in the new tree for comparison (not need each new item compared).
- Finding exact and complete frequent item sets.

2 Related Work

FPTree data structure which was used in FP-growth algorithm [2] to represent a compact tree structure of transaction DB into main memory. FP-growth algorithm used a divide-and-conquer method which produces frequent itemsets, and needs only two DB scans. This algorithm works very efficient memory usage and processing cost.

Most successful data mining method has led to many FP-growth-based algorithms which used for a static data set to reduce the DB scans times. But, FPTree based FP-growth algorithm cannot be applied to stream data mining environment.

However, FP-growth algorithm has become used on many stream data mining methods, such as FP-stream [5], DSTree [21], CanTree [13], FUFP-Tree [7, 8], CPSTree [20], and so on [16, 17]. Furthermore, many researcher study based on apriori-based algorithms, such as SWF [13], SWFI stream [12], and MFI-TransSW [15]. These incremental mining methods show good performance and mining results for several applications, but basically which has a limitation on dealing with data streams.

There are several studies [9, 10, 20], more processing cost and memory is a considerably which commonly needed to generate and test the candidate itemsets. The output results of candidate itemsets generation have huge processing cost, especially if there are a huge amount of items (or the length of candidate itemsets is long).

3 Background Knowledge of GM-Tree

This section described tree structure which is called GM-Tree. GM-Tree is being maintained frequent patterns. The data structure of GM-Tree improved functionality and its performance by combining canonical ordering and batch increment techniques. The proposed method is used by incremental mining technique to maintain frequent item sets that are discovered in transaction databases. GM-Tree updates its data structure efficiently When databases are frequently changed by additions, deletions and modifications of transactions. Batch Incremental technique mentions merging into two datasets (here in the form of a tree) to obtain a new data set that is equivalent to the entire database formed by the two sets. The tree structure is made by more compact, avoids sequential increment of transaction, canonically ordered of nodes and gives a scalable algorithm with minimum overheads of modifying the tree structure of update operations. This algorithm makes a single scan through the first database to build a tree structure. The tree so formed has items arranged from root to the leaves in a lexicographical order way, hence ordering of the items is unaffected by their frequencies.

The support frequency is taken into account while mining the tree. Now to deal with the dynamic environment, a new similar tree is constructed from new transactions in the database. Once created, the new tree is merged with the last updated tree forming the corresponding tree structure for the whole updated database, avoiding a re-scan of the entire updated database and thus providing a better efficiency. The above statements can be summarized into two important properties of the GM-Tree described below:

Property 1: Nodes in GM-Tree are ordered lexicographically and thus the ordering is unaffected by changing item frequency.

Property 2: New transactions are used to generate another tree which is then merged with the last updated tree, preventing re-scan of the entire updated database.

Figure 1 shows the GM-Tree generated from Table 1. Improved GM-Tree solves the limitations of the above invoked trees as follows:

(1) In case of the GM-Tree, nodes are arranged in a canonical (i.e. lexicographic) order and hence while merging two trees, it is not necessary to check and swap nodes with the changing frequency.
(2) GM-Tree is not affected by frequency count and thus swapping or bubbling and re-scanning of the nodes can be completely avoided which makes it more time efficient.
(3) During GM-Tree construction, when a new tree is formed, only the nodes of this new tree needs to be compared with the last updated tree nodes. Thus, when the data size is increased and the size of the tree increases, a large number of unnecessary comparisons can be avoided. This indicates that GM-Tree is well suited for extremely large database.
(4) GM-Tree needs more memory while merging two trees but reduces the computational time drastically as there is no swapping or re-scanning of nodes required. Moreover, in this modern world, space requirement (i.e. main memory) is no more a big concern [3, 21].

Algorithm steps for improved GM Trees are as follow:

(1) Create a Tree (T) from the content of the original database ($db_{original}$) and order the nodes of T lexicographically.
(2) Create another new Tree (t) from the database (db_{new}) having new transactions (considering those transactions which are entered into the database within a predefined time) and order the nodes of t lexicographically.
(3) Merge the tree t to the last updated tree T. The tree thus formed after merging t to T will have the content of the entire updated database ($db_{original}$) $\cup$ (db_{new}) with tree nodes ordered lexicographically (i.e. $T \leftarrow T \cup t$).
(4) Continue with Step 2 when new transactions entered into (db_{new}).

Table 1. Transaction database

Original database		New entries (1st Group)	
Trans Id	Transactions	Trans Id	Transactions
t1	{p, a, s, t}	t5	{a, t, c}
t2	{b, s, c, a, t}	t6	{s, t, a, c}
t3	{a, b, q, s, t}	New entries (2nd Group)	
t4	{t, a, s}	t7 {b, c, q, t}	

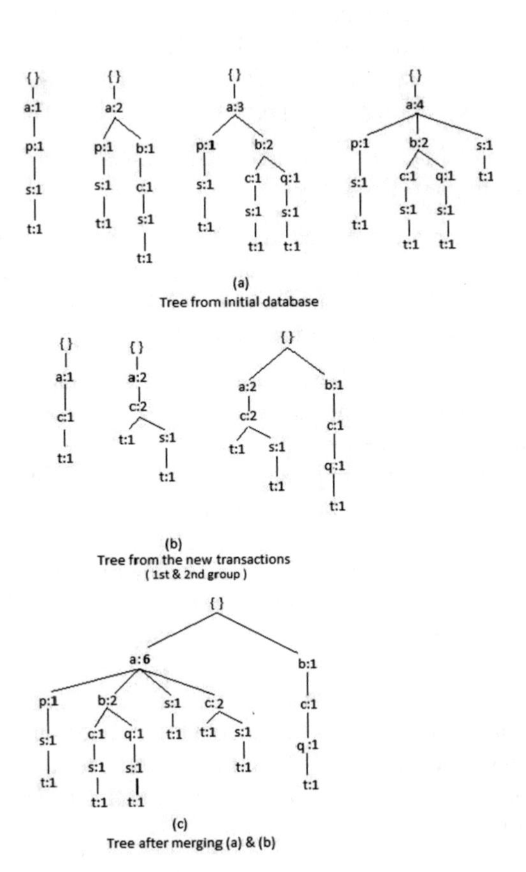

Fig. 1. GMTree after adding each transaction

4 Proposed Methodology

The proposed system used GMTree-GTree data structures with sliding-window method. This data structures are more efficient than CanTree-GTree data structures. This data structures of new tree for data nodes need compared on similar item generally form single node. In CanTree, it may generate a skewed tree with too many branches and hence with too many nodes. It considers one transaction from database at a time, which drops its time efficiency. It produces concise tree only if majority of transactions have a common pattern.

4.1 GMTree-GTree Algorithm

The GMTree-GTree algorithm [4] makes use of the sliding window technique [2]. A batch contains a group of transactions that is treated as a single entity. A sliding-window consists of 'k' groups of transactions, where 'k' is the window size. When a window becomes full, the earliest batch is removed and fresh batch is inserted.

The following data structures are used in this method:

(1) GMTree [14, 20] is a base tree that efficiently stores the transactions in the current window. A GMTree will have an item-table (iTable) and last-node-of-transaction-table (lTable) associated with it. Each row of iTable consists of the item id, the item's support count, and a list of nodes with this item in GMTree. Each row of lTable consists of the index of the batch, and a list of the last nodes of transactions in GMTree.
(2) GTree is a projection-tree, built from the GMTree and it is used to mine frequent patterns. A GTree also has an iTable associated with it.
(3) Original tree nodes are comparing with similar items generally forms a single node in the new tree.

ALGORITHM 1. Tree Generation from transaction database (DB)
Data: Transaction Database(DB)
Result: lexiographic order of Tree nodes
1. Set the Transaction Pattern base empty
2. Declare head node of GM-Tree, head = null
3. Scan the transaction database (DB), sorting each transaction T_1 lexicographically until the last transaction reached
for every $T_1 \in DB$ **do**
  if the first item does not exist as the child of the head **then**
    add the transaction to the child of head with count value 1
    add the transaction to *iTable*
  end
  if first item exists as a child of the head **then**
  while items $I \in T_1$, equals to the node of the Tree **do**
    increment their count value by 1
  end
    add the remaining items $I \in T1$ to the tree with a count value as 1
  end
end

ALGORITHM 2.Merge (T, t)
Data: Tree (T) $\in$ ($db_{original}$) and Tree (t) $\in$ (db_{new}) , new batch
Result: New Merged Tree ($T \leftarrow T \cup t$) having the lexicographic order of nodes
intialization: $head_1 \leftarrow$ T and $head_2 \leftarrow$ t
for all child nodes, $N_2 \in head_2$ **do**
  if N_2 exists as a child node of $N_1 \in head_1$ **then**
    add the count of the two nodes
    Merge (N_1;N_2)
    update N_1 to iTable
  end
  else
    add N_2 as a child of $head_1$
  end
end
Add T to *lTable* for a new batch;

ALGORITHM 3. GMTree-GTree

Input: a sequence of batches
Output: frequent itemsets window, *listOfItemsets*
Method:

```
GMTree-Gtree( ) {
    While (true) {
        Get a new batch;
        If a window is full of batches,
            Delete the oldest batch from GMTree;
            Add the new batch to GMTree;
        For each data item D in GMTree, construct GTree_D from GMTree;
        itemset = null;
        listOfItemsets = null;
        call PreorderTraverse(GTree_D, itemset);
    }
}

PreorderTraverse(GTree, itemset) {
    If (GTree is null) return;
    newItem = itemset + item(root(GTree)); add newItem to listOfItemsets;
     For each data item X in (I(GTree)-{root of GTree}) {
        If the support of X >= minimum support { construct GTree_X from
        GTree;
        call PreorderTraverse(GTree_X, newItem);
        }
    }
}
```

All frequent or infrequent data items of transactions are stored in the base-tree. The GMTree is a little different from FPTree. In GMTree the data items of a transaction are sorted in canonical order before adding them to the GMTree whereas in FP Tree the ordering is based on frequency.

Two DB scans are required in the case of FP Tree algorithm, whereas only one DB scan will be required by the GMTree. Hence for real-time applications, the GMTree-GTree algorithm works better and is more suitable than the FP-growth algorithm [6, 18, 20].

4.2 Mining Closed Frequent Item Sets, Association Rules and Implementation on Hadoop

Modify GMTree-GTree algorithm [4] is used to mine for closed frequent itemsets [4, 11]. Alogrithms are both closed and their support is more than a minimum threshold. Consider an item set for which there does not exist any superset which has equal support count, then that item set is closed in the specifies data set.

Closed frequent item sets are suitable for used more than maximal frequent item set because its efficiency is of more importance that space, the support of the subsets is provided. Hence this methods, an additional pass is not required to find this information.

Knowledge discovery process is important in data mining and which is usually obtained by mining association rules [1, 2, 19]. These are basically used if/then statements that support us to discover relationships between data which seems unassociated in a relational database, data warehouses or repositories also. For example, consider two frequent item sets {a, c, e} with support count 2 (Fig. 2).

The dataset is illustrated in Table 2 as follows:

Table 2. Sample dataset

TID	Items
100	a c d
200	b c e
300	a b c e
400	b e
500	a c e

For every non empty subset s of I, output the rule: s -> (I-s), if supportcount (I)/supportcount(s) >= minimum confidence.

Let us suppose the minimum confidence be 60.
For R_1: a, c -> e
Confidence = 2/3 = 66.66%. Rule is selected. For R_2: c -> a, e
Confidence = 2/4 = 50%. Rule is rejected.

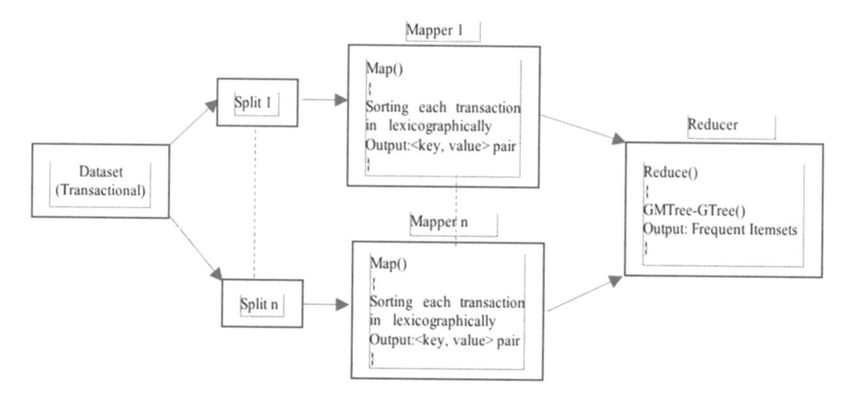

Fig. 2. Hadoop GMTee-GTree flowchart

5 Results and Discussion

The GMTree-GTree algorithm was first implemented in single node using Java and was tested using the kosarak (.gz) dataset which is a real world dataset from an hun`garian news portal and consists of click stream data of a newspapers web portal. Then, this algorithm was implemented on the Hadoop framework after making some modifications and the time taken was compared to that of single node using Java. The dataset used for comparison was a web documents dataset "webdocs" from the FIMI repository. It is a transactional dataset that contains the main characteristics of a spidered collection of web html documents. The size of the dataset is about 1.48 GB and contains approximately 17 lakh transactions. The experiments were performed on the following system: Single node using Java: Hardware: Intel core i5, 8 GB RAM, CPU 2.4 GHz Software: Windows 8.1, Java with JDK 1.8 Hadoop implementation: Hardware: Intel core i5, 8 GB RAM, CPU 2.4 GHz Software: Ubuntu 16.10, Java with JDK 1.8, Hadoop 2.7.2 (pseudo distributed mode).

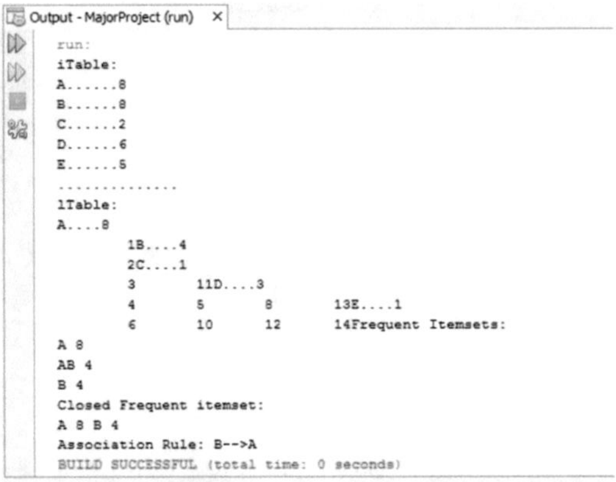

```
Output - MajorProject (run)
run:
iTable:
A......8
B......8
C......2
D......6
E......5
..............
lTable:
A....8
          1B....4
          2C....1
          3         11D....3
          4         5         8         13E....1
          6         10        12        14Frequent Itemsets:
A 8
AB 4
B 4
Closed Frequent itemset:
A 8 B 4
Association Rule: B-->A
BUILD SUCCESSFUL (total time: 0 seconds)
```

Fig. 3. iTable and lTable along with complete and closed frequent item sets

Figure 3 shows the iTable and lTable of the constructed GMTree along with the complete frequent item sets, closed frequent item sets along with their support count and the association rules with 60% confidence and minimum support set to 4. This result is for the small dataset, as shown in Table 1. As we can see, the iTable contains the items A, B, C, etc. and their frequencies. The lTable contains the item, its maximum frequency and the last nodes in which it is present, which helps us to track the transactions easily. For example, here for C, the max frequency is 1 and it is present in the nodes 3 and 11. So, from this we get the frequent item sets by eliminating the ones below the minimum support count. Also, the closed frequent item sets and the association rules are obtained by the method as discussed in Sect. 4.2.

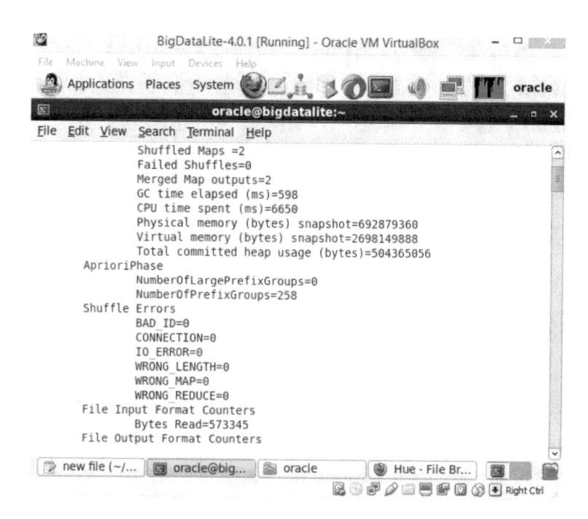

Fig. 4. Hadoop implementation

Figure 4 is a screenshot when the Hadoop implementation of the GMtree-Gtree algorithm was completed successfully. The GMTree-GTree algorithm was executed in the Hadoop framework with 50 input splits, 2 maps and 2 reducer. The total execution time taken is almost 2 h using Hadoop while it took almost 10 h to complete execution using single node Java program.

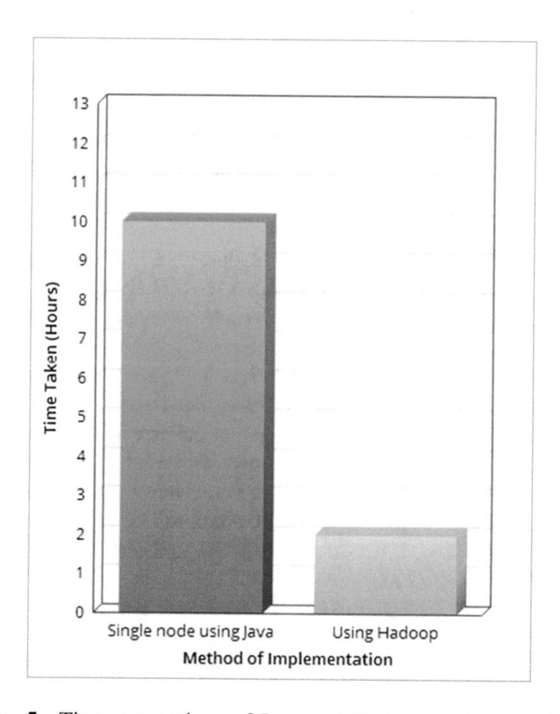

Fig. 5. Time comparison of Java and Hadoop implementation

Figure 5 shows a graph that compares the execution time taken by the implementation in single node Java and Hadoop framework. It can be seen that it takes much lesser time using Hadoop as compared to simple Java execution as Hadoop is designed to handle big data efficiently and splits the given input and feeds each input split into different mappers that execute parallely. This leads to the significant time reduction in Hadoop framework as compared to a sequential execution in single node using Java.

6 Conclusion and Future Work

From the results obtained, we can observe that the Hadoop GMTree-GTree algorithm works well for mining frequently occurring patterns in real-time streaming data. As the window slides, the algorithm adds new transactions to the GMTree without any need for restructuring. Here, GTree is used for constructing the projection-tree in order to discover the frequently occurring item sets. So, this algorithm would be more time-efficient for mining the complete frequent item sets from dynamic, streams of data also. Using the same proposed algorithm in a MapReduce framework significantly lowers the execution time taken.

As future work, other tree algorithms for data stream mining considering the real-time conditions will be implemented in Hadoop or any other similar frameworks like Apache Spark, Storm etc.

References

1. Agrawal, R., Imielinski, T., Swami, A.N.: Mining association rules between sets of items in large databases. In: Proceedings of the ACM SIGMOD Conference on Management of Data, pp. 207–216 (1993)
2. Chang, J., Lee, W.: A sliding window method for finding recently frequent itemsets over online data streams. J. Inf. Sci. Eng. **24**(4), 753–762 (2004)
3. Cheung, W., Zaiane, O.R.: Incremental mining of frequent patterns without candidate generation or support constraint. In: Proceedings of Seventh International Database Engineering and Applications Symposium, pp. 111–116. IEEE (2003)
4. Chi, Y., Wang, H., Yu, P.S., Muntz, R.R.: Catch the moment: maintaining closed frequent item sets over a data stream sliding window. Knowl. Inf. Syst. **10**(3), 265–294 (2006)
5. Giannella, C., Han, J., Pei, J., Yan, X., Yu, P.S.: Mining frequent patterns in data streams at multiple time granularities, pp. 192–209, 1 November 2016
6. Han, J., Pei, J., Yin, Y., Mao, R.: Mining frequent patterns without candidate generation: a frequent-pattern tree approach. Data Min. Knowl. Discov. **8**, 53–87 (2004)
7. Hong, T.-P., Lin, C.W., Wu, Y.L.: An efficient FUFP-tree maintenance algorithm for record modification. Int. J. Innovative Comput. Inf. Control **4**(11), 2875–2887 (2008)
8. Hong, T.P., Lin, C.W., Wu, Y.L.: Incremental fast updated frequent pattern trees. Expert Syst. Appl. **34**, 2424–2435 (2008)
9. Jiang, N., Gruenwald, L.: Research issues in data stream association rule mining. SIGMOD Rec. **35**(1), 14–19 (2006)
10. Krempl, G., Zliobaite, I., et al.: Open challenges for data stream mining research. ACM SIGKDD Explor. Newslett. **16**(1), 1–10 (2014)

11. Kumar, V., Satapathy, S.R.; A novel technique for mining closed frequent item sets using variable sliding window. In: IEEE International Advance Computing Conference (IACC) 2014, pp. 504–510 (2014)
12. Lee, C.H., Lin, C.R., Chen, M.S.: Sliding window filtering: an efficient method for incremental mining on a time-variant database. Inf. Syst. **30**, 227–244 (2005)
13. Lee, C., Lin, C., Chen, M.: Sliding window filtering: An efficient method for incremental mining on a time-variant database. In: Information Systems. Sliding Window-Based Frequent Pattern Mining, pp. 227–244 (2005)
14. Leung, C.K.S., Khan, Q.I., Hoque, T.: CanTree: a tree structure for efficient incremental mining of frequent patterns. In: Proceedings of the Fifth IEEE International Conference on Data Mining (2005)
15. Li, H., Lee, S.: Mining requent itemsets over data streams using efficient window sliding techniques. Expert Syst. Appl. **36**, 1466–1477 (2009)
16. Li, H., Zhang, N., Chen, Z.: A simple but effective maximal frequent itemset mining algorithm over streams. J. Softw. **7**(1), 25–32 (2012)
17. Mao, G., Wu, X., Zhu, X., Chen, G.: Mining maximal frequent itemsets from data streams. J. Inf. Sci. **33**(3), 251–262 (2007)
18. Nguyen, T.T.: A Compact FP-tree for Fast Frequent Pattern Retrieval. PACLIC 2013, vol. 27 (2013)
19. Shen, L., Shen, H., Cheng, L.: New algorithms for efficient mining of association rules. Inf. Sci. **118**, 254–268 (1999)
20. Tanbeer, S.K., Ahmed, C.F., Jeong, B.S., Lee, Y.K.: Sliding window-based frequent pattern mining over data streams. Inf. Sci. **179**, 3843–3865 (2009)
21. Zaki, M.J., Hsiao, C.-J.: Charm: an efficient algorithm for closed itemset mining. In: SDM, vol. 2, pp. 457–473. SIAM (2002)

Investigation of the Use of Learning Management System (Moodle) in University of Computer Studies, Mandalay

Thinzar Saw(✉), Kyu Kyu Win(✉), Zan Mo Mo Aung(✉), and Myat Su Oo(✉)

University of Computer Studies, Mandalay, Myanmar
thinzarsaw@gmail.com, kyukyu567@gmail.com, zanmomoaung@gmail.com, Myathsuoo90@gmail.com

Abstract. Nowadays, there is a substantial growth in the use of e-learning platforms in higher education around the world. Currently, the use of e-learning technology has become necessity within the Higher Education Institutions (HEIs) in Myanmar. Learning Management System (LMS) emerged as a tool in teaching and learning environment. LMS such as Modular Object Oriented term Developmental Learning Environment (Moodle) is a course management system through the Internet. This paper aims to investigate the undergraduate students' and teachers' perceptions on LMS (Moodle) usage in University of Computer Studies, Mandalay (UCSM). Quantitative data are collected through a questionnaire that is answered by undergraduate students and teachers in UCSM. The collected data was analyzed using the SPSS (Statistical Package for the Social Sciences). A total of 318 respondents answered the questionnaire. The results of this study indicate that Moodle is a real support for teaching and assessment evoluation in UCSM. However, there is no significant effect on improving students' semester grades and the interaction between students and teachers before and after using Moodle.

Keywords: E-learning · Learning Management System (LMS) Moodle

1 Introduction

Since the last few years, there has been increasingly interesting to the use of e-learning for society and educational environments. During the time, online learning, also called e-learning, along with various integrations of e-learning with traditional classes is improving rapidly. Nowadays, the use of e-leaning platforms in higher education over the world is growing substantially and it has gradually been as an important facilitator in teaching and learning process. Higher education institutions (HEIs) play a crucial role in undertaking research and incubating the innovative and creative thinking for globally and economically competitive society [5].

In other countries, Learning Management Systems (LMS) has been adopted by many HEIs. Because of various advantages including flexible learning times and effective learning, LMS has been widely used in higher education [7]. Modular Object

T. T. Zin and J. C.-W. Lin (Eds.): ICBDL 2018, AISC 744, pp. 160–168, 2019.
https://doi.org/10.1007/978-981-13-0869-7_18

Oriented term Developmental Learning Environment (Moodle), also recognized as a LMS, and is a course management system through the Internet. There are so many advantages using Moodle, for example, student-teacher interactions, foster student independence, and allow students more flexible time for learning, etc. Moodle can be installed at any number of servers without any cost and it is no need to pay maintenance costs for enhancing the system [9]. From the students' perspective, Moodle provide them with the ability to access the course materials, delivered by the instructors, and use communication and interactive features in their learning activities [1]. Educational institutions widely use this system to enhance their traditional teaching. The users around the world, such as universities, schools, teachers, instructors, courses, and societies, use this learning platform.

Being a developing technologically, Myanmar has a long way to go to reach high level growth rate for e-learning. However, the rate of LMS use among HEIs in Myanmar is very still low. It is obvious that the lack of knowledge about the LMS user acceptance, technology, infrastructure, human resources and afraid of IT. Although several studies have focused on the student's perception of LMS among other countries, there has been very little research reported on lecturers' perception of LMS.

Therefore, this research aims to investigate the use of Moodle Learning Management System as part of the teaching process related to undergraduate students' and teachers' perceptions in UCSM, to improve the teaching and learning techniques in computer universities and to know the usefulness of Moodle usage. This paper intends to achieve the improvements of teaching-learning process in other computer universities.

2 Related Works

In the past decades, the effectiveness of LMS has been studied on students' assessments as a more frequently used teaching technique. There are many advantages derived from e-learning. A well designed e-learning system could provide advantages like timely access to resources, up-to-date learning materials [3], quickness access, cost effective [4], interactive and collaborative [7]. Holley claimed that students in HEIs who participate in online learning achieve better performance than students who engage in traditional face-to-face learning [1]. Moreover, it was shown that the students' grades and qualifications increased when using the mixed technology of e-learning in Moodle and face-to-face lectures.

From the another point of view, Russell (2001), it was also stated that there was no statistically obvious differences between classical and online learning. By using LMS, teachers can create and manage educational courses more quickly and more easily. In addition, they can also exchange information with students over the network, engage students in online discussion through forum and student performance can also be accessed via online [4]. The students are provided with ability to access lecture notes, and use communication and interactive features in their learning activities by using LMS [3].

Some researchers pointed out that Moodle e-learning system is an important guideline for university management when further investigating how to enhance students' performance on different levels in the teaching process. Beside, Wiburg (2003) argues that the integration of LMS in teaching and learning is associated with several learning opportunities, such as enhancing students' critical thinking and problem-solving skills development.

Apparently, the success of LMS in any educational institutions can be defined by students' acceptance and utilization of LMS in classes. Understanding students' perception towards an e-learning system is a crucial issue for improving e-learning usage and effects. It was described that user acceptance is often the pivotal factor determining the success or failure of an information system project.

The use of e-learning materials increases as a fundamental resource for institutions among other countries. In the last decade, in Myanmar, there was an effort to apply LMSs in all universities. All of the functionalities provided by Moodle e-learning system tend to motivate students, enhance their efficiency and cost-savings [12] and especially tend to accelerate their learning processes [13]. However, till now, there are a small number of computer universities in Myanamr using LMS. The obvious facts for not using LMS in computer universities are the lack of knowledge about the LMS, technology, infrastructure and human resources.

This study focuses on the undergraduate student's and teachers' perception of LMS usage in Computer Universities, especially UCSM. Nevertheless, this paper tried to fill above mentioned obvious facts and highlight the vital role of Moodle usage on undergraduate students' and teachers' perception in the learning environment at UCSM.

3 Research Methodology

3.1 Data Collection Procedures

Survey questionnaire are the main methods of data collection. The survey data are collected using the paper survey collector. A total of 240 students and 78 of teachers responded to the questionnaire. The 22 questions used in this study examine the perceptions on students and teachers based on three categories (general conveniences, emotions and technological barriers). Most of the statements are constructed by referencing previously questionnaire (e.g., Faculty survey sample spring-2010). Specifically, this questionnaire aims to establish students' perceptions of how interesting, how useful the Moodle and examine teachers' perceptions how to get more comprehensive view of the Moodle.

In this questionnaire, four types of questions such as 5-point Likert scales, multiple choices, closed and open type. Likert scales are coded to 1–5, where 1 = Strongly disagree, 2 = Disagree, 3 = Neutral, 4 = Agree and 5 = Strongly agree. Multiple choice questions are coded to 1-5 according to rank order, Closed type (Yes and No) questions are coded to 1–2, where 2 = 'Yes' and 1 = 'No'.

The data for all 318 respondents in 2017-2018 academic years in UCSM are presented in Moodle survey data. Each of the questions is known as a variable number (Q1, Q2, etc.) Each variable number corresponds to the question number in survey data (i.e. Q1 is question 1; Q2 is question 2, etc.) An important issue arises in the management of data as to how to handle 'missing data'. Missing data arise when respondents fail to reply to a question either by accident or because they do not want to answer the question. The missing data has been treated as a zero (0).

3.2 Statistical Analysis

To analyze the data, two statistical methods are used. In the first analysis, factor analysis is run to obtain reliability and validity measures for the items of the questionnaire using SPSS. Testing the validity of the questionnaire was conducted using Pearson Correlations Coefficient in this study. In the validity test, significance level of 1% and the total of survey respondents are used. If sig. (2-tailed) value is less than significant value (0.01) and the value obtained from analysis is greater than r table product moment 0.165, it can be concluded that the question is valid.

After completing the validity test of the questionnaire, the consistency and reliability of a questionnaire is needed to test. To determine reliability, the internal consistency of the items representing each questionnaire is evaluated by using Cronbach's alpha. Cronbach's alpha method is most commonly used and popular to assess the internal consistency of a questionnaire (or survey) that is made up of multiple Likert-type scales and items.

Cronbach's alpha is calculated based on the respondents' answers. The range of values for the reliability method is from 0 to 1. Values near 1 show good reliability. Usually, if the value is more than 0.7, the questionnaires are reliable. The Cronbach's alpha values for teachers and students are shown in Table 1.

Table 1. Reliability statistics for the perception of students and teachers on Moodle

	Cronbach's alpha	No. of respondents
Teachers	0.764	78
Students	0.714	240

In second, descriptive statistics is used to describe the numerical data where data are presented in the form of tables, charts or summarization by means of percentiles and standard deviation.

4 Findings

This study examines students' and teachers' perception on LMS (Moodle) usage in UCSM. The LMS (Moodle) has been introduced to 2014–2015 academic years in UCSM. The results of descriptive analysis on Moodle usage based on the survey data of only first semester at 2017–2018 academic years in UCSM are shown as follows Tables 2 and 3 with three categories:

According to the students' and teachers' perceptions, Moodle is easy and user-friendly. It is a real support for teaching, assessment evaluation and technology. And also time table, course information and students' semester grade results are accessed by using Moodle. Thus, most students and teachers prefer to make tutorials and assignment on Moodle to traditional. But only half of the teachers and students would like to make tutorials and lab test because of internet access. 44% of teachers in UCSM agree that teachers' workload can be reduced by using Moodle. But, there is no significant effect on improving students' semester grades and the interaction between students and teachers.

Table 2. The Survey Questionnaire: Responses and Percentages for students

General convenience	Strongly disagree	Disagree	Neutral	Agree	Strongly agree
It is easy to download the course materials	2 (0.8%)	6 (2.5%)	20 (8.4%)	117 (49%)	94 (39%)
It is easy to submit and download assignments	1 (0.4%)	10 (4.2%)	30 (12.6%)	154 (64.4%)	44 (18.4%)
Moodle enables to access time tables and course information	1 (0.4%)	19 (8%)	67 (28.2%)	134 (56.3%)	17 (7.1%)
Moodle enables to make online discussions and forums for students	20 (8.4%)	58 (24.5%)	116 (48.9%)	36 (15.2)	7 (3%)
Moodle enables to get easily interaction between students and teachers	9 (3.8%)	51 (21.5%)	87 (36.7%)	82 (34.6%)	8 (3.4%)
Moodle enables to achieve the teaching and learning improvement	1 (0.8%)	14 (5.9%)	71 (29.7%)	131 (54.8%)	21 (8.8%)
Moodle enables to get grade results for students' assessments	5 (2.1%)	13 (5.5%)	63 (26.6%)	126 (53.2%)	30 (12.7%)
Emotions				**Yes**	**No**
Do you think Moodle is a real support for learning process?				202 (85.6%)	34 (14.4%)
Do you have enough time to use Moodle during weekdays?				93 (38.9%)	146 (61.1%)
Do you want to use Moodle in 24 h during weekdays?				154 (64.4%)	85 (35.6%)
Would you like to use chat and forum functions of Moodle for knowledge sharing?				192 (80.3%)	47 (19.7%)
Do you like learning with Moodle?				213 (91.4%)	20 (8.6%)
Do you think Moodle is user-friendly?				191 (82%)	42 (18%)
Overall do you satisfy with Moodle service?				184 (78%)	45 (22%)
Technology barriers				**Yes**	**No**
Do you think this university can support the 24 h electricity?				180 (75.3%)	59 (24.7%)
Do you think this university needs the extra technical supports such as computers, lab room and internet access, etc.?				179 (75.5%)	58 (24.5%)
Do you need additional time and training to learn with the Moodle				114 (48.3%)	122 (51.7%)
In addition to Moodle, have you ever been used other Learning Management System?				69 (29.7%)	163 (70.3%)
Do you have difficulties when using Moodle?				71 (30%)	166 (70%)

Table 3. The Survey Questionnaire: Responses and Percentages for teachers

General convenience	Strongly disagree	Disagree	Neutral	Agree	Strongly agree
It is easy to upload the course materials	2 (2.6%)	1 (1.3%)	4 (5.1%)	49 (62.8%)	22 (28.2%)
It is easy to download the answers of tutorials and assignments in Moodle	2 (2.6%)	1 (1.3%)	15 (19.2%)	46 (59.0%)	14 (17.9%)
Moodle enables to access time tables and course information	2 (2.6%)	2 (2.6)%	14 (18.4%)	47 (61.8%)	11 (14.5%)
Moodle enables to get easily interaction between students and teachers	2 (2.6%)	1 (1.3%)	20 (26.0%)	51 (66.2%)	3 (3.9%)
Moodle enables to view semester grade results for students' assessments	3 (4.0%)	4 (5.3%)	15 (20.0%)	46 (62.3%)	7 (9.3%)
I prefer to make the Software and Hardware Lab test on Moodle than traditional test	2 (2.7%)	15 (20.0%)	21 (28.0%)	31 (41.3%)	6 (8.0%)
I prefer to make tutorial and assignments on Moodle than traditional.	3 (3.8%)	6 (7.7%)	13 (16.7%)	47 (60.3%)	9 (11.5%)
Teachers' workload can be reduced by using Moodle	3(4.0%)	10 (13.3%)	29 (38.7%)	30 (40.0%)	3(4.0%)
Emotions				Yes	No
Do you think Moodle is a real support for teaching-learning process?				70 (92.1%)	6 (7.9%)
Would you like to use chat and forum functions of Moodle for knowledge sharing?				41 (56.2%)	32 (43.8%)
Do you have enough time to use Moodle during weekdays?				60 (81.1%)	14 (18.9%)
Do you think that Moodle enables to improve the students' semester grade?				35 (50.7%)	34 (49.3%)
Do you want to use Moodle in 24 h during weekdays?				38 (50%)	38 (50%)
Do you like teaching with Moodle?				61 (87.1%)	9 (12.9%)
Do you think Moodle is user-friendly?				67 (88.2%)	9 (11.8%)
Overall do you satisfy with Moodle service?				64 (82%)	10 (18%)
Technology barriers				Yes	No
Do you think this university can support the 24 h electricity?				63 (84.0%)	11 (14.7%)
Do you need additional time and training to learn with the Moodle				25 (67.7%)	24 (33.3%)
Do you think this university needs the extra technical supports such as computers, lab rooms and internet access, etc.?				55 (78.6%)	15 (21.4%)
In addition to Moodle, have you ever been used other Learning Management System?				12 (16.9%)	59 (83.1%)
Do you have difficulties when uploading and downloading files?				17 (22.4%)	59 (77.6%)

Most students would like to use chat and forum functions of Moodle for knowledge sharing. Almost all teachers and students apply Moodle for teaching and learning at least three times per week. Therefore, the frequency of using LMS has been good. Teachers have enough time to use Moodle during weekdays. Nevertheless, there isn't enough for students. The desire ratios of the students and teachers to use Moodle in 24 h during weekdays are about 2:1 and 1:1.

Some difficulties are faced when using Moodle because of low internet speed. Thus, internet access should be speed up. Although ppt and pdf files are easy to upload and download, mp3 and mp4 files are difficult. They prefer to use Wi-Fi access the whole place in UCSM. They want to use Moodle not only local network but also other network.

To use Moodle, there must be technical resources (computers, lab rooms, faster internet access and 24 h electricity). These resources have been completely supported in this university. Teaching and learning with Moodle is comfortable for students and teachers. Overall, they are satisfied with Moodle service.

5 Discussion

The success of LMS in any educational institutions starts by students' acceptance and promotes students' utilization of LMS in classes. Understanding students' perception towards an e-learning system is a crucial issue for improving e-learning usage and effects (Davis 1993). In this work, the perception of teachers and students on the use of Moodle are examined for teaching and learning process.

This study shows the positive impact LMS on students and teachers. According to the response from the undergraduate students of academic year 2017-2018 in UCSM, the frequency of using LMS has been good. The students satisfied that course materials have become more convenient. Students learning skills have been improved. Overall the students and teachers responses suggest that satisfaction as "positive teaching and learning experience" in course management. Novo-Corti, Varela-Candamio and Ramil-Diaz (2013) reported an increase in the performance of students (grades and qualifications) when using the mixed technology of e-learning in Moodle and face-to-face lectures. In this study, there is no significant effect on improving students' semester grades and the interaction between students and teachers.

Moreover, LMS provided the students with ability to access lecture notes, and interactive features in their learning activities [8]. Similarly, this study finds that it is a real support for teaching-learning process, assessment evaluation and technical resources. The findings described that if the internet speed is better than now, teachers and students' interaction can be increased in assessing with Moodle. The study could be conducted to examine the perceptions of students and teachers from the other Computer Universities in Myanmar to get more comprehensive view of perception the Moodle.

In addition to currently used Moodle functions, most of the students and teachers want to utilize other functions to Moodle such as video lecture files, user recent activity, weekly lessons schedule, chat and forums. And they would like to get the feedback of their assignments. When they forget the password, they want to recover themselves. Most of the undergraduate students have not been used other open source learning management system.

By conducting pilot study, the research finds out which of the three categories have the most influence on usefulness, the results indicated that use of learning management system has the positive impact, good and finally fair which shows the students' and teachers' awareness of the benefits. Moodle is the best communication tools with user friendly interface and easily accessible.

6 Conclusion

Learning Management System (LMS) has become a very important tool in education. LMS systems also provide tools for interactive teaching and learning between teachers and students and upgrade their teaching techniques. This research study the use of LMS (Moodle) by analyzing students' and teacher's perceptions in UCSM to improve the teaching and learning techniques in computer universities and to know the awareness of its benefits. The results indicate that Moodle is a real support for teaching and assessment evaluation in UCSM. But, there is no significant effect on improving students' semester grades and the interaction between students and teachers. This may be weakness of questionnaire in this research. Thus, it may be the better result by using more related questions.

This study has a few limitations to examine the use of Moodle LMS in UCSM. The population was limited to the undergraduate students and teachers of 2017-2018 academic years (only first semester) in UCSM. In consequence, the result of the study may not reflect the general use of information technology at other universities.

Future studies could be conducted to examine the perceptions of undergraduate students and teachers from the other Universities in Myanmar that has been used LMS to get more comprehensive view of perception the Moodle. Mobile devices such as tablets, personal digital assistants (PDAs), and smart phones, and the use of different modes of communication such as SMS messaging, are the way of the future. The LMS might need to adapt to catch up popular technology to become more usable for future students.

References

1. Fidani, A., Idrizi, F.: Investigating students' acceptance of a learning management system in university education: a structural equation modeling approach. In: ICT Innovations 2012 Web Proceedings, pp. 311–320 (2012)
2. Al-Assaf, N., Almarabeh, T., Eddin, L.N.: A study on the impact of learning management system on students of the University of Jordan. J. Softw. Eng. Appl. **8**(11), 590 (2015)
3. Sen, T.K.: Application of blended and traditional class teaching approach in higher education and the student learning experience. Int. J. Innovation Manag. Technol. **2**(2), 107 (2011)
4. Hwa, S.P., Hwei, O.S., Peck, W.K.: An investigation of University Students' acceptance towards a learning management system using technology acceptance model. J. Educ. Soc. Sci. **2**(5), (2016)

5. Umek, L., Keržic, D., Tomaževic, N., Aristovnik, A.: moodle e-learning system and students' performance in higher education: the case of public administration programmes. In: International Association for Development of the Information Society. International Conference e-Learning (2015)
6. Thuseethan, S., Achchuthan, S., Kuhanesan, S.: Usability evaluation of learning management systems in Sri Lankan universities. arXiv preprint arXiv:1412.0197 (2014)
7. Mafata, M. P.: Investigation of the use of Learning Management Systems in Educational Technology Modules: A Case Study. University of KwaZulu-Natal (2009)
8. Alaofi, S.: How do Teachers and Students Perceive the Utility of Blackboard as a Distance Learning Platform? Case study. Taibah University, Saudi Arabia (2016)
9. Al-sarrayrih, H.S., Knipping, L., Zorn, E.: Evaluation of a Moodle based learning management system applied at berlin institute of technology based on ISO-9126. In: Proceedings of ICL 2010 Conference Hasselt Belgium, pp. 880–887 (2010)
10. Lonn, S., Teasley, S., Krumm, A.: Investigating undergraduates' perceptions and use of a learning management system: A tale of two campuses. In: Annual Meeting of the American Educational Research Association, 16 April, San Diego, California, Retrieved June, vol. 6, p. 2014 (2009)
11. Lin, C.H.: Using moodle in general education English as a second language program: Taiwanese college student experiences and perspectives. J. Educ. Soc. Res. **3**(3), 97 (2013)
12. Al-Busaidi, K.A., Al-Shihi, H.: Instructors' acceptance of learning management systems: a theoretical framework. In: Communications of the IBIMA (2010)
13. Cavus, N., Momani, A.M.: Computer aided evaluation of learning management systems. Procedia Soc. Behav. Sci. **1**, 426–430 (2009)
14. Landau, S., Everitt, B.S.: A Handbook of Statistical Analyses Using SPSS. Chapman & Hall/CRC Press LLC, Boca Raton (2004)

User Preference Information Retrieval by Using Multiplicative Adaptive Refinement Search Algorithm

Nan Yu Hlaing(✉) and Myintzu Phyo Aung

Myanmar Institute of Information Technology, Mandalay, Myanmar
nanyu.man@gmail.com

Abstract. One of the experiments of user preference information retrieval is to rank the most relevant documents. This system used MARS (Multiplicative Adaptive Refinement Search) algorithm for user preference information. User can give a query as an input. The input is the query that contains song name, song types, singer names that are part of the information which the user wants to know. The song information is retrieved from the web pages. The system returns the song information that is related to the input query that similarity values of web pages and query by using Cosine Similarity. From the initial result, user's preference can mark by clicking the corresponding check box at interface. When the user marked the relevant document, and then refined by using MARS algorithm. The end result is the collection of documents list at a central location. The relevant documents are showed at the top of the system as output. This is called for choosing the proper methods to evaluate the system performance. The traditional performance measures, this system used precision and recall, are relies on user's relevance judgment.

Keywords: IR · MARS · Cosine similarity

1 Introduction

Nowadays, assessment of information retrieval systems is a critical part of information retrieval research. Information retrieval (IR) is retrieving the relevance information from a set of information resources. The objective of any IR system is to identify relevant documents to a user's query. Most of information retrieval system requires high performance, for example speed, consistency, accuracy and ease of use in retrieving relevant documents to satisfy user queries. This paper describes how to work the approach taken to retrieve song information retrieval system and demonstrates how to measures the performance of information retrieval system.

2 Related Work

Users can search the information with some iterations of relevance feedback for some search system to gain certain important search high performance. This system used an algorithm, MA query expansion algorithm. It is used query modification methods to

T. T. Zin and J. C.-W. Lin (Eds.): ICBDL 2018, AISC 744, pp. 169–178, 2019.
https://doi.org/10.1007/978-981-13-0869-7_19

progress the query vector adaptively. It is implemented for MARS to show effectiveness and efficiency of their performance. The performance of this algorithm is better than other similarity-based relevance feedback algorithms that is query reformation methods in information retrieval system, is basically used in an adaptive directed learning algorithm for searching documents by linear classifier. The system's experiments show that MA algorithm considerably increases the search performance in information retrieval system [1].

3 Background Theory

3.1 Information Retrieval System

Information retrieval (IR) is the activity of finding relevant information to the user's need from a set of information resources. User can be searched with text or other content-based indexing. Information retrieval is the identification and efficient use of searching for information in a document. User can search document of texts, document itself, metadata, songs and images.

An information retrieval process begins when a user enters a query into the system. Queries are user defined statements of needed information. In information retrieval, a query does not match any object in the group; instead it matches with numerous objects from which the most relevant documents are taken into consideration of more evaluation.

Most IR system computes ranking of relevant documents that are related document to the user's query. After entering the query to the system, IR access to the content management modules that are directly connected with large collection of data objects. The returned results are generated by the system that is shown to the user by some GUIs. The process repeats and returned results are changed until the user satisfied for what he is actually seeking.

3.2 Information Retrieval Models

Information Retrieval (IR) models are a main part of IR research and IR systems. Firstly, the relevance means to be associated with the similarity between a query and a document. In the second, it used a binary random variable to assessment the value of relevance variable by using relevance and probabilistic models. In the third section, the uncertainty of gathering queries of the documents is displayed to the relevance ambiguity. There are many kinds of IR models such as Boolean, VSM, BIRM, Probability, Query reformulation, Relevance feedback, Page rank etc.

3.2.1 Vector Space Model

This model is most well known in IR. It represents the documents and queries as vectors in a multidimensional space, their dimensions are the vector of index terms with related weights which represent the importance of the index terms in the document and within the whole document collection. A query is also displayed as a list of index terms with related weights that represent the importance of the index terms in the query. The relevance document for user's query is estimated by computing some distance or similarity metric between the two vectors [5].

The term weight of the song document can be calculated by:

$$w_{td} = tf_{t,d}.log\frac{|D|}{|\{d' \in D|t \in d\}|} \quad (1)$$

Where,

$tf_{t,d} = term\ frequency\ of\ the\ termt\ and\ song\ document\ d,$
$log\frac{|D|}{|\{d' \in D|t \in d\}|}$ = inverse song document frequency,
$|D|$ = the total number of song documents in the document set,
$|\{d' \in D|t \in d'\}| = number\ of\ song\ documents\ containing\ the\ term\ t,$

3.2.2 Cosine Similarity

Cosine similarity is the similarity function that is often used in information retrieval. It is a measure of similarity between two vectors and in the IR it measures the angle between two documents. For two vectors of attributes, a and b, the angle of cosine similarity, θ, is represented using a dot product:

$$Similarity = cos\theta = \frac{a.b}{||a||||b||} \quad (2)$$

The term frequency vectors of documents are represented by attribute vectors a and b. The range of cosine similarity is usually [− 1,1]. The resulting ranges of similarity −1 is exactly opposite, 1 is exactly the same, 0 is indicating independence and similarity and dissimilarity is indicated by in between values. In IR, document vector are usually non-negative, so that the resulting ranges will be 0 to 1. The angle between two documents can never be greater than 90° [6].

Cosine Similarity for Documents

The similarity of two documents d_1 and d_2 is:

$$\text{Similarity}(\text{d}_1, d_2) = \cos\left(\overrightarrow{d_1}, \overrightarrow{d_2}\right) = \frac{\overrightarrow{d_1}\overrightarrow{d_2}}{\left|\overrightarrow{d_1}\right|\left|\overrightarrow{d_2}\right|} \quad (3)$$

$\left|\overrightarrow{d_1}\right| = length\ of\ the\ vector\ d_1$
$\left|\overrightarrow{d_2}\right| = length\ of\ the\ vector\ d_2$
$\overrightarrow{d_1}\overrightarrow{d_2} = doc\ product\ of\ two\ vectors$

This equation is shown for document with term weight:

$$Similarity(d_1, d_2) = \frac{\sum_{a=1}^{n} W_{a,1} W_{a,2}}{\sqrt{\sum_{a=1}^{n} W_{a,1}^2} \sqrt{\sum_{a=1}^{n} W_{a,2}^2}} \quad (4)$$

This equation is using the cosine similarity between song document d_2 and query d_3:

$$Similarity(d_2, d_3) = \frac{\vec{d_1} . \vec{d_2}}{\left|\vec{d_1}\right|\left|\vec{d_2}\right|} = \frac{\sum_{a=1}^{n} W_{a,1} W_{a,3}}{\sqrt{\sum_{a=1}^{n} W_{a,1}^2} \sqrt{\sum_{a=1}^{n} W_{a,2}^2}} \quad (5)$$

3.3 Relevance Feedback

The relevance feedback mostly used in informational retrieval system so as to progress the final results of retrieving. When the user enters the query to the system, the relevance feedback is to show the results that are firstly returned of the user's query to the system. The process of relevance feedback is

- Firstly, user's query
- The system return of initial results that are set of documents
- User check returned documents that are relevance or not
- The system shows most of the relevance information of the user's feedback by computing
- The system runs with new query
- The system shows a reviewed collection of results that are related with user's query. The result is most relevance for user by computing the precision and recall.

The process of relevance feedback can be one or more iterations for user's need. When the user didn't know about what he/she wants, it is difficult to formulate the query but it is easy to know when he/she sees the set of documents and can do many iteration of refinement process at that time. Relevance feedback can be most useful and has proven very effective for improving accuracy, performance of song information retrieval system [3].

4 Multiplicative Adaptive Refinement Search (MARS)

A multiplicative refinement technique is used to identify a user preference information expression. The query vector will be promoted by using multiplicative factor, when an index term is judged by the user. If a document is judged as relevant, its terms of the song document are promoted by a factor or if not the terms will be demoted.

In this system, the function f(x): [0,1] $\rightarrow$ R^+ are used as a non-decreasing updating functions, because the proportional value of the index term may be updating function of the index term [1].

4.1 Multiplicative Adaptive Refinement Search (MARS) Algorithm

```
Algorithm MA(q0, f, Θ )
(i) Inputs:
sq0: the non negative initial vector
f(x):[0,1] R , the updating function
Θ ≥ 0, the classification threshold
(ii) Set k=0;
(iii) Classify and rank document with the linear
classifier (qk , Θ )
(iv) While {
rq (relevant query )
for (i =1,...,n){
        if (di ≠0){
        if (qi,k ≠ 0) set qi,k+1 = qi,k
        else set qi,k+1 = 1
        if ( d is relevant ) {
        set qi,k+1 = (1 + f(di))qi,k+1
        set rqi = di;
      }else {
      set  qi,k+1 / (1+f(d )) set q I,k+1 = qi,k+1
di))
      set rqi = null
    }else set qi,k+1 = qi,k
        }queryexpansion(rq,q,k)
          }
(v) It the user has no judged any document in the
k-th step, then stop. Otherwise, let k=k+1 and go to
step (iv).
```

4.2 Query Expansion

Query expansion is used as a process of the information retrieval during which a user's initial query statement is enhanced by adding search terms to improve retrieval performance. Query expansion is efficient by the fact that initial query formulation does not always return the exact information needed by the user. The application of thesauri to query expansion and reformulation has most popular and widely used.

There are three different methods of query expanding: Manual, Interactive and Automatic. Manual and Interactive query expansion require user involvement. Automatic query expansion is the process of supplementing and adding terms or phrases to the original query to improve the retrieval performance without user's intervention [4].

4.2.1 Query Expansion Algorithm

```
    queryexpansion (rq q k)
      {
        set newquery = Same Keyword in rq;

newquery=(newquery1,…………,newquerym)
        for(i=1,……,n){
        set count = 0;
        for (j=1,….,m){
        if(newqueryj contains in rqi)
        set count = count+1;
        }qi,k=qi,k+count;
        }
    }
```

5 System Design of the Proposed System

In this section, the system flow diagram is shown in Fig. 1. When the system start, user can give a query as an input. This query contains keywords such as song's title, song's type, singer's name, song's series title, etc.

This system is implemented in Web Base application. When the user enters the query, the system will find the documents by using Cosine Similarity method. From the

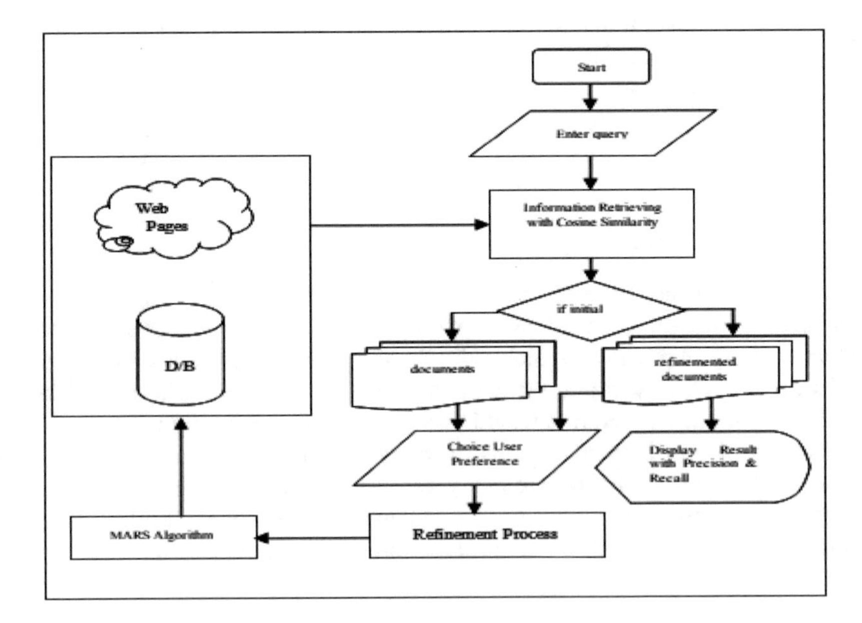

Fig. 1. System flow diagram

initial result, the system will allow the user to do the refinement process if the user wishes. When the user selected the relevant document, and then refine by using MARS algorithm. Once Refinement process is done, the query may be an expansion. Finally, the documents are almost same as the user desires.

6 Refinement Process

Refinement relations are defined for many different specification techniques and for different semantic frameworks. The classification is scored the pages, when the user takes a refinement process. By observing the classification of various scores that result of the refinement is in improved performance [7].

In this function, the main major of this function is to make relevant up or down according to the users relevant by using "MARS Refinement algorithm". This algorithm will promote the relevant documents and demote the irrelevant documents. (e.g.: If the user preferences the three documents, and then the function will search the similarity keywords from these document and update the query. By using this new query, the system will refine documents and then display again to the user. These documents will be on top of the list in the web page.

6.1 Sample Result of Refinement Process

Documents of song information
Age 40 Moe Moe Sunday rainning Pop
Kin Sar Big Bag Ve Lane Alternative Pop
Allow me to see you Aung Phyo Unknown Pop
Yin Mhar A Shi Tine Ait Za Ni Unknown pop
Htar khae Mae Chit thu Mie Mie Khae Khan sat ka,chay yar myar Pop
Ar Nar Light Tar Examplez JoJar Unknown Pop
Min Ko Thati Ya Yin Kaung, Kaung Moe Moe The best Melody World Pop
Su, Taung Mel Zaw Paing Kyi Nu Kwint Pop

Initial rank of the documents
document 1 : 6.92
document 2 : 6.182
document 3 : 6.059
document 4 : 5.936
document 5 : 5.813
document 6 : 5.69
document 7 : 5.567
document 8 : 5.444

For irrelevant document

Original rank value = 6.92

Updating function = 0.0021591

New rank value = original rank value/(1 + updating function)

= 6.925/1 + 0.0021591

= 6.9050913

For irrelevant document

Original rank value = 6.059

Updating function = 0.2493198

New rank value = original rank value ×(1 + updating function)

= 6.0591 + 1.2493198

= 7.5696287

Relevant documents
document 3
document 4

New Ranks for documents
document 1 : 6.9050913
document 2 : 4.9115114
document 3 : 7.5696287
document 4 : 8.3134846
document 5 : 3.3415453
document 6 : 3.1838346
document 7 : 2.8255713
document 8 : 2.9971855

New Query
Allow me to see you Aung Phyo Unknown Pop
Yin Mhar A Shi Tine Ait Za Ni Unknown pop

new query: [Pop, Unknown]

Result of Refinement Process
A Lwan Win Ga Bar Ait Za Ni Aung Thwal Pop
Allow me to see you Aung Phyo Unknown Pop
Yin Mhar A Shi Tine Ait Za Ni Unknown Pop
Complicated Avril Levenge Unknown Punk
Maryar Lay Phyu Unknown Rock
Ar Nar Light Tar Examplez + JoJar Unknown Pop
Min Ko Thati Ya Yin Kaung Kaung + Moe Moe The best Melody World Pop
Hamster never be replaced Sai Sai Kham Hlaing Unknown Hip Hop

7 Precision and Recall

The system can measure the performance of song information by using precision and recall. The measures require a set of song documents that are related with user's query [2].

7.1 Precision

The precision is the fraction of retrieved song documents that are relevant to the user's query:

$$Precision = \frac{number\ of\ relevant\ song\ document\ returned}{total\ number\ of\ song\ document\ returned} \tag{9}$$

7.2 Recall

The recall is the fraction of the relevant song documents that are successfully retrieved.

$$Recall = \frac{number\ of\ song\ document\ returned}{total\ number\ of\ relevant\ song\ document} \tag{10}$$

8 Conclusion

In the conclusion, the system accepts query and then find the similar documents by using cosine similarity. This system is intended to use in the field of song group. User can get song information by entering only query. Instead of interacting with song types, this system can give information immediately. From the resulting pages, the user gives relevant feedback. According to this feedback, the system uses MARS algorithm which promotes or demotes the ranks by the documents' status (relevant or irrelevant). For more user preference, the precision will be more increased and the recall will be more decreased. Finally, system's results are more user preference song information.

References

1. https://pdfs.semanticscholar.org/09db/f0df802c1601076ae9b7f55b1149892a0706.pdf
2. Information Retrieval. https://en.wikipedia.org/wiki/Information_retrieval
3. Relevance Feedback and Query Expansion. Cambridge University Press (2009). https://nlp.stanford.edu/IR-book/pdf/09expand.pdf
4. Query Expansion. https://en.wikipedia.org/wiki/Query_expansion
5. Lewandowski, D.: Web searching, search engines and Information Retrieval (2005)
6. Cosine Similarity. https://en.wikipedia.org/wiki/Cosine_similarity
7. Refinement Process. https://en.wikipedia.org/wiki/Refinement

Proposed Framework for Stochastic Parsing of Myanmar Language

Myintzu Phyo Aung(✉), Ohnmar Aung, and Nan Yu Hlaing

Myanmar Institute of Information Technology, Mandalay, Myanmar
myitzu.mm@gmail.com, mamalay2009@gmail.com,
nanyuhlaing@gmail.com

Abstract. Parsing is breaking a sentence into its constituent nonterminal. Parsing is useful in the study of artificial intelligence for various reasons, such as, for an index-term generation in an information retrieval; for the extraction of collocation knowledge from large corpora; development of computational tools for language analysis. In this paper, a framework for stochastic parsing of Myanmar Language is presented. This parsing system will use the context free grammar and stochastic context free grammar. Myanmar sentence will be accepted as input. And then this input sentence will be tokenized, segmented and assigned part of speech tags. Finally, stochastic parse tree will be generated by using stochastic context free grammar as the output of stochastic parsing system.

Keywords: Parsing · Stochastic parsing · Stochastic context free grammar Parse tree

1 Introduction

Classification or categorization of texts is the classifying documents into predefined categories automatically. There are many applications in text classification such as spam filtering, email routing, language identification, genre classification, readability assessment etc. Because mediating between linguistic expression and meaning is important and the syntactic structure of one language is differing form one another, syntactic parsing is a central task in natural language processing. For example, English has SVO order (subject-verb-object) Myanmar Language has SOV (Subject Object Verb) order and so on. The order of other syntactic elements, such as the order of a head noun and its modifiers can also be processed beyond the order of subject, verb and object.

Since language can be used not only for expressing thoughts but also for exchanging information, it plays an important role in human communication. Processing of natural language which is the branch of linguistics, artificial intelligence and computer science, aims to have the interaction among natural language of human beings and computers. Examples of language processing tasks include part-of-speech (POS) tagging, named entity recognition and word sense disambiguation. The task which will be focused in this paper is shallow parsing which identifies the non-recursive cores of various phrase types in text, possibly as a precursor to full parsing or information extraction.

T. T. Zin and J. C.-W. Lin (Eds.): ICBDL 2018, AISC 744, pp. 179–187, 2019.
https://doi.org/10.1007/978-981-13-0869-7_20

2 Related Work

In recent years, the artificial intelligence community has studied various stochastic behaviors of natural language (NL) to carry out successful information extraction processes. This section describes the related work of the system. Automatic summarization of Myanmar text was presented in [17]. In their approach, Myanmar document is accepted as input. Firstly, sentence boundary is identified, and word segmentation, stop word removal and part of speech tagging are performed as preprocessing steps. Features such as location, similarity to the title and numerical data in sentence are extracted from sentence by using a training corpus of document summary pair.

Chunking can be used in several tasks and also a previous step providing input to the next steps. A chunk is basically the identification of parts of speech like short phrases and consists of single content word surrounded by a constellation of function words. Myanmar Phrase identification system by using Conditional Random Fields was presented in [20]. The system was firstly trained with training data and tested with testing data. Although there were some error identifications, the system got the best accuracy.

Morphological analysis is an important first step in many natural language processing tasks such as parsing, machine translation, information retrieval, part of speech tagging among others. A probabilistic language model for Joint Myanmar Morphological Segmentation and Part of Speech Tagging was presented in [1]. In their experiments, three testing corpora are used for evaluation in order to calculate the accuracy of the word segmentation and tagging. Each corpus contains 150 sentences these are from news websites and Myanmar Grammar books.

Construction of Finite State Machine (FSM) for Myanmar noun phrase structure in Context Free Grammar is presented by [9] to apply in Myanmar noun phrase identification and translation system which is part of Myanmar to English machine translation system. New phrase chunking algorithm for Myanmar Natural Language processing has been reported in previous study [10].

Phrase Structure Grammar for Myanmar Noun Phrase Extraction is presented in [11]. In their approach, Context Free Grammar is generated and used for extraction of Myanmar Noun Phrase. The system is tested on 1500 testing sentences: 1000 simple sentences and 500 compound sentences. The evaluation measures are defined in terms of the precision, recall and f-measure and shown that the system got the high accuracy.

3 Lexical Categories of Myanmar Language

Myanmar (Burmese) language, the official language of the Union of Myanmar is spoken by 32 million as a first language and as a second language by ethnic minorities in Myanmar. A basic unit of every language is its sentence. Sentences in Myanmar language consist of grammatically structured words. These words are grouped into classes, known as part of speech (POS) which have the similar syntactic behavior and semantic type. There are eight basic parts-of-speech in the Myanmar language. The two most important POS are noun and verb. These word classes or POS are shown in Table 1.

Table 1. Lexical categories of Myanmar language

Lexical Category	**Definition**	**Examples**
Noun	names of persons, places, things or concepts	မောင်မောင်၊ ကျောင်းသား၊ စာအုပ်
Verb	expression of actions or states in a sentence	စားသောက်
Pronoun	replace nouns or can be used instead of nouns	ကျွန်တော်၊ကျွန်မ
Adjective	describing the properties of nouns	တော် ၊ထို၊ဤ
Adverb	modifiers to verbs, adjectives and other adverbs	မြန်မြန်၊ခင်ခင်မင်မင်
Preposition	indication of the relationships between the nouns or pronouns and some other parts of speech	သည်၊က၊မှာ
Conjunction	join clauses, parallel nouns, and adjectives	နှင့်၊သို့မဟုတ်၊မှတစ်ပါး
Particles	indicating words which is subjects, objects, place, or time etc	များ၊တို့

4 Different Levels of Language

In order to understand language, language is studied at different levels. There are seven levels for linguistic analysis. These levels of language study are presented in Table 2 [8].

Table 2. Different level of language study

Level	Description	Example usage/System
Phonological	The interpretation of speech sounds within and across words	Speech-recognizing systems
Morphology	The study of the meaningful parts of words	Automatic stemming, truncation or masking of words
Lexicology	The study of words	Parts-of-speech tagging or the use of lexicons

(*continued*)

Table 2. (*continued*)

Level	Description	Example usage/System
Syntactic	The study of the rules, or "patterned relations", that govern the way the words in a sentence are arranged	Parsing algorithms
Semantics	The study of the meaning of word more complex level of linguistic analysis	Identification of automatically phrases of two or more words that when looked at separately have quit e different meanings
Discourse analysis	Works with units of text longer than a sentence	By understanding the structure of a document, Natural Language Processing systems can make certain assumptions
Pragmatics	The study of how the context influences meaning	

The above levels of linguistic processing reflect an increasing size of unit of analysis as well as increasing complexity and difficulty as we move from phonological level to pragmatics.

5 Proposed Stochastic Parsing System

Every language has a grammar. Strings of symbols for a language can be derived from the grammar of that language. If there is an input text and a grammar, parsing or syntactic analysis can be done. Which is the process of determining that the text was generated by the grammar, and how it was generated, which grammar rules and which alternatives were used in the derivation. In the process of top down derivation, the process starts from the starting symbol down towards the given string. Bottom up parsing works in the opposite direction from top-down. A bottom-up parser starts with the sting of terminals itself and builds from the leaves upward, working backwards to the start symbol by applying the productions in reverse. This section describes the proposed stochastic parsing system for Myanmar Language which is depicted in Fig. 1.

5.1 Preprocessing

Myanmar Language likes other South Asia languages do not place spaces between words but spaces are usually used to separate phrases. Besides, in Myanmar sentences, some are written with propositions or particles and some are written without propositions or particles.

The language is classified into two categories. One is formal, used in literary works, official publication, radio broadcasts, and formal speeches. The other is colloquial, used in daily conversation and spoken. In Myanmar script, sentences are clearly delimited by a sentence boundary marker but words are not always delimited by spaces. Therefore some preprocessing steps such as tokenization, segmentation, and part of speech tagging are required before the parse tree of the sentence is generated.

5.1.1 Tokenization

Tokenization is the process of breaking up the given text into units called tokens. Tokens may be words or number or punctuation mark. Tokenization does this task by locating word boundaries. So tokenization can also be said to be the identification of syllable boundaries. Ending point of a word and beginning of the next word is called word boundaries. There are many challenges in tokenization which depends on the type of language [21].

The tokenization process of the system uses Myanmar 3 based tokenizer. It is based on Myanmar writing format, start with one of consonants, vowel, number, special character. For example, for the input Myanmar sentence "နွေဥတုသည်ပူပြင်းသည်", the tokenization process of the system tokenizes the sentence as နွေ/ဥ/တု/သည်/ပူ/ပြင်း/သည်.

5.1.2 Segmentation

In Myanmar script, sentences are clearly delimited by a sentence boundary marker, but words are not always delimited by spaces. Therefore, segmenting Myanmar sentences into word is a challenging task. Segmentation process constructs the Myanmar words form syllables. In the segmentation process of the system, input Myanmar sentences are separated into words by using longest matching approach with the aid of Myanmar to English bilingual lexicon. As an example, for the input sentence, "နွေဥတုသည်ပူပြင်းသည်", the process will generate နွေဥတု/သည်/ပူပြင်း/သည်.

5.1.3 Part of Speech (POS) Tagging

In Myanmar, there are eight parts of speech: noun, pronoun, verb, adverb, adjective, preposition, conjunction, and particle as described in Table 1. Part of speech tagging process of the system assigns each word of the sentence with these tags [19]. The part of speech tagging process of the system will tag the segmented sentence နွေဥတု/သည်/ပူပြင်း/သည် as follows:

- နွေဥတု-common countable singular
- သည်-subject preposition
- ပူပြင်း-verb
- သည်-verb preposition

5.2 Stochastic Parsing

Analysis of a sentence, similar to the lexical analysis for computer languages, which is known as parsing identifies the constituent parts of sentences (nouns, verbs, adjectives, etc.) and then links them to higher order units that have discrete grammatical meanings (noun groups or phrases, verb groups, etc.).

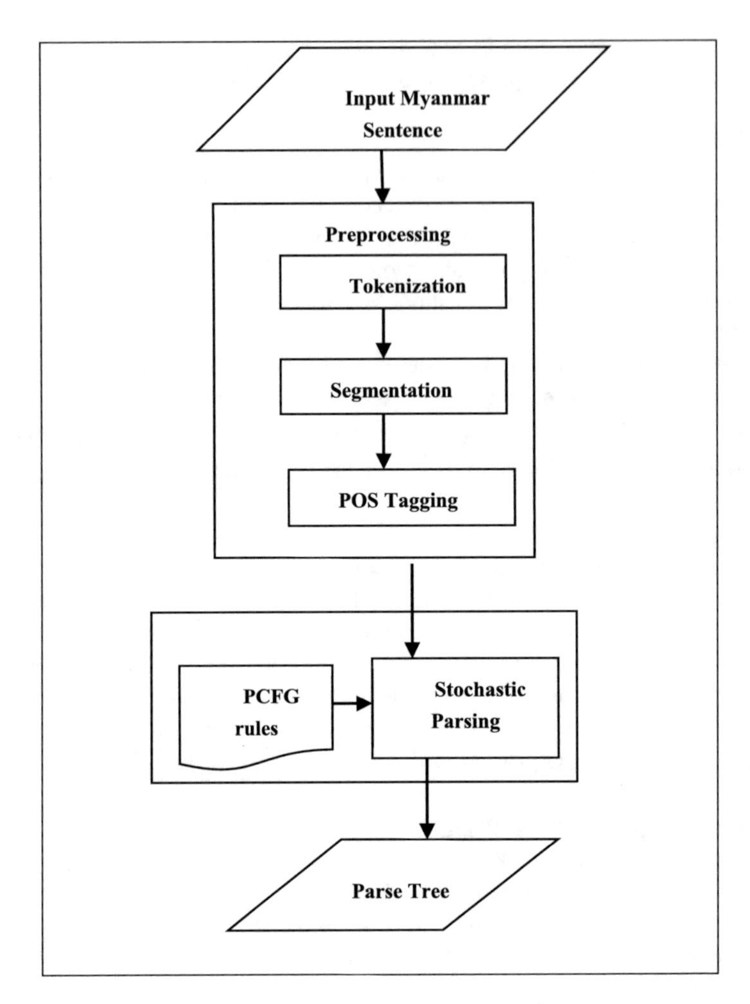

Fig. 1. Proposed stochastic parsing system

5.2.1 Context Free Grammar

A context free grammar (CFG) can be defined as the following 4 tuple:

G = (N, Σ, R, S), where

1. N is a set of nonterminal symbols.
2. Σ is a set of terminal symbols.
3. R is a set of rules or productions in the form of $A \rightarrow \beta$, where $A \in N$ and β is an ordered list of symbols drawn from $N \cup \Sigma$.
4. S is the starting symbol.

Sentences can be generated from a CFG with the derivation process: the derivation starts with S, rewrites a nonterminal A by replacing it with the right-hand side of a rule with A on the left-hand side. This process is repeated until a string of terminals is obtained. Parsing can also be defined as the obtaining of this derivational process for a target output sentence. Because, natural language is highly ambiguous, there may be many CFG parse trees for a single sentence [5]. For Myanmar Language, these Context Free Grammar rules are defined with the Part of Speech (POS) sequences of Myanmar language [12]. The example CFG rules are presented in Fig. 2.

NP→Noun POPOS NCCS PAIDNUM
NP→PP Noun PAIDNUM
NP→PP NP
NP→ Noun POPOS NP
PP→NounPOPOS
NP→NounPAIDNUM
Noun→ NCCS
Noun→NCPS

Fig. 2. Example CFG rules

5.2.2 Stochastic Context Free Grammar

Probabilistic context free grammar is an extension of CFGs in which each rule has a probability $p \in [0, 1]$. In a consistent PCFG, the sum of the probabilities of all the rules which have the same nonterminal on the left-hand side from a probability distribution must be one.

NP→Noun POPOS NCCS
PAIDNUM ,0.053
NP→PP Noun PAIDNUM ,0.043
NP→PP NP, 0.02
NP→ Noun POPOS NP, 0.05
PP→ Noun POPOS ,
NP→ Noun PAIDNUM
Noun→ NCCS , 0.083

Fig. 3. Example PCFG rules

The PCFG is mainly used in selection of the best parse for a sentence among multiple parse, according to its probability model. The probability of a parse tree in a PCFG can be calculated by multiplying the probabilities of each of the rules in the parse tree [5]. Some of these PCFG rules are described in Fig. 3.

6 Conclusions

Chunking or shallow parsing segments a sentence into a sequence of syntactic constituents. In this paper, stochastic parsing system for Myanmar language is presented. In this system, Context Free Grammar and Stochastic Context Free Grammar (SCFG) will be used for parsing of Myanmar sentence. There are two types of Myanmar Language, written language and spoken Language. The system presented in this paper is based on the written form Myanmar Language. For the generation of Grammar rules (i.e. CFG and PCFG), example sentences from Myanmar Grammar book were used as training sentences. This CFG and SCFG will be applied in bottom up parser and then the parse trees of Myanmar sentences will be generated and the accuracy of two grammars, CFG and PCFG, will be compared. Parsing of a sentence can help many other NLP applications such as text summarization, machine translation, phrase chunking and information retrieval systems.

References

1. Cing, D.L., Htwe, T.M.: A probabilistic model for joint myanmar morphological segmentation and part of speech tagging. In: International Conference on Computer Applications, Yangon, Myanmar, February 2017
2. Keryszig, E.: Advanced Engineering Mathematics, 7th edn.
3. Feng, F., Bruce Croft, W.: Probabilistic techniques for phrase extraction. J. Inf. Process. Manage. Int. J. **37**, 199–220 (2001)
4. Oliveraet, F., et al.: Systematic noun phrase chunking by parsing constraint synchronous grammar in application to Portuguese Chinese Machine Translation. In: Proceedings of the Sixth International Conference on Information Technology Applications (2009)
5. Cheung, J.C.K.: Parsing German topological fields with probabilistic context-free grammars. M.Sc. thesis, Graduate Department of Computer Science, University of Toronto
6. Okell, J., Allott, A.: Burmese/Myanmar Dictionary of Grammatical Forms. Cruzan Press (2001)
7. Batra, K.K., Lehal, G.S.: Rule based machine translation of noun phrase from Punjabi to English. Int. J. Comput. Sci. Issues (2010)
8. Karoo, K., Katkar, G.: Analysis of probabilistic parsing in NLP. Int. J. Innovative Res. Comput. Commun. Eng. **4**(10) (2016)
9. Aung, M.P., Lynn, K.T.: Construction of finite state machine for myanmar noun phrase. In: Proceedings of MJIIT_JUC Joint International Symposium, Hiratsuka, Japan (2013)
10. Aung, M.P., Moe, A.L.: New phrase chunking algorithm for myanmar natural language processing. Int. J. Appl. Mech. Mater **695**, 548–552 (2015)
11. Aung, M.P.: Constructing Myanmar phrase structure grammar for myanmar noun phrase extraction. In: The Seventh International Conference on Science and Engineering, Yangon, Myanmar, December 2016
12. Aunget, M.P., et al.: Stochastic context free grammar for statistical parsing of Myanmar. In: International Conference on Computer Applications, Yangon, Myanmar, February 2018
13. Win, M.T., et al.: Burmese phrase segmentation. In: Proceedings of Conference on Human Language Technology for Development, Alexandria, Egypt, pp. 27–33, May 2011
14. Hopple, P.M.: The Structure of Nominalization in Burmese
15. Nugue, P.M.: An Introduction to Language Processing with Pearl and Prolog

16. Soe, S.P.: Aspects of Burmese Language. Department of Myanmar, UDE University of Yangon (2010)
17. Yuzana, S.S.L., Nwet, K.T.: Framework of the extractive myanmar text summarization with Naïve Bayes classifier. In: International Conference on Computer Applications, Yangon, Myanmar, February 2017
18. Myint, S.T.Y., Khin, M.M.: Lexicon based word segmentation and part of speech tagging for Myanmar text. Int. J. Comput. Linguist. Nat. Lang. Process. **2**(6), 394–403 (2013)
19. Myint, S., Khin, M.M.: Lexicon based work segmentation and part of speech tagging for written Myanmar text. Int. J. Comput. Linguist. Nat. Lang. Process. **2**(6), 396–493 (2013)
20. Yin, Y.M.S., Soe, K.M.: Identifying Myanmar phrases using conditional random field
21. www.language.worldofcomputing.net/category/tokenization

Information Communication Systems and Applications

FDTD Based Numerical Calculation of Electromagnetic Wave Radiation in Multi-layer Circular Cylindrical Human Head

Z. M. Lwin[1(✉)] and M. Yokota[2]

[1] Interdisciplinary Graduate School of Agriculture and Engineering, University of Miyazaki, 1-1 Gakuen Kibanadai Nishi, Miyazaki 889-2192, Japan
z3t1601@student.miyazaki-u.ac.jp

[2] Department of Electrical and Systems Engineering, University of Miyazaki, 1-1 Gakuen Kibanadai Nishi, Miyazaki 889-2192, Japan
m.yokota@m.ieice.org

Abstract. Electromagnetic radiation rate in the multi-layer circular cylindrical human head model are studied using FDTD (finite different time domain) method in two different frequencies 3.35 GHz and 4.5 GHz. Human head is designed as the layered structures such as skin, fat, bone, Dura, CSF (Cerebrospinal fluid) and brain. The electrical and magnetic fields are numerically examined by FDTD method of two dimensional Maxwell's equation. For 2D FDTD, the electric field is polarized along the cylindrical axis and FDTD method is implemented with Matlab programming. The electric field point source is kept near cylindrical human head and then analyze the nature of wave propagation to the tissue layer by layer using the impedance mismatch and specific absorption rate (SAR). Only some part of head that located near the source has the high level of E field absorption rate and then it sharply decreases. The electric field absorption rate reach peak in Skin and CSF layer. This paper also mentions the relationship between the rising of SAR rate and the reflection rate at each tissue layer interface.

Keywords: FDTD · Human head · SAR

1 Introduction

The environment of people is overwhelming with radio wave and most of the people cannot stand without the modernize luxurious device. Every day, people use computer, cell phone, microwave and other different kinds of devices. Even if we don't use these things, there are so many wireless transmissions tower or antenna station in our surrounding. And also, modern society, people spend their spare time using the cell phone for internet browsing, conversation, playing game, reading, etc. It may have some unwanted effects by using the electronic devices especially mobile phone for long time in every day. Generally, Human body are radiated in many ways from various source. According to the nature of Human body, it can be able for absorbing the electromagnetic field. In this way, it leads to increase the temperature of the cells and tissue of human body also called heating effects or thermal effects. The human metabolism of cells and

T. T. Zin and J. C.-W. Lin (Eds.): ICBDL 2018, AISC 744, pp. 191–198, 2019.
https://doi.org/10.1007/978-981-13-0869-7_21

tissue may be able to destroy by that effects. So studying the effects of electric field propagation is important. Specific absorption rate is important step to study the thermal effect in numerical way.

Nishizawa *et al.* [1] studied propagation of electromagnetic plane wave to the 2D elliptic shape human body in 300 MHz and 1300 MHz is calculated using Method of Moment. Human body is considered as three-layers structured such as Skin, Fat and Muscle. Local SAR and whole body average SAR value is mentioned as a result. The rectangular shape three-layer human tissue with air interface is examined using 2D FDTD in TM mode by Alisoy *et al.* [2]. It is intended to simulate as the biomedical device. So point source is kept not only near body but also inside of the human skin and fat. The propagation characterize is presented using the reflection properties of electromagnetic wave in human tissue layers. Seyfi *et al.* [3] computed 1 g and 10 g SAR by 2D FDTD method in 900 MHz on the homogeneous circular cylindrical human head and also studied the shielding effect of reducing SAR rate from the head using point source. Elliptic cylindrical 2D human body model is studied by Caorsi *et al.* [4] using not only homogeneous human model but also multi-layer structure via Mathieu's functions in 900 MHz. SAR is calculated as a result of plane wave incident in TM mode. Two dimensional human body model is investigated in both TE and TM mode using FEM method in Siriwitpreecha *et al.* [5]. They used MRI based 2D slice of human body model to investigate the SAR and temperature distribution due to the leakage electromagnetic wave in 300 MHz, 915 HHz, 1300 MHz and 2450 MHz. Using MRI image, nine layers of human model are analysis and then their results are validated with simple three layers' elliptical human body model.

In this study, radiation and propagation of electromagnetic wave to human head is numerically examined using the well-known method called Finite different time domain (FDTD) method. FDTD is one of the most usable method for electromagnetic problem. FDTD was applied in biomedical application in Taflove et al. [6]. Practically, the thickness of Tissue layer and dielectric properties of human tissues are varied from one person to other. It is difficult to get the suitable and stable exposure rate results of computing in numerical way. But FDTD has ability to be adaptable in calculating efficiently in heterogeneous human body. So it is a popular method to implement in biomedical electromagnetic filed.

In this paper, the propagation characteristic of the electromagnetic wave to the human head model is examined in frequency 3.35 GHz and 4.5 GHz, which are assuming of the 5G frequencies. Human head is designed as the layered structures such as skin, fat, bone, Dura, CSF tissue and brain. EM wave Radiation and Electric field propagation are numerically examined by FDTD method of two dimensional Maxwell's equation. E polarization is assumed and FDTD codes are implemented with Matlab programming. The electric field point source is located at near cylindrical human head. Electric field propagation to the tissue layer by layer using impedance mismatch and SAR distribution are studied.

2 Modeling and Method

Human Head are designed as six different layers namely, Skin, Fat, Bone, Dura, CSF and Brain. Each layer has different dielectric properties and these value depend on the frequency. Table 1 mentions the dielectric permittivity and conductivity of human tissues at frequency of 3.35 GHz and 4.5 GHz. The dielectric properties of tissue are from online dielectric properties database [7]. These parameters of tissue layers of human head are present in Table 1. Calculation area is 30 cm square and the circular cylindrical human head is place inside that square. Point source is added 2 cm apart from the human head. The radius of human head is 9 cm. The geometrical presentation of the system is as shown in Fig. 1. The wide Blue Area is free space and then is small point source. And then, Circle with Colorful layers indicated each tissue layer. The outermost cyan color layer mean skin, Blue for Fat, light blue for Bone, green for Dura, yellow for CSF and the innermost green layer for brain.

Table 1. Dielectric properties of human head

Layer	3.35 GHz		4.5 GHz			Thickness
	σ (S/m)	ε	σ (S/m)	ε	ρ (kg/m^3)	(cm)
Skin	1.9363	37.135	2.6861	36.18	1010	0.1
Fat	0.14767	5.1888	0.21187	5.0766	920	0.14
Bone	0.58152	10.874	0.8438	10.28	1810	0.41
Dura	2.2569	40.903	3.1523	39.477	1010	0.05
CSF	4.3952	64.825	5.8729	62.854	1010	0.2
Brain	2.11205	41.344	3.0332	39.9105	1040	16.2

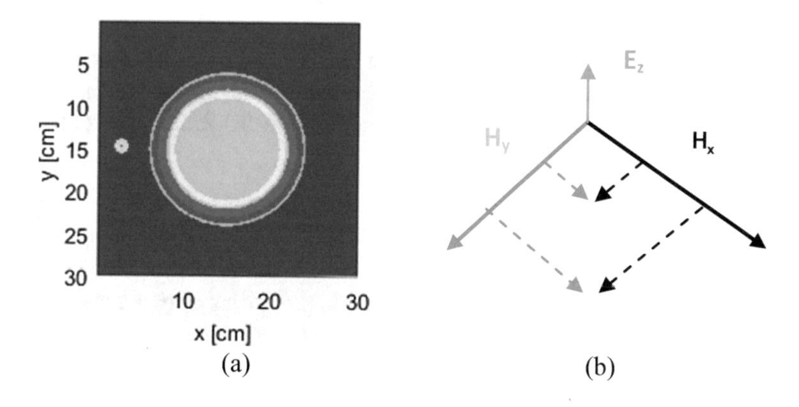

Fig. 1. (a) Cylindrical human head model and point source, (b) 2D TM mode in Yee grids (E_z, H_y, H_x)

2.1 FDTD and SAR Equations

Using 2D FDTD method, the intensity of Electric field is determined using E polarized wave (Ez, Hx, Hy) as in Eqs. (1) to (3).

$$\frac{\partial E_z}{\partial y} = -\mu \frac{\partial H_x}{\partial t} \quad (1)$$

$$\frac{\partial E_z}{\partial x} = \mu \frac{\partial H_y}{\partial t} \quad (2)$$

$$\frac{\partial H_y}{\partial x} - \frac{\partial H_x}{\partial y} = \varepsilon \frac{\partial E_z}{\partial t} + \sigma E_z \quad (3)$$

where μ, ε and σ are permeability, permittivity and conductivity of dielectric tissue layer's properties [6]. In the FDTD calculation, the parameters are as follows: For 3.35 GHz implementation, FDTD cell size ($\Delta x = \Delta y$) is 0.22 mm, time resolution (Δt) is 0.47 ps and wavelength is 89.6 mm with point 50 grid resolution. The calculation area is 1349 × 1349 and total FDTD iteration times is 17067. The highest permittivity value is existed in CSF layer as 64.825 and the minimum wavelength is 11.1 mm. For 4.5 GHz, FDTD cell size ($\Delta x = \Delta y$) is 0.17 mm, time resolution (Δt) is 0.35 ps and wavelength is 66.7 mm with point 50 grid resolution. The calculation area is 1784 × 1784 and total FDTD iteration times is 22225. The highest permittivity value is also existed in CSF layer as 62.845 and the minimum wavelength is 8.4 mm.

For the absorbing boundary of FDTD calculation, Mur 2nd ABC is used to absorb the electromagnetic wave at the border of the studied area. SAR is calculated using the following equation

$$SAR = \frac{\sigma E^2}{2\rho} [W/kg] \quad (4)$$

where ρ is the density of tissue [8].

When the electromagnetic wave is incident to the surface of dielectric medium, the intrinsic impedance is occurred at that interface where the incidence waves partly transmitted and partly reflected [9]. The reflection coefficient (R) for each layer interface is calculated from impedance (η).

$$R = \frac{\eta_{i+1} - \eta_i}{\eta_{i+1} + \eta_i} \quad (5)$$

The impedance is computed from the dielectric properties of that layer.

$$\eta_i = \sqrt{\frac{\mu_i}{\varepsilon_i}}, \text{ where } i = 0, 1, 2, 3, 4, 5, 6 \text{(layers)} \quad (6)$$

In this study, it has six layer of human head and air or free space is also counted. It can be clearly seen as in Fig. 2.

Air	Skin	Fat	Bone	Dura	CSF	Brain
ε_0	ε_1	ε_2	ε_3	ε_4	ε_5	ε_6
μ_0	μ_1	μ_2	μ_3	μ_4	μ_5	μ_6
η_0	η_1	η_2	η_3	η_4	η_5	η_6

R_0 R_1 R_2 R_3 R_4 R_5

Fig. 2. Layered structure of system

3 Analysis and Results

In simulation, the field distribution and SAR are examined in order to compare the effect of the frequencies. Figure 3(a) and (b) show the electric field distribution of last time step of FDTD iteration for the frequency 3.35 GHz and 4.5 GHz, respectively.

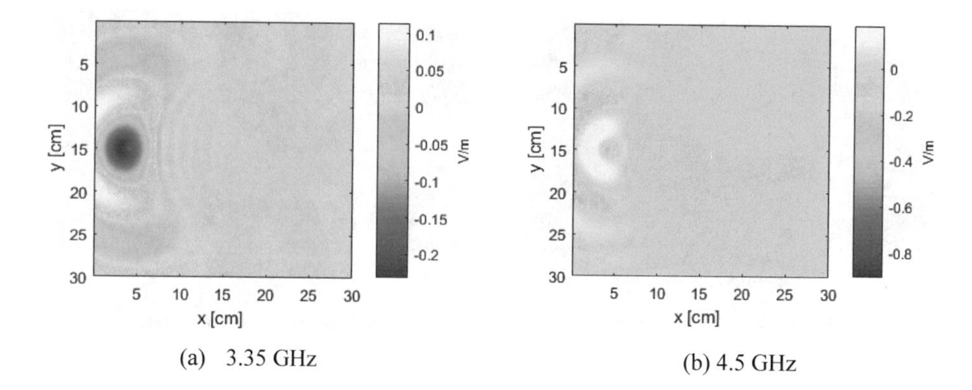

(a) 3.35 GHz

(b) 4.5 GHz

Fig. 3. E filed distribution to head for different frequency (a) 3.35 GHz and (b) 4.5 GHz

Local SAR distribution also shows in Fig. 4. SAR rate is high in near of electric field source. And, the electric field abortion rate in human head can be clearly seen in the map of Logarithmic Local SAR Distribution in Fig. 5. The region near the electric source is the highest absorption rate.

The nature of E field distribution of the given two frequencies is compared in Fig. 6 (a). The Electric field intensity is gradually decrease inside the human head in both frequencies but 3.35 GHz has the higher intensity rate than 4.5 GHz. In both case, over 90% of incident electric field reflected at the air-skin interface and only few percent transmitted into the head. That transmitted power is also absorbed by skin and CSF layers, so not much E field transmitted to brain. So the highest level of electric field distribution can be seen about 1 cm of the outermost part of the head and then it decreases significantly.

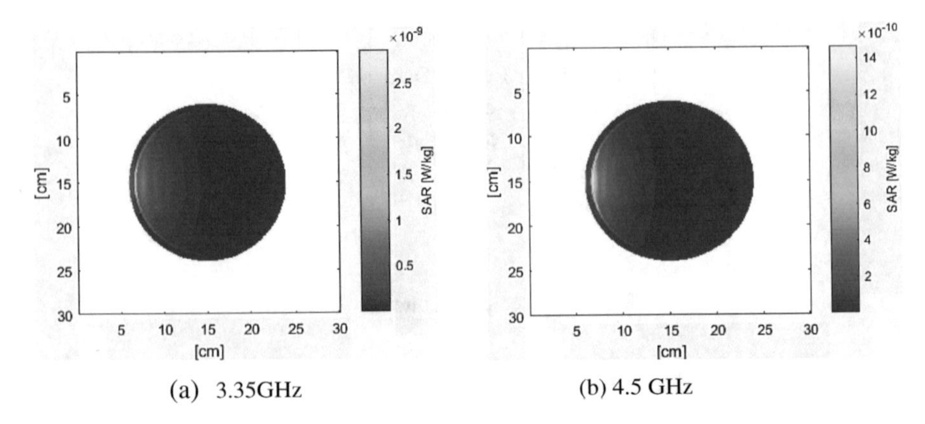

Fig. 4. Local SAR distribution: (a) 3.35 GHz and (b) 4.5 GHz

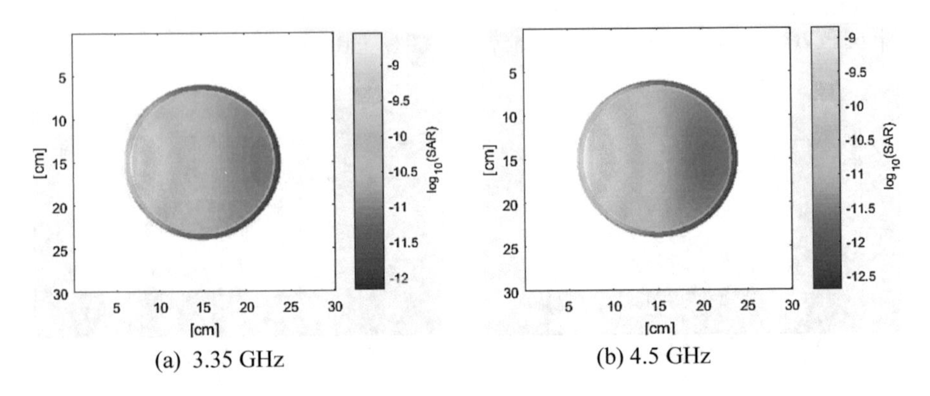

Fig. 5. Logarithmic local SAR distribution: (a) 3.35 GHz and (b) 4.5 GHz

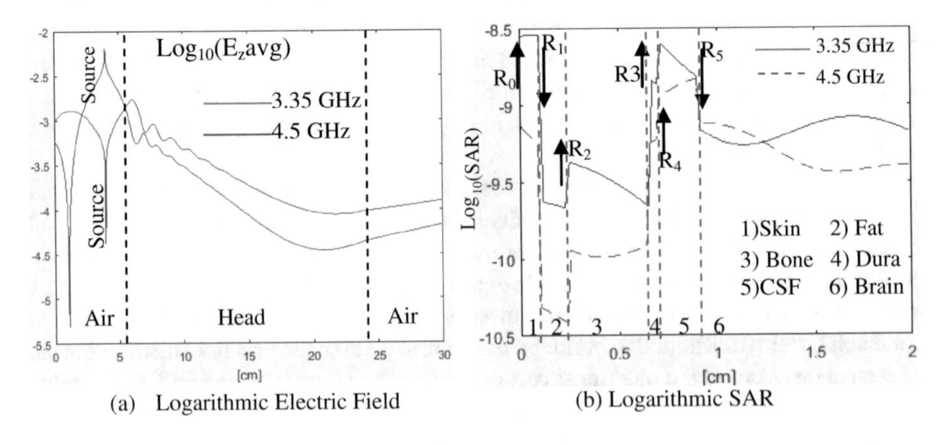

Fig. 6. Cross sessional view: (a) Logarithmic electric field and (b) Logarithmic SAR

The Local SAR in human head in cross section view is shown as in Fig. 6(b). Consequence of E field transmission, SAR value is sharp rise in Skin and CSF layer. The absorption rate in Fat and Bone Layer is not much compared with other layers because of the nature of permittivity and conductivity of that tissue.

The absorption rate at each layer interface occurs many changes as a result of the dissimilarity of the permittivity and conductivity value of consecutive tissue layers. It can be seen by using the impedance mismatch of reflection rate. The Reflection (Rx) of layer x (as in Fig. 2 and Fig. 6(b)) has the relation with increasing and decreasing of absorption rate. The direction of the arrow in Fig. 6(b) means the increasing and decreasing of SAR rate at that interface. Table 2 shows the R value of each interface from R0 to R5.

Table 2. Reflection R at each layer (Fig. 2)

Frequency	R_0	R_1	R_2	R_3	R_4	R_5
3.35 GHz	−0.99	0.46	−0.18	−0.32	−0.11	0.11
4.5 GHz	−0.99	0.45	−0.17	−0.32	−0.12	0.11

According to Fig. 6(b) and Table 2, it can be concluded that if R is less than zero, SAR rate increment occurs at that interface. Otherwise, the absorption rate decrease.

4 Conclusion

Multi-layer based human head model is examined with numerical approach using FDTD 2D TM method including with the antenna configuration of near field. As the dielectric properties form layer by layer are not same, the absorption and propagation of E field have many changes. These changes are validated with impedance and reflection properties of layer interface. It can be seen the agreement with other previous researches, CSF and Skin is the highest SAR rate.

Two types of frequency are investigated in different types of tissue layers. The results show that the nature of propagation is almost same. The increase and decrease of SAR rate in each tissue layer interface are also happen in same direction. But the higher frequency 4.5 GHz is lower absorption rate than 3.35 GHz.

Using of Matlab programming, computation time is too long in practical usage. To get the faster computation time, Matlab Mex function is considered to use. This paper forecast on two dimensional case, it will be extended to examine in three dimensional structure.

References

1. Nishizawa, S., Hashimoto, O.: Effectiveness analysis of lossy dielectric shields for a three-layered human model. IEEE Trans. Microwave Theor. Tech. **47**(3), 277–286 (1999)
2. Alisoy, H.Z.: An FDTD based numerical analysis of microwave propagation properties in a skin-fat tissue layers. Optik J. **124**(21), 5218–5224 (2013)

3. Seyfi, L., Yaldiz, E.: Numerical computing of reduction of SAR values in a homogenous head model using copper shield. In: Proceedings of the World Congress on Engineering (WCE 2010), UK, vol. 2, pp. 1–5 (2010)
4. Caorsi, S., Pastorino, M., Raffetto, M.: Analytic SAR computation in a multilayer elliptic cylinder for bioelectromagnetic applications. Bioelectromagnetic **20**(6), 365–371 (1999)
5. Siriwitpreecha, A., Rattanadecho, P., Wessapan, T.: The influence of wave propagation mode on specific absorption rate and heat transfer in human body exposed to electromagnetic wave. Int. J. Heat Mass Transfer **65**, 423–434 (2013)
6. Taflove, A., Hangess, S.C.: Computational Electrodynamices: The Finite-Difference Time-Domain Method. Artech House, Boston (2000)
7. Andreuccetti, D., Fossi, R., Petrucci, C.: An internet resource for the calculation of the dielectric properties of body tissues in the frequency range 10 Hz–100 GHz, IFAC-CNR, Florence, Italy (1997). Based on data published by Gabriel, C., et al. 1996. http://niremf.ifac.cnr.it/tissprop/. Accessed 11 Feb 2018
8. Ishimaru, A.: Electromagnetic Wave Propagation, Radiation, and Scattering, 2nd edn. IEEE Press, UK (2017)
9. Cheng, D.K.: Field and Wave Electromagnetics, 2nd edn. Addison-Wesley Publishing Company, New York (1983)

Improved Convergence in Eddy-Current Analysis by Singular Value Decomposition of Subdomain Problem

Takehito Mizuma(✉) and Amane Takei

University of Miyazaki, 1-1, Gakuen Kibanadai Nishi, Miyazaki-shi, Miyazaki 889-2192, Japan
tcl5037@student.miyazaki-u.ac.jp

Abstract. The purpose of this study is to improve the convergence of the iterative domain decomposition method for the Interface Problem in time-harmonic eddy-current analysis. The solver applied is the A-ϕ method, which consists of the magnetic vector potential A and an unknown function of the electric scalar potential ϕ. However, it is known that the convergence of the iterative domain decomposition method deteriorates for the interface problem in analyses with large-scale numerical models. In addition, the equation obtained by the A-ϕ method is a singular linear equations. In general, iterative methods are applied to solve this equation, however it is difficult to achieve high-precision because of the truncation error. In this research, to solve this problem, a direct method using a generalized inverse matrix based on a singular value decomposition method is introduced to solve the subdomain problems. Although this increases the computational cost, high-precision arithmetic becomes possible. Here, we investigate the improvement in the convergence of the interface problem by comparing our proposed method with previous method, when applied to the standard time-harmonic eddy-current problem.

Keywords: Iterative domain decomposition method
Time-Harmonic Eddy-Current analysis · Singular value decomposition

1 Introduction

Transformers, rotating machines, linear motors and other devices are becoming more sophisticated and are used for a variety of applications. To improve the accuracy of their design, it is important to accurately estimate the electromagnetic characteristics inside electrical equipments. Electromagnetic field analysis based on numerical methods such as the finite element method, performed on computers is an effective for this purpose.

To realize high-accuracy electromagnetic field analysis, it is desirable to use a high-density numerical model that precisely reproduces the target system. Parallelization methods for numerical analysis based on the finite element method have been studied to enable more accurate results. The iterative domain decomposition method [1] is one of these methods. In this method, the entire analysis region is first divided into smaller non-overlapping regions called subdomains. Next, finite element analysis is

T. T. Zin and J. C.-W. Lin (Eds.): ICBDL 2018, AISC 744, pp. 199–205, 2019.
https://doi.org/10.1007/978-981-13-0869-7_22

performed the subdomain (called the subdomain problem) for each iteration of the subdomain boundary problem (called the interface problem). The solution for the entire analysis region is obtained by an iterative method. The subdomain problem is highly parallelizable because the solutions for the subdomains can be calculated independently.

In general, it is known that the convergence of the solver in the interface problem deteriorates as the degrees of freedom increases. In addition, the convergence can be improved by improving the accuracy of the solutions in the subdomain problem. The nature of the coefficient matrix of the simultaneous linear equations to be solved in the subdomains depends on the problem under consideration. For singular problems such as eddy-current problems, an iterative method is generally used as a solver in subdomain problems.

However, as solutions obtained by iterative methods usually contain truncation errors, it is difficult to obtain high-accuracy solutions in the subdomain problem with double precision floating point numbers. In [2], an improvement in the convergence of the iterative method is realized with high-precision arithmetic using pseudo quadruple precision. However, this method has a much higher calculation time than conventional methods.

On the other hand, to solve simultaneous linear indeterminate equations, it is necessary to obtain a generalized inverse matrix. In general, memory requirement in an iterative method is less than a case of using direct methods, however the accuracy is poor. Direct methods using singular value decomposition can be applied instead of iterative ones. Singular value decomposition methods decompose the coefficient matrix of the equation to be solved into a diagonal matrix consisting of singular values and two unitary matrices. By using these three decomposed matrices, the generalized inverse matrix can be easily obtained. This method has no truncation error, unlike the iterative method. In addition, if the generalized inverse matrix is kept, the solution can be found by performing the matrix-vector product with the known vector only once.

In this research, we improve the convergence for the interface problem by introducing a direct method using singular value decomposition in the subdomain problems. The performance of the proposed method is compared with a previous method using a standard model.

2 Numerical Analysis

2.1 Time-Harmonic Eddy Current Analysis

The A-ϕ method with for the entire analysis region as Ω consists of, a function of the magnetic vector potential as ϕ, and unknown function the electric scalar potential ϕ [V] for the electric scalar potential, as an unknown function and is known as the fundamental equation of for the time-harmonic eddy-current problem. The region Ω is composed of a conductor region R and a non-conductor region S, which are not overlapping. The finite element method in Eqs. (1) and (2) is obtained by approximating A as a linear Nedelec tetrahedron and ϕ as a regular tetrahedron [1]. The element under consideration, A, is placed on the edge of the element, and ϕ is placed on the node.

$$\left(\nu rot A_h, rot A_h^*\right) - \left(j\omega\sigma A_h, A_h^*\right) + \left(\sigma grad \phi_h, A_h^*\right) = \left(\tilde{J}_h, A_h^*\right) \tag{1}$$

$$\left(\sigma grad \phi_h, grad \phi_h^*\right) - \left(j\omega\sigma A_h, grad \phi_h^*\right) = 0 \tag{2}$$

Here, (•,•) is the L^2 inner product in the region Ω. A_h, ϕ_h and J_h are finite element approximations of A, ϕ and J respectively, and A_h^* and ϕ_h^* are arbitrary test functions corresponding to A and ϕ, respectively. ν [m/H] is the magnetic reluctivity, σ [S/m] is conductivity, ω [rad/s] the angular frequency, j is an imaginary unit, ν is a piecewise positive constant, σ is a positive constant in the R region, and zero in the S region. J_h is the current density corrected so that J satisfies the equation of continuity (3) [3].

$$div J = 0 \text{ in } \Omega \tag{3}$$

2.2 Iterative Domain Decomposition Method

In this section, we describe the iterative domain decomposition method which applied to the eddy current analysis using code developed by the authors. The simultaneous linear Eq. (4) to be solved is obtained by performing finite element division on the region Ω. This region is divided into N subdomains without overlap (Eq. (5)).

$$Ku = f \tag{4}$$

$$\Omega = \cup_{i=1}^{N} \Omega^{(i)} \tag{5}$$

K is a coefficient matrix, u is an unknown vector, f is a known vector, and subscript (i) is the data on $\Omega^{(i)}$. u_B is the degrees of freedom of the new boundary region. Equation (4) becomes Eq. (6) when the domain decomposition is applied with $u_I^{(i)}$ as the degrees of freedom of the internal nodes of the subdomain.

$$\begin{bmatrix} K_{II}^{(1)} & 0 & 0 & K_{IB}^{(1)} R_B^{(1)\mathrm{T}} \\ 0 & \ddots & 0 & \vdots \\ 0 & 0 & K_{II}^{(N)} & R_{IB}^{(N)} R_B^{(N)\mathrm{T}} \\ R_B^{(1)} K_{IB}^{(1)\mathrm{T}} & \cdots & R_B^{(N)} K_{IB}^{(N)\mathrm{T}} & \sum_{i=1}^{N} R_B^{(i)} K_{BB}^{(i)} R_B^{(i)\mathrm{T}} \end{bmatrix} \begin{bmatrix} u_I^{(1)} \\ \vdots \\ u_I^{(N)} \\ u_B \end{bmatrix} = \begin{bmatrix} f_I^{(1)} \\ \vdots \\ f_I^{(N)} \\ f_B \end{bmatrix} \tag{6}$$

$R_B^{(i)\mathrm{T}}$ is a 0–1 matrix for limiting u_B to the internal degrees of freedom $u_B^{(i)}$ of subdomain Ω. From this, Eqs. (7) and (8) are obtained. $f_B^{(i)}$ is the right-hand vector of the equation for u_B. Equation (8) is an interface problem, which is an expression for satisfying the continuity between subdomain areas [1].

$$K_{II}^{(i)} u_I^{(i)} = f_I^{(i)} - K_{IB}^{(i)} u_B^{(i)} \quad (i = 1, \ldots, N) \tag{7}$$

$$\left\{\sum_{i=1}^{N} R_B^{(i)}\left\{K_{BB}^{(i)} - K_{IB}^{(i)T}\left(K_{II}^{(i)}\right)^{\dagger} K_{IB}^{(i)}\right\}R_B^{(i)T}\right\}u_B$$
$$= \sum_{i=1}^{N} R_B^{(i)}\left\{f_B^{(i)} - K_{IB}^{(i)T}\left(K_{II}^{(i)}\right)^{\dagger} f_I^{(i)}\right\} \tag{8}$$

The eddy-current problem involves specificity. † of $\left(K_{II}^{(i)}\right)^{\dagger}$ represents a generalized inverse matrix. Equation (8) can be rewritten as follows.

$$Su_B = g \tag{9}$$

$$S = \sum_{i=1}^{N} R_B^{(i)} S^{(i)} R_B^{(i)\mathrm{T}} \tag{10}$$

$$S^{(i)} = K_{BB}^{(i)} - K_{IB}^{(i)\mathrm{T}}\left(K_{II}^{(i)}\right)^{\dagger} K_{IB}^{(i)} \tag{11}$$

Here, S is the Schur complement matrix, and $S^{(i)}$ is the local Schur complement, matrix in the subdomain $\Omega^{(i)}$. Equation (9) can be solved by the conjugate orthogonal conjugate gradient (called the COCG) method with diagonal scaling preconditioning. In implementation, the finite element analysis is performed on internal degrees of freedom in each subdomain when the degrees of freedom on the boundary of the subdomains are fixed with the Dirichlet boundary condition. The largest computational load in this algorithm is the vector product operation of the Schur complement matrix in the iterative method of the interface problem. The finite element analysis performed in each subdomain is called the subdomain problem. With the iterative domain decomposition method this can be divided into two problems, the interface and subdomain problems.

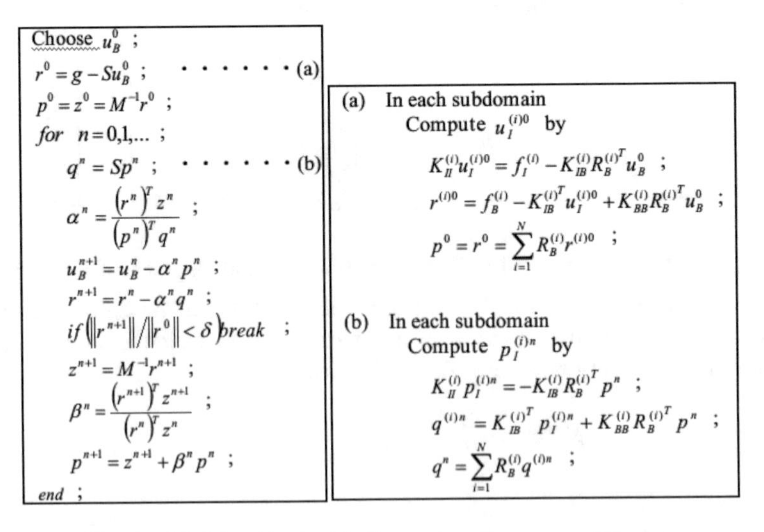

Fig. 1. COCG method with diagonal scaling preconditioning.

For the interface problem in Eq. (9), the algorithm shown in Fig. 1 is applied. First, calculate the internal boundary degrees of freedom u_B. δ is the convergence value, which is a positive constant. $\|\cdot\|$ represents the 2-norm. It is necessary to perform a vector multiplication operation of the Schur complement matrix S in (a) and (b) of Fig. 1 for the COCG step when calculating the initial residual. However, the construction of S is very computationally expensive compared with the coefficient matrix K. Therefore, substituting (a) and (b) in the algorithm with the finite element analysis of the subdomain problem using Eqs. (10) and (11), S is calculated indirectly [1]. Finally, $u_I(i)$ for each subdomain is calculated by Eq. (7) to obtain a solution for the whole region.

3 Improvement of Convergence in the Interface Problem

In this research, the singular linear equations in the subdomain problem are solved by the generalized inverse matrix based on singular value decomposition. Various methods for obtaining the singular value decomposition have been proposed. In this research, the extended version of Golub-Reinsch's singular value decomposition algorithm [4], which is widely used, is applied. The matrix K is decomposed into the following form:

$$K = U\Sigma V^{\mathrm{H}} \tag{12}$$

Here, Σ is a diagonal matrix composed of singular values of K, U and V are unitary matrices, and H is Hermitian. The solution of the equation $Kx = b$, for unknown vector x and known vector b, is obtained by the following equation.

$$x = V\Sigma^{\dagger}U^{\mathrm{H}}b \tag{13}$$

This method is adopted for the subdomain solver in the iterative domain decomposition method. An advantage of this approach is that here is no need to solve the regular matrix, so it is possible to use a suitable solver for each problem, such as the A-ϕ method for eddy current analysis.

4 Analysis

4.1 Analysis Model

To examine the proposed method, we use the TEAM Workshop Problem 7 [5] (Fig. 2) which is a standard problem in eddy-current analysis. The model considered here is shown in Fig. 2. A thick aluminum plate with a hole, which is placed eccentrically, is set un-symmetrically in a non-uniform magnet. A field is produced by the exciting current which varies sinusoidally with time. Physical properties are set as in [5]. This model is split into tetrahedral elements and divided into 13,453,974 elements, 2,332,695 vertexes, and 15,997,971 edges.

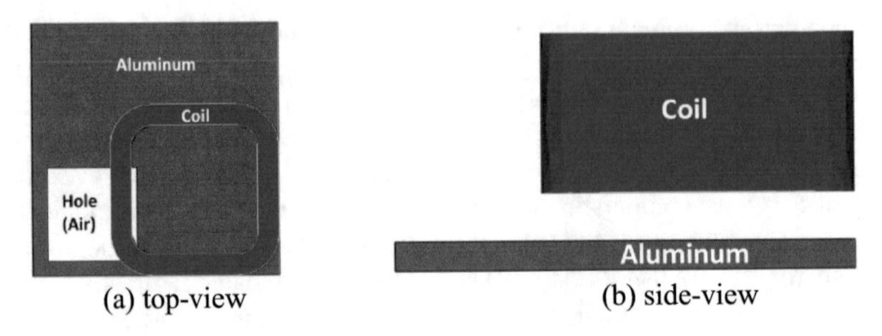

Fig. 2. TEAM workshop problem 7 model.

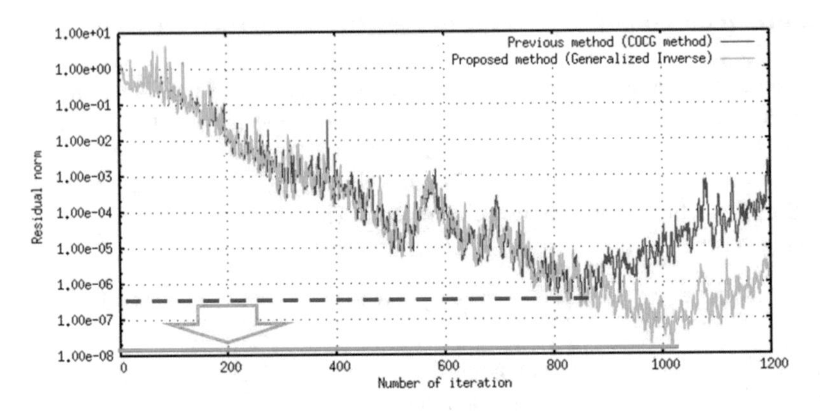

Fig. 3. Convergence history for the interface problem.

Figure 3 shows the convergence results of analyses under the above conditions. The horizontal axis is the number of iterations, and the vertical axis is the relative residual norm of the interface problem. The analysis method with the highest accuracy is obtained when the calculation is stopped when the residual norm shows the smallest value. In addition, the smaller the residual norm, the more accurate the solution is obtained. As shown in Fig. 3, we have succeeded in reducing the residual norm by 1–2 orders compared with the previous iterative method, and it is thus confirmed that a high-accuracy can be realized with our method.

5 Conclusion

We tried the improvement of the convergence in the interface problem when introducing a direct method using singular value decomposition to solve the subdomains in the eddy-current problem based on the iterative domain decomposition method. In the subdomain problem, typically iterative methods are used. However, the introduction of the generalized inverse matrix has been shown as an effective option.

References

1. Kanayama, H., Sugimoto, S.: Effectiveness of A-phi method in a parallel computing with an iterative domain decomposition method. IEEE Trans. Magn. **42**(4), 539–542 (2006)
2. Mizuma, T., Takei, A.: Improvement of convergence properties of an interface problem in iterative domain decomposition method using double-double precision. In: Proceedings of The 34rd JSST Annual Conference (2016)
3. Kanayama, H., Shioya, R., Tagami, D., Matsumoto, S.: 3-D eddy current computation for a transformer tank. COMPEL **21**(4), 554–562 (2002)
4. Wilkinson, J.H., Reinsch, C. (eds.): Handbook for Automatic Computation, Linear Algebra. Springer-Verlag, New York (1971)
5. Fujiwara, K., Nakata, T.: Results for benchmark problem 7 (asymmetrical conductor with a hole). COMPEL **9**(3), 137–154 (1990)

Development and Validation of Parallel Acoustic Analysis Method for the Sound Field Design of a Large Space

Yuya Murakami(✉), Kota Yamamoto, and Amane Takei

University of Miyazaki, 1-1 Gakuenkibanadainisi, Miyazakis-shi, Miyazaki, Japan
hl13049@student.miyazaki-u.ac.jp

Abstract. Recently, acoustic analysis techniques are used in the design of acoustic environments both inside and outside of rooms containing a sound source. However, when the analysis area is expanded or the frequency is increased, unknowns of a solving liner equation increase, and large-scale analyses becomes necessary. Therefore, in this study, a large-scale acoustic analysis method has been developed based on a parallel finite element method. In the proposed large-scale analysis method, an iterative method is applied to the interface problem as the iterative domain decomposition method, which is known to be an effective parallelization method. To improve the convergence of the iterative method, balancing domain decomposition has been applied as a preconditioner step, and its effectiveness in large-scale acoustic analysis is demonstrated. The acoustic analysis code is developed in this study that is shown to be capable of the large-scale analysis of finite element models on the order of tens of millions of elements with high accuracy.

Keywords: Acoustic analysis
Balancing domain decomposition preconditioning
Parallel finite element method

1 Introduction

In recent years, acoustic analysis technology has been increasingly used in the design of acoustic environments, such as concert halls and live houses [1], with the aid of the drastic performance improvement of computers, as symbolized by the emergence of the supercomputer called the K computer [2]. In the conventional design of acoustic environments, scaled acoustic model experiments using prototypical miniature models are conducted before developing the actual design, which is very costly and time-consuming. To solve this problem, the application of acoustic analysis technology has been attempted. With such technology, the factors hindering the sound are determined in advance, the quality of the acoustic space is raised in the design, and the time and cost required for construction are dramatically reduced. However, because of the large scale of the analysis domain and the expansion of the analysis frequency band, analysis of large-scale problems is a problem in acoustic wave analysis.

T. T. Zin and J. C.-W. Lin (Eds.): ICBDL 2018, AISC 744, pp. 206–214, 2019.
https://doi.org/10.1007/978-981-13-0869-7_23

Therefore, in this research, we can use the unstructured grid and examine the large scale finite element acoustic analysis method with domain decomposition method known as parallelization method suitable for parallel memory parallel computer.

Presently, the use of Beowulf-type personal computer (PC) clusters, in which many PCs are connected via a network, and advanced distributed memory parallel computers, such as supercomputers, is becoming widespread. The use of such parallel computers and the parallel analysis code operating on them opens up the possibility of the analysis of tens of millions to hundreds of millions of elements. In the present study, parallel computing algorithms for finite element analysis and other libraries, including domain segmentation, were improved, and numerical examples for structural and electromagnetic field analysis on the scale of tens to hundreds of billions of finite elements were developed [3, 4].

Based on the above background, in this study, a new three-dimensional acoustic analysis method was developed using a parallel finite element method based on the iterative domain decomposition method (IDDM) on the premise of using parallel computers. Firstly, as a basic study, an acoustic analysis code using the finite element method was developed, and a test code was written with a maximum error of approximately 1.4% compared with the reference solution in the benchmark problem [5]. Based on this test code, a large-scale acoustic analysis method based on the parallel finite element method was developed by introducing an iterative domain decomposition method for parallelization.

In the iterative domain decomposition method, it is known that the convergence of the iterative method applied as a solution of the interface problem worsens as the number of regions increases. Balancing domain decomposition (BDD) preconditioner has been proposed as a method of improving the convergence of interface problems [6, 7]. The BDD preconditioner is a pretreatment method that improves the convergence of the interface problem and achieves high-speed analysis, and its effect has been confirmed by structural analysis and application to thermal problems [8]. In addition, the basic equation of the development code is a modified version of the d'Alembert's equation to the Helmholtz equation. There are almost no examples of parallel calculation using the BDD preconditioning for the Helmholtz equation, and there are no reports on the effectiveness of BDD preconditioner in large-scale problems. Therefore, in this study, parallel finite element acoustic analysis code based on the iterative domain decomposition method with BDD preconditioner was applied to a large-scale problem involving tens of millions of elements, and the effectiveness of the method was confirmed.

It was then demonstrated that the analysis code developed in this study has reasonable accuracy as a numerical acoustic analysis method and that the convergence of the iterative method of solving the interface problem is improved by BDD preconditioner. In addition, as a real environmental problem, acoustic analysis was conducted in a simple live house model, and the effectiveness of the method was confirmed.

2 Formulation

2.1 Formulation of Equations for the Velocity Potential

Consider the three-dimensional region V. The boundary S of V is given by the sound source S_1 and the sound absorption boundary S_2 Here, for a medium with density ρ [kg/m^3] and a sound velocity c [m/s], the equation governing the region V is given by d'Alembert's equation for the velocity potential ϕ as

$$\frac{\partial^2\phi}{\partial x^2}+\frac{\partial^2\phi}{\partial y^2}+\frac{\partial^2\phi}{\partial z^2}=\frac{1}{c^2}\frac{\partial^2\phi}{\partial t^2}\ \text{in}\ V, \tag{1}$$

Here, when the velocity potential ϕ is defined as $\phi = \Phi\exp(-j\omega t)$ Eq. (1) becomes

$$\frac{\partial^2\Phi}{\partial x^2}+\frac{\partial^2\Phi}{\partial y^2}+\frac{\partial^2\Phi}{\partial z^2}-k^2\Phi=0\ \text{in}\ V. \tag{2}$$

This makes it possible to obtain the shape of the Helmholtz equation in steady state. Here, j is the imaginary unit, $k = \omega/c$ is a wave number, and ω is an angular frequency.

2.2 Finite Element Model

From Eq. (2), a weak formulation is derived using the Galerkin method. Consider finite element division of region V. When discretizing the velocity potential Φ using tetrahedral primary elements, Eq. (2) yields the following discretization equation:

$$[A]\{\Phi\}-k^2[M]\{\Phi\}+j\rho\omega[C]\{\Phi\}=\{q\}, \tag{3}$$

Here, [·] is a matrix, { · } is a vector, and q is point sound source.

3 Parallelization Method

An iterative domain decomposition method was introduced as a method of efficiently solving large-scale linear simultaneous equations. In an iterative domain decomposition method, the analysis region is divided into arbitrary partial regions, and analysis in each partial region is executed in each step of the iterative calculation of the balance calculation between the partial regions. When the balance calculation is complete, the solution can be obtained. This method has been applied as an effective parallelization method for large-scale analysis [8]. The finite element equation (Eq. (3)) given in the previous chapter can be written as

$$[K]\{u\}=\{f\} \tag{4}$$

where K is a coefficient matrix, u is an unknown vector, and f is a known vector. The analysis region V is divided into N pieces such that the minimum unit is an element and the boundaries between the regions do not overlap. Let u_B be the degree of freedom newly generated on the boundary between the regions and $u_I^{(i)}$ the degree of freedom occurring on the partial region, and if we apply the region partitioning method to Eq. (4)

$$\begin{bmatrix} K_{II}^{(1)} & \cdots & 0 & K_{IB}^{(1)} R_B^{(1)T} \\ 0 & \ddots & \vdots & \vdots \\ 0 & \cdots & K_{II}^{(N)} & K_{IB}^{(N)} R_B^{(N)T} \\ R_B^{(1)} K_{IB}^{(1)T} & \cdots & R_B^{(N)} K_{IB}^{(N)T} & \sum_{i=1}^{N} R_B^{(i)} K_{BB}^{(i)} R_B^{(i)T} \end{bmatrix} \begin{Bmatrix} u_I^{(1)} \\ \vdots \\ u_I^{(N)} \\ u_B \end{Bmatrix} = \begin{Bmatrix} f_I^{(1)} \\ \vdots \\ f_I^{(N)} \\ f_B \end{Bmatrix} \quad (5)$$

4 Validation of Analysis Method

4.1 Benchmark Model

In the accuracy validation and performance evaluation of the analysis method, the Code_Aster benchmark model AHLV 100 [9] Fig. 1 was used. This test model is an acoustic tube with a sound-absorbing end with a length of 1 m, a height of 0.1 m, and a width of 0.2 m. and At the left end, the vibration boundary condition $v_n = 0.014$ [m/s] corresponding to a sound source was applied, and the sound absorption boundary condition $Z_n = 445.9\ [\mathrm{kg/m^3 \cdot s}]$ was applied at the right end. In addition, $f = 500$ Hz was used as the analysis frequency, and air $\rho = 1.3\ [\mathrm{kg/m^3}]$ was used as the medium of the sound field. The speed of sound was considered to be $c = 343$ m/s. All surfaces excluding the sound source and the sound-absorbing end were assumed to be perfectly rigid (total reflection). The number of elements was 140,604, the number of nodes was 26,927, and discretization was performed using a tetrahedral primary element.

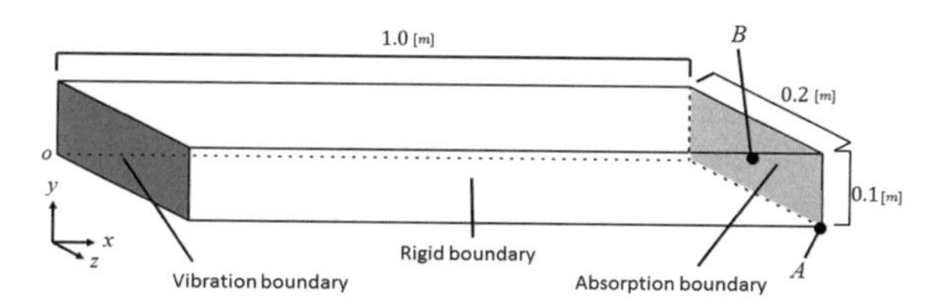

Fig. 1. Benchmark model: AHLV100.

The points considered in the accuracy validation were the two nodes labeled A and B on the sound-absorbing surface in Fig. 1. These points respectively correspond to the node at the vertex of the sound-absorbing surface and that located approximately at the center of the sound-absorbing surface.

The finite element formulation considered in this study and the accuracy validation used to confirm the appropriateness of the boundary conditions were then implemented. To validate the accuracy of the proposed method, solutions obtained using Theoretical value [10] and a test code [5] were used as reference. In the convergence assessment, the calculation was terminated when the relative residual norm in the conjugate orthogonal conjugate gradient (COCG) method reached a value of 10^{-6} or less [11].

4.2 Accuracy Validation

The analysis results using AHLV 100 are given in Table 1. The results for the real part of the sound pressure obtained using the developed code and the test code are shown in Fig. 2(a) and (b), respectively. A comparison of the visualization results obtained using the developed code and the test code confirms that they are in good agreement. Comparing the values of points A and B with The theoretical value and development code, the error is within ±1.4 [%].

Table 1. Analysis results obtained using AHLV 100

Point	A		B	
	Real	Imag.	Real	Imag.
Theoretical	6.0237	1.6387	6.0237	1.6837
Development code	6.0136	1.6587	6.0210	1.6613
Error [%]	−0.1677	1.2205	−0.0448	−1.3304

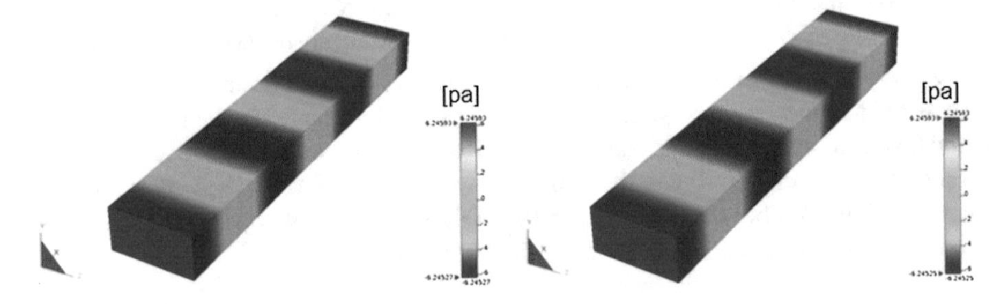

Fig. 2. Visualization of the results obtained using (a) the developed code and (b) the test code. The color scale corresponds to the real part of sound pressure.

From the above results, the error rate of the parallel finite element analysis code developed in this research was found to be smaller than the error rate of 5% generally accepted in numerical analysis, confirming that the proposed method has a high accuracy. Also, because the results obtained using the developed code show good agreement with those using the test code, the proposed method is considered reasonable.

5 Validation Using a Real Environment Model

5.1 Live House Model

To confirm the effectiveness of the parallel acoustic analysis code developed in this research, large-scale analysis based on a real acoustic environment was carried out. A small acoustic environment (live house) was considered as the analysis object. The shape model of the live house is shown in Fig. 3. This model was built based on live houses currently in operation [12]. The dimensions of the shape model are given in Fig. 3(b). It is 6 [m] in width, 6 [m] in height, 15 [m] in depth and contains shapes simulating a sound source, a stage, a human body, and a structure inside the room. One yz surface of the rectangular parallelepiped on the stage was defined as a sound source with a sound pressure of 0.2 Pa, a frequency of 150 Hz, and the sound characteristics of a classical guitar. The specific acoustic impedance was set as the sound-absorbing boundary condition the floor was modeled as wood. The surfaces of the other walls and structures were modeled as glass wool, and the human body structure was assumed to be cylindrical in shape. The medium was assumed to be air ($\rho = 1.3$ kg/m^3), and the speed of sound was set to $c = 340$ [m/s].

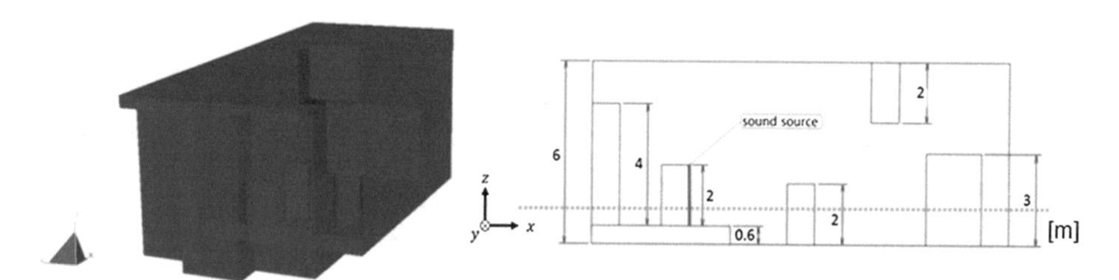

Fig. 3. (a) Appearance of the analyzed shape model. (b) Cross-sectional view of the analyzed shape model in the xz plane and relevant model dimensions.

The computer-aided design (CAD) model was divided using a tetrahedral primary element with a maximum element side length of 0.047 m, and the obtained numerical model had 51,527,590 elements and 8,484,771 nodes. The selected maximum element side length is approximately 1/48 of the wavelength of the analysis frequency of 150 Hz, which provides sufficient resolution with respect to the wavelength.

The convergence of the problem can be greatly improved by performing BDD preconditioner, which adds a corrected solution obtained using a course grid to the solution when solving the Neumann–Neumann pretreatment known to be well-suited to the preconditioner of the interface problem. BDD preconditioner is based on the multigrid method, and its effectiveness in structural and thermal problems has been demonstrated [8]. BDD preconditioner was applied prior to the application of the COCG method to the regional equilibrium problem. The performance evaluation of BDD preconditioner introduced for the iterative domain decomposition method developed in this research is described.

5.2 Analysis Environment

The parallel computing environment used in this study is a PC cluster composed of 22 multicore of Intel Core i7 2600 central processing units (CPUs) (3.4 GHz, L2 256 kB $\times$ 4, L3 8 MB) and 22 PCs (88 cores) equipped with 24 GB memory. In addition, MPICH 3.2 was used as a communication library.

5.3 Analysis Results

Figure 4 shown Indicates the convergence history of the residual norm in the iterative method (COCG method) of the interface problem in the case of preconditioner, diagonal scaling preconditioner, BDD preconditioner. Figure 5 shows a heat map of the real part of the obtained sound pressure in the cross section corresponding to the location of the broken line in Fig. 3(b). Table 2 shown Comparison of iteration count, calculation time, memory used.

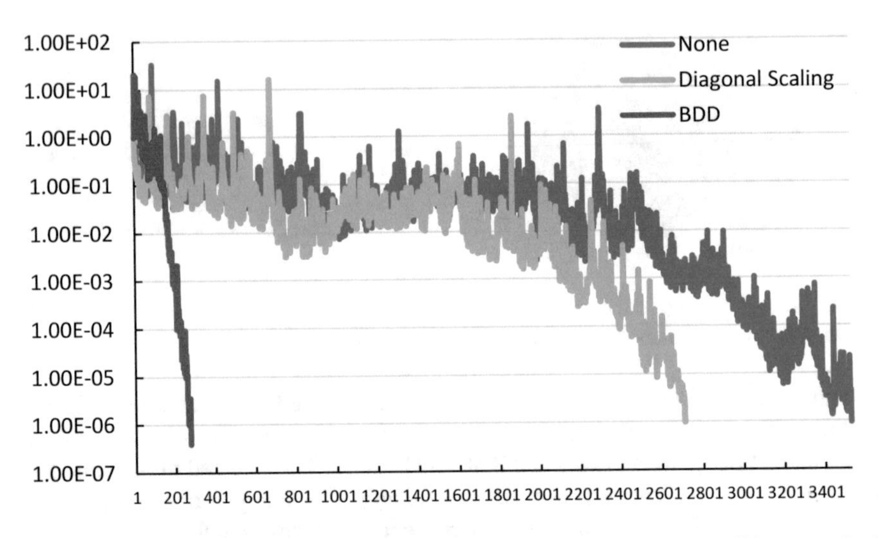

Fig. 4. Convergence history of the residual norm obtained using each preconditioner method.

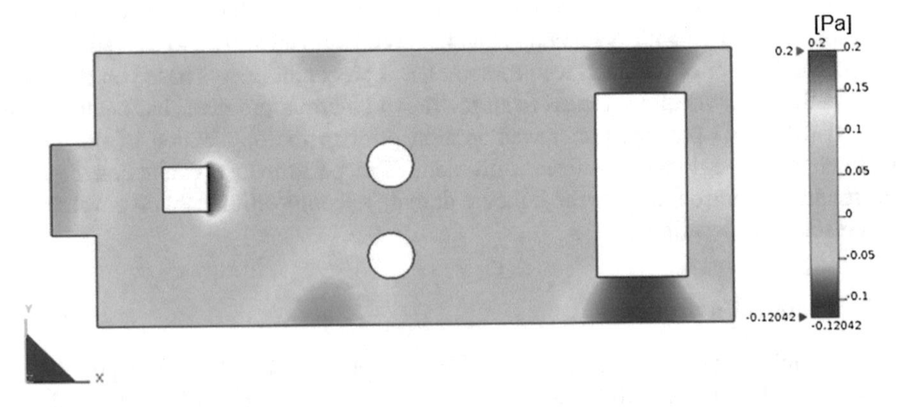

Fig. 5. Visualization of the real part of the sound pressure in a cross section in the *xy* plane.

Table 2. Comparison of iteration count, calculation time, memory used

	CGloop [time]	Time [s]	Mem [GB]
None	3528	14634.3	0.37
Diagonal scaling	2710	14313.7	0.37
BDD	275	13863.8	0.46

Figure 5 can be confirmed that the sound field in the live house has high and low sound pressure, and has wave mobility. From the Fig. 4, Fig. 5, Table 2, the proposed method can be applied to a large-scale model with a complex shape based on a real environment; furthermore, BDD pretreatment greatly contributes to enabling large-scale analysis using a real environment model.

6 Conclusion

This paper has described the development and examination of a large-scale acoustic analysis method using the parallel finite element method based on the iterative domain decomposition. In the accuracy validation of the analysis code using the AHLV 100 model, the error rate of the code has been approximately 1.4%, which is smaller than the error rate of 5% generally allowed in numerical analysis. Moreover, the sound pressure results obtained using our analysis code in this study showed good agreement with the test code, thereby validating the analysis code. In addition, BDD preconditioner was introduced to obtain the iterative method of the interface problem in the iterative domain decomposition method. It was confirmed that applying BDD preconditioner greatly improved the convergence of the solution, and it has been shown that the proposed method can be applied to real world problems.

References

1. Otsuru, T., et al.: Large scale sound field analysis by finite element method applied onto rooms with temperature distribution. In: Proceedings of ICSV 9/CD-ROM (2002)
2. Institute of Physical and Chemical Research Computational Science Research. Homepage. http://www.aics.riken.jp/jp/. Accessed 24 Feb 2018
3. Takei, A., Muroya, K., Yoshimura, S., Kanayama, H.: Finite element analysis for microwave frequency electromagnetic fields using numerical human models. Jpn. Simul. Soc. J. **4**(3), 81–95 (2012)
4. Takei, A., Yoshimura, S., Kanayama, H.: Large scale parallel finite element analyses of high frequency electromagnetic field in a commuter train environment, vol. 1, pp. 1–10 (2009)
5. Yamamoto, K., Kudou, A., Takei, A.: Research and development large-scale acoustic analysis based on the parallel finite element method. Jpn Soc. Mech. Eng. Kanto Branch No. 23, OS0701-04 (2017)
6. Mandel, J.: Balancing domain decomposition. Commun. Numer. Methods Eng. **9**, 233–241 (1993)
7. Mandel, J., Brezina, M.: Balancing domain decomposition: theory and computations in two and three dimensions, computational mathematics group. Technical report 7 (1993)

8. Ogino, M., Shioya, R., Kanayama, H., Tagami, D., Yoshimura, S.: Parallel elastic finite element analysis using the balancing domain decomposition. Trans. Jpn. Soc. Mech. Eng. A **69**(685), 1360–1367 (2003)
9. Takei, A., Sugimoto, S., Ogino, M., Yoshimura, S., Kanayama, H.: EMC analysis by large-scale full-wave analysis with an iterative domain decomposition method. In: 62nd National Congress of Theoretical and Applied Mechanics, ID: OS22-09
10. Delmas, J.: AHLV100 – Guide wave at anechoic exit, Code_Aster, Version 11. https://www.codeaster.org/V2/doc/v13/en/man_v/v8/v8.22.100.pdf. Accessed 2018
11. Okamoto, N., Yasuda, Y., Oturu, T., Tomiku, R.: Application of Krylov Subspa-CE methods to finite element sound field analysis: numerical analysis of large-scale sound fields using iterative solvers part II, vol. 71, no. 610, pp. 11–18 (2006)
12. Star Pone's Café. http://mandala.gr.jp/SPC/home. Accessed 28 Feb 2018

Secret Audio Messages Hiding in Images

Saw Win Naing(✉) and Tin Myint Naing(✉)

Faculty of Information and Communication Technology,
University of Technology (Yatanarpon Cyber City), Pyin Oo Lwin, Myanmar
sawwinnaing321@gmail.com, utinmyintnaing08@gmail.com

Abstract. Today, security of data transmission is a big problem in society. Information security fields play a vital role in network communication. To secure information, cryptography and steganography are widely used. Image steganography become popular because images are frequently used on internet. In this paper, the combination of cryptography and steganography is used to provide more secure which mean that it provides two layers protection. The secret audios or important speeches are recorded and this secret data is encrypted using Advance Encryption Standard (AES). This cipher data is embedded into cover images using Random-Based Least Significant Bit method. The random positions of image for LSB are also generated by the use of AES key. The resulting stego images are sent to the receiver and the receiver needed to perform the reverse processes to obtain the secret data. The image quality is analyzed in terms of both Mean Square Error (MSE) and Peak Signal to Noise Ratio (PSNR) measured between the cover image and stego image.

Keywords: Information security · Advance Encryption Standard (AES)
Least Significant Bit method
Mean Square Error (MSE) and Peak Signal to Noise Ratio (PSNR)

1 Introduction

Information security is defined as the protection of information in one system that store and transmit that information. For an organization, information is valuable and should be appropriately protected. Security is especially important in a business environment increasingly interconnected, in which information is now exposed to growing number and a wider variety of threats and vulnerabilities. Cause damage such as malicious code, computer hacking, and denial of service attacks have become more common, more ambitious, and more sophisticated. So, by implemented the information security in an organization, it can protect the technology assets in use at the organization [1].

Cryptography is a science that applies complex mathematics and logic to design strong encryption methods. It provides data security and many cryptography techniques are employed, such as symmetric and asymmetric techniques. Symmetric techniques mean that both the sender and receiver use the same cryptographic key where asymmetric methods mean different keys are used in both sender and receiver. Different types of asymmetric cryptography techniques, such as RSA (Rivest Shami and Adleman), DiffieHellman, DSA (Digital Signature Algorithm), ECC (Elliptic curve cryptography) are analyzed. Different types of Symmetric encryption algorithms such as

T. T. Zin and J. C.-W. Lin (Eds.): ICBDL 2018, AISC 744, pp. 215–223, 2019.
https://doi.org/10.1007/978-981-13-0869-7_24

Digital Encryption Standard (DES), Triple Digital Encryption Standard (TDES) and Advanced Encryption Standard (AES) Algorithm [2].

Steganography is widely used in digital fields and its techniques are developed day by day [4]. The information security fields include watermarking, cryptography and steganography. The purpose of both Steganography and cryptography are intended to protect information from unauthorized parties. The main difference between them is that steganography is concerned with hiding data and hiding the existence of the data, where cryptography is concerned with transforming data into an unreadable form. In this paper, Data is encrypted (cryptography) and then this is hidden within a carrier image (steganography) to provide another layer of protection.

This paper presents that embedding data into images. The goal of this work is to provide high level of security, maximum embedding capacity, efficiency and reliability for secret communication using image processing and stenographic techniques. In this study, the audio file is taken as secret information. According to the size of the secret audio file, the number of required images is calculated. For enhancing the security, Advance encryption standard (AES) is used for the encryption of audio file and this encryption technique is more secure than other encryption techniques today. Finally, secret message will be embedded in the images by Least Significant Bit (LSB).

The remainder of this paper is organized as follows: Sect. 2 describes proposed method of paper in which overview of the system design is provided. The details of AES and LSB are also explained in Sect. 2. Section 3 illustrates the experimental results, MSE and PSNR calculations and performance analysis and conclusion is explained in Sect. 4.

2 Proposed Methods

The main processes of this system are secret audio and cover images acquisition, applying AES algorithm, embedding process using LSB technique to obtain stego images.

Firstly, Secret audio is loaded into the system. According to the size of the audio files, the numbers of required images are calculated. The secret audio is encrypted using AES (advance encryption algorithm); it is the most used algorithm for encryption for its security. Finally, encrypted audio file is embedded in images using LSB (least significant bit) technique and stego images are obtained. This stego images are send to the receiver in which the receiver has the pre-shared key. The receiver need to extract the encrypted audio from each stego image using LSB extraction. And then the receiver will be obtained the cipher data. This cipher data is decrypted with AES to obtain the original secret audio. The receiver needs to do reverse processes by using this key. The propose solution is analyzed on the basis of PSNR (peak signal to noise ratio) and MSE (mean square error) frequency between stego image and original image (Fig. 1).

2.1 Secret Audio and Cover Image Acquisition

Sampling rate or sampling frequency defines the number of samples per second taken from a continuous signal to make a discrete or digital signal. The sampling rate are

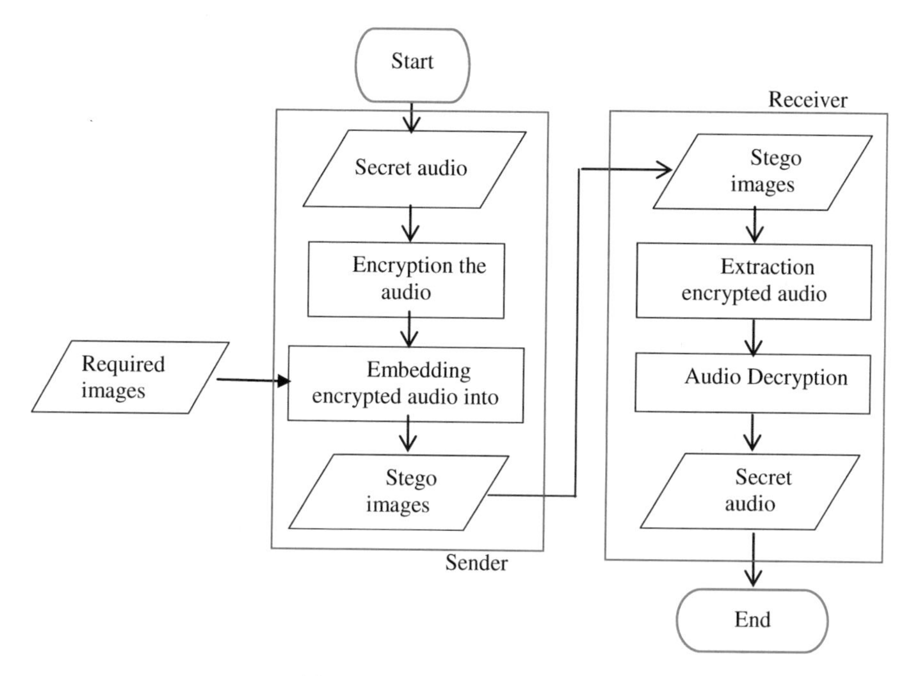

Fig. 1. Proposed system design

measured in hertz (Hz). The range of human hearing is generally considered to be 20 Hz to 20 kHz. In this system, sampling rate 4 kHz is used which is enough to hear the voice messages. According to the duration of the audio file, the sample data may be varied.

Calculations of number of samples in audio:

- For 3 s audio with 4000 sampling rate—3 × 4000 = 12,000 samples
- For 30 s audio with 4000 sampling rate—30 × 4000 = 120,000 samples
- For 1 min audio with 4000 sampling rate—60 × 4000 = 240,000 samples
- For 3 min audio with 4000 sampling rate—3 × 60 × 4000 = 720,000 samples
- For 5 min audio with 4000 sampling rate—5 × 60 × 4000 = 1,200,000 samples

In this paper, the cover images with 1280 × 720 resolution are used. In 24-bit image, all red, green and blue color components can be used. Therefore, each pixel has 3 bits of secret message.

Total Number of Pixel in each images:

- For 1280 × 720 images—1280 × 720 = 921600 pixels (1 channel)
  = 921600 × 3 = 2764800 pixels (RGB channel)

Calculation of number of required images:

- Number of bits in 3 s audio—12000 × 8 = 96000 bits
  Number of required images = 96000/2764800 = 0.0347 ~ 1 image
- Number of bits in 5 min audio—1200000 × 8 = 9600000 bits
  Number of required images = 9600000/2764800 = 3.4722 ~ 4 images

2.2 Advance Encryption Standard (AES)

The AES Algorithm is a symmetric-key cipher, in which both the sender and the receiver use a single key for encryption and decryption. There are three types of AES i.e. AES-128, AES-192 and AES-256 according key sizes: 128, 192 and 256 bits [7]. This key size determines the security level as the size of key increases the level of security increases. The data block length is fixed to be 128 bits. In this paper, AES-128 is used for encryption and decryption processes.

AES consists two main phases:

- Key Expansion
- AES Encryption and Decryption

2.2.1 Key Expansion

- Rotword performs a one-byte circular left shift on a word.
- Subword performs a byte substitution on each byte of its input word, using S-box.
- The result of steps 1 and 2 is XORed with a round constant [8].

2.2.2 AES Encryption and Decryption

The AES algorithm is an iterative algorithm. Each iteration can be called a round, and the total number of rounds is 10, 12, or 14, when key length is 128,192, or 256, respectively. In this paper, key length AES-128 is used. So it has 10 rounds and each round has the following four stages. For encryption process, the last round has only three stages: add round key, substitute bytes and shift rows.

- Add Round Key: Add Round Key stage is a simple XOR operation between the State and the Round Key [5].
- Substitute Bytes: S-box is represented as a 16×16 array, rows and columns indexed by hexadecimal bits [6]. Each byte is replaced by byte indexed by row (left 4-bits) and column (right 4-bits) of a 16×16 table [8].
- Shift Rows: The Shift Rows step operates on the rows of the state; it cyclically shifts the bytes in each row by a certain offset. For AES, the first row is left unchanged. Each byte of the second row is shifted one to the left. Similarly, the third and fourth rows are shifted by offsets of two and three respectively [5].
- Mixed Columns: The transformation can be defined by following matrix multiplication on State array. Each element in the product matrix is the sum of products of elements of one row and one column [5] (Fig. 2).

2.3 Least Significant Bits (LSB)

In computing, Least Significant Bit (LSB) substitution method is one of the earliest and simplest methods for hiding information. In LSB encoding, the least significant bits of the cover media are altered to include the secret message. In this approach, random based LSB will be used; mean that secret bit is embedded into random pixel of images.

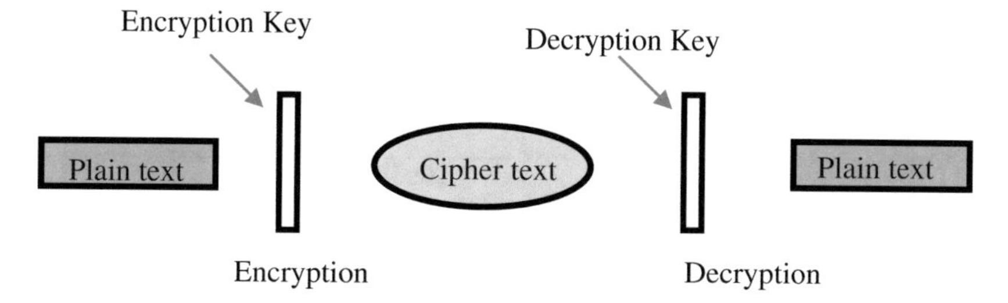

Fig. 2. Process of advanced encryption standard

For Example: to hide letter A whose binary value is 01000001, which will replace the least significant values of the pixels of images. The random sequence is obtained by pre-shared key.

Random Pixel Positions for embedding:

63	38	2	31	27	50	25	7

In Table 1, for 8 × 8 resolution image, there are total pixel positions of cover image but only random pixel positions are used in this approach.

Table 1. Total pixel positions of cover

1	2	3	4	5	6	7	8
9	10	11	12	13	14	15	16
17	18	19	20	21	22	23	24
25	26	27	28	29	30	31	32
33	34	35	36	37	38	39	40
41	42	43	44	45	46	47	48
49	50	51	52	53	54	55	56
57	58	59	60	61	62	63	64

According to the Table 2, there are some advantages of using LSB. They are it does not affect the size of the cover data because it does not increase or decrease the number of the cover data bytes too much. It just replaces some last data bits of cover images with secret bits without affecting the size. It does not make noticeable changes to the cover data. LSB works in the least bit (rightmost) which has the least weight between all the bits in a byte. This amount of change is neither noticeable by human eye nor ear. Some of the bits are replaced with its same value as shown in Table 2.

Table 2. Random-based LSB substitution

A. The series of cover bytes before embedding process			
63	38	2	31
11010010	01001010	10010111	10001100
27	50	25	7
00010101	01010111	00100110	01000011
B. The resulting bytes after embedding process			
63	38	2	31
1101001(0)	0100101(1)	1001011(0)	1000110(0)
27	50	25	7
0001010(0)	0101011(0)	0010011(0)	0100001(1)
(-) The bit new value is identical to its original value (-) The bit new value is different from its original value			

3 Experimental Results of the System

Firstly, Secret audios are loaded into the system. And the number of images is determined by the size of audio file. In which, only '.mp3' or '.wav' formats can be used. The secret audio is encrypted using AES algorithm. The resulting cipher audio is embedded into the images using random based LSB. The key can be typed alphabets, number or special characters. This is used in AES encryption, decryption and LSB Algorithm in both sender and receiver.

3.1 Mean Square Error and Peak Signal to Noise Ratio

One of the objectives of this paper is to analyze the image quality before and after embedding process in which Mean Square Error (MSE) and Peak signal to Noise Ratio (PSNR) are used. MSE is the cumulative square error between the original and embedded image, whereas PSNR represents the peak error. The lower the MSE and higher the PSNR values indicates that the embedded image quality is better than original image quality. Generally higher value of PSNR indicates that reconstructed image is high quality. The PSNR between two images having 8 bits per pixel or sample in terms of decibels (dBs) is given by:

$$\mathrm{PSNR} = 20\log_{10}\left[\frac{\mathrm{B}^2 - 1}{\sqrt{\mathrm{MSE}}}\right] \tag{1}$$

Where MSE represents the mean square error and B represents the bits per sample and MSE is calculated as:

$$\mathrm{MSE} = \frac{1}{\mathrm{MN}} \sum\nolimits_{x=0}^{M-1} \sum\nolimits_{y=0}^{N-1} (C(x, y) - S(x, y))^2 \quad (2)$$

Where (x, y) are the two coordinates of image, (M, N) are the two dimensions. So, S (x, y) creates steganography image and C (x, y) creates cover image [11].

Table 3 shows the MSE and PSNR calculations on images. In sender, the size of secret audio file is needed to send to the receiver. This size is embedded to the last image. In table, 5 min audio just only needs the four images to embed it. This amount of images is enough to embed 5 min audio files that required images calculation is shown in previous calculations. In receiver, the size of audio file is extracted in last image. All the encrypted samples are embedded into these 4 images. In above table, the MSE and PSNR are calculated on 4 images.

Table 3. MSE and PSNR calculations

Image no.	MSE	PSNR
1	0.0124	60.7295
2	0.0176	58.6843
3	0.0145	59.5937
4	0.0085	66.5937

3.2 Performance Analysis of AES and LSB

Table 4 presents the performance analysis of AES and LSB. Unlike image and text, audio has the large number of samples. Besides, AES works only 16 bytes at a time. So, AES takes a lot of time to encrypt these samples. For LSB, pixels positions are randomly extracted from images. Since LSB is one-bit substitution method, it also takes a lot of time to embed. And there are various types of audio sample with sampling rate 4000 Hz that are tested on AES and LSB methods. Sampling frequency 4000 Hz is enough to hear the voice message. All these results are tested with Intel PENTIUM processor.

Table 4. Performance analysis of AES and LSB substitution

Sampling rate (Hz)	Audio name	Length	No. of images requirement	Encryption time	Embedding time
4000	Call	3 s	$0.0347 \sim 1$	10 s	20 s
4000	Report	30 s	$0.3472 \sim 1$	1 min 20 s	2 min 10 s
4000	Secret news	1 min	$0.6944 \sim 1$	3 min	5 min 20 s
4000	Mail	3 min	$2.0833 \sim 3$	9 min 37 s	18 min
4000	Reporter	5 min	$3.4722 \sim 4$	15 min 25 s	28 min 10 s

4 Conclusions

The main aim of this paper is to provide secure secret audio transmission over network. In this paper, the combination of cryptography and steganography are used to provide more secure for transmission. The stego images are sent to the receiver without noticing which has secret audio. Since AES works 16 data block to encrypt, it takes too much time when the large audio file is loaded into the system. In AES, the key maintains the highest possible security level for the proposed system. Random based LSB is more secure than sequential LSB because the attackers have to know the random sequences to get the secret bits. The experimental results are analyzed with the help of MATLAB. In this paper, the images are used as cover objects. The number of required images will be varied according to the size of audio. This is because it is needed to hide large audio files. As the result of this, there takes amount of time. For future work, the texts and images can be used as secret audio instead of audio. And the system can be implemented using AES-192 and AES-256 to provide more security. Various types of LSB techniques can be used. The system is implemented by the use of MATLAB programming.

Acknowledgments. First of all, I would like to express my special thanks to Dr. Aung Win, Rector, University of Technology (Yatanarpon Cyber City). I would like to express my respectful gratitude to Dr. Soe Soe Khaing, Pro-rector, University of Technology (Yatanarpon Cyber City), for her kind permission to complete this thesis. I am thankful to her for giving enough consideration to her ideas and views. I would like to express grateful thank to Dr. Cho Me Me Maung, Professor and Head and Dr. Hnin Hnin Htun, Professor and Course Coordinator, Department of Computer Engineering, for her kind guidance, encouragement in making my thesis to complete successfully. I am very grateful to my supervisor, Dr. Tin Myint Naing, Associate Professor, University of Technology (Yatanarpon Cyber City), for his kind guidance and encouragement. He has been very supportive in this thesis, and also guides a lot, particularly, at the level of quality of presentation.

References

1. Jirwan, N., Singh, A., Vijay, S.: Review and analysis of cryptography techniques. Int. J. Sci. Eng. Res. **4**(3), 1–6 (2013)
2. Dilawar, A.: Image Steganography: for hiding audio messages within gray scale images. Department of Computer Science and Software Engineering, International Islamic University (2014)
3. Sakthisudhan, K., Prabhu, P., Thangaraj, P.: Secure audio steganography for hiding secret information. In: International Conference on Recent Trends in Computational Methods, Communication and Controls (ICON3C 2012). Bannari Amman Institute of Technology, Erode – India (2012)
4. Sherly, A.P., Amritha, P.P.: A compressed video steganography using TPVD. Int. J. Database Manag. Syst. (IJDMS) **2**(3), August 2010. TIFAC CORE in Cyber Security, Amrita Vishwa Vidyapeetham, Coimbatore
5. Announcing the ADVANCED ENCRYPTION (AES): Federal Information Processing Standards Publication 197, United States National Institute of Standards and Technology (NIST), 26 November 2001

6. Stallings, W.: Cryptography and Network Security Principles and Practices, 4th edn., 16 November 2005
7. Lin, E.T., Dell, E.J.: A review of data hiding in digital images. In: Video and Image Processing Laboratory School of Electrical and Computer Engineering
8. Dilawar, A.: Image Steganography: For Hiding audio messages within Gray scale images. Department of Computer Science and Software Engineering, International Islamic University (2014)
9. Bhagyashri, A.P. et al.: Review of an improved audio steganographic technique over LSB through random based approach. IOSR J. Comput. Eng. (IOSR-JCE). Government Engineering College Aurangabad MS, India

Location Based Personal Task Reminder System Using GPS Technology

Thwet Hmue Nyein(✉) and Aye Mon Yi

Department of Information Science, University of Technology (Yatanarpon Cyber City), Pyin Oo Lwin, Myanmar
thwethmuenyein.thn@gmail.com, ayemonyi@gmail.com

Abstract. Most of the reminders in mobile phones are purely time specific i.e. this will provide alarm only on that particular time. Some tasks are only meaningful to be performed at a specific location. So, this location-based task reminder system is developed. It reminds the user about the location when he/she enters within alert range of the predefined location. All the user need to have is the android mobile phone. The main objective of the research is to develop a GPS (Global Positioning System) based application to handle the following requirements: to provide the alarm to the user when he/she reaches near the preselected destination, to retrieve the users' current location geographical coordinates, to allow the users to create the reminder for their target location and save that reminder to the list, to allow the users to delete and edit their reminders in list. Based on the saved location, the alarm will ring automatically and the reminding task detail will display when the user reaches within the target location. The distance between the user current position and the destination can get by using Haversine Formula. This location-based reminder system will act as an assistance for frequent travelers to visit new places with the help of LBS (Location Based Services).

Keywords: GPS (Global Positioning System) · Haversine formula
LBS (Location Based Services)

1 Introduction

Nowadays, LBS services and the concept about location learning are attracted by people. With a short period of time, these services became very popular by supporting to the users as mobile system. Recently, smart phones are most popular equipment between users and it plays major role as part of their daily lives. So, this system is provided as a mobile application. As the human are busy more and more and they have to perform various tasks, it is very hard to remember and perform those tasks. So, this system will help the user to perform the task without worrying to remember.

There are three basic segments to develop this application. The absence of any of these segments means the impossibility of developing this application. These basic segments are:

T. T. Zin and J. C.-W. Lin (Eds.): ICBDL 2018, AISC 744, pp. 224–230, 2019.
https://doi.org/10.1007/978-981-13-0869-7_25

User – a person who uses the mobile phone and install this application on his/her mobile phone;
Mobile phone – a hardware equipped terminal which can install this application;
GPS system – system of satellites and receivers intended for positioning [1].

The rest of the paper is organized as follows: Sect. 2 discusses related work, Sect. 3 gives background theory, Sect. 4 shows system overviews, Sect. 5 describes system implementation and result, Sect. 6 describes accuracy analysis and finally we conclude the paper in Sect. 7.

2 Related Work

Marmasse and Schmandt (2003) developed a system that used GPS technology called commotion [4], which to learn about the users frequently visited locations, and allowed users to attach reminders to these locations. The reminders could be voice- or text-based lists.

Vasileios Zeimpekis, George M. Giaglis and George Lekakos [7] developed a technique based on LBS for indoor and outdoor positioning system to get the user's location. This system is based on wireless communication system with mobility. The different types of location-based services are used to get the actual location by using three different positioning techniques such as self-positioning, remote positioning and wireless positioning technique.

An application called Place_Its [6], developed by Sohn et al (2005), uses GSM cells and GSM phones for place detection and can be used on many cell phones. Many users minimized their usage of other reminder tools by using this system.

PlaceMail [3] was developed by Ludford et al. (2005). In this system, GPS was used when available. When GPS signal was not available, the coordinates of the serving GSM cell was used. Reminders were triggered both the proximity of the defined places to the current location and the speed and direction of the user's phone [2].

3 Background Theory

This system is one of the example of Location Based Services (LBS). Almost all LBS systems use GPS to get geographical coordinates of the locations.

3.1 Location Based Services (LBSs)

LBSs are information services accessible with mobile devices through the mobile network and utilizing the ability to make use of the location of the mobile device. These definitions describe LBS as an intersection of three technologies: Internet, GIS (geographic information system), Mobile devices (see Fig. 1) [5].

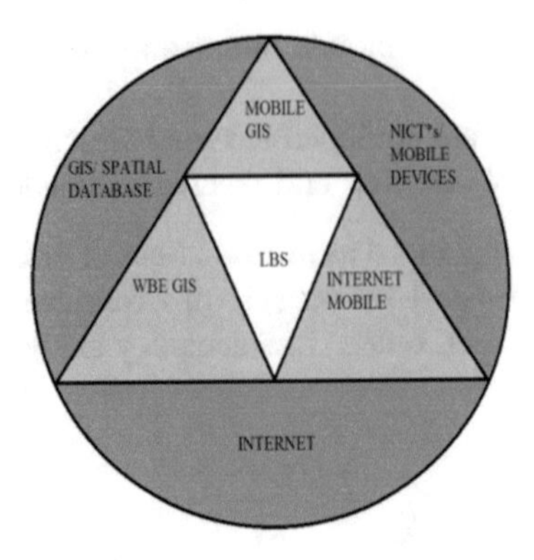

Fig. 1. LBS as an intersection of technologies

3.2 GPS

The global positioning system (**GPS**) is a 24-satellite based navigation system that uses multiple satellite signals to find a receiver's position on earth. The U.S. Department of Defense (DoD) developed GPS technology. The technology was originally used for military purposes. Since 1980, when GPS technology was available to the consumer market, it has become common in cars, boats, cell phones, mobile devices and even personal heads-up display (HUD) glasses. GPS receivers find user's location by getting coordinate information from three or four satellite signals. That information includes the position of the satellite and the precise time of transmission. With three signals, any 2D position can be found on earth; additional satellite signals make it possible to find altitude. GPS technology works in almost any condition and is accurate to within 3–15 m, depending on the number of signals received [11].

4 System Overview

The proposed system is implemented as the reminder system based on location. The user can get his/her current location by using GPS. When the user reaches the location, the system will check the task reminder's specified location for its latitude and longitude. And then, the distance between the task's reminder location and user's current location compare with the user's predefined alert radius. If the distance between is smaller than the alert radius, the alarm of the task reminder will be generated and an alert will be given to the user via notification about the task detail the user has set before. All these activities will be performed by using GPS service.

This system requires the user to set the location at which he/she has to complete the task and the tasks which the user wants to remind him/her are also specified. Whenever the user reaches within the alert radius of that location, the system reminds the user of the task through notification, therefore enabling the user to complete the task as promptly as possible. In this proposed system, it doesn't provide a continuous reminder to the user instead it just simply displays the task reminder until the user declines that notification.

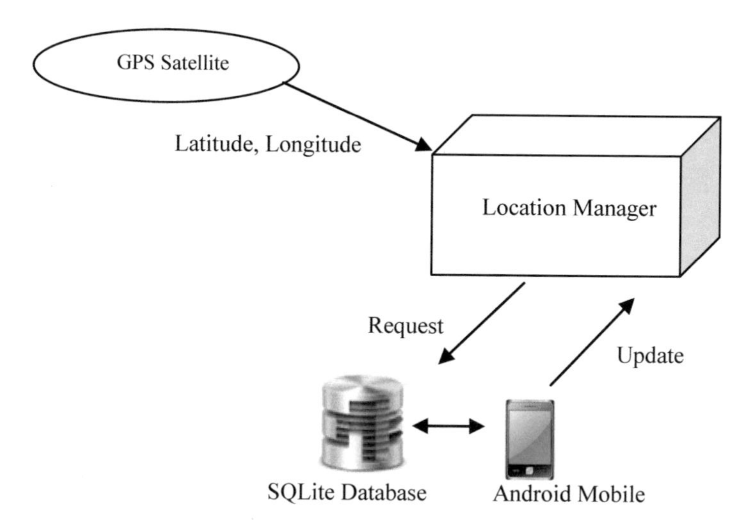

Fig. 2. Architecture of proposed system

This system also allows the user to set multiple task reminders at the same location. And when the user reaches the specified location all the tasks that have to be performed are notified to the user so that no jobs are missed or left incomplete. All the tasks that have to be performed are notified to the user so that no jobs are missed or left incomplete (Fig. 2).

5 System Implementation and Result

The proposed system can be implemented as follows:

1. Install the application and find user location.
2. Enable GPS provider
3. Enable Background service

Input: Task Detail (task title, task description, location, alert radius, alarm and notification)

Processing:

1. Track current location using GPS.
2. Calculate distance between user's current location and task reminder's location in background services.

Output: Notify the user with task detail.

Figure 3 shows creation of new reminder with task details.

Figure 4 shows notifying the user with task detail.

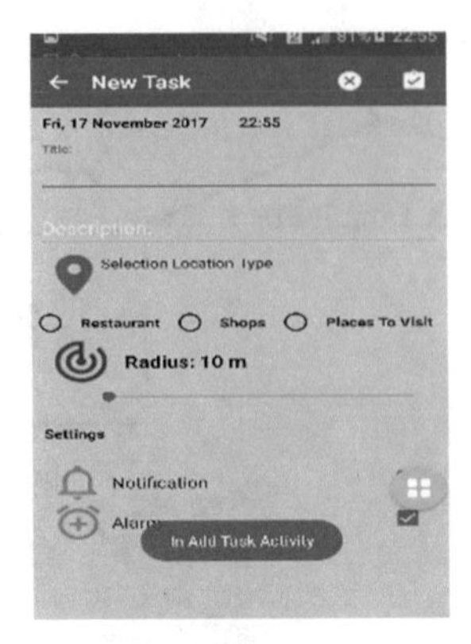

Fig. 3. Set reminder

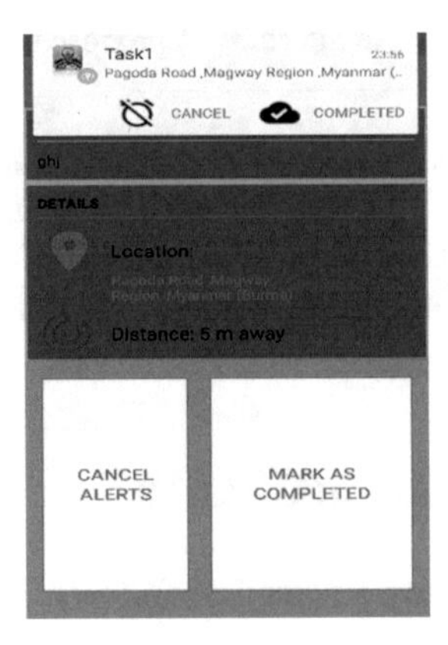

Fig. 4. Notifying user with alarm

6 Accuracy Analysis

To access the accuracy of our distance calculation, we compared the distances between GPS coordinates of different Myanmar cities using Haversine Formula, Vincenty Formula to take account the flattened shape of the earth [10] and great circle distance form distance calculator website. For the Vincenty Formula, the distances were computed using GPS Visualizer tool [9]. The GPS coordinates of different Myanmar cities [8] were used as inputs to the algorithm.

Table 1. Great circle distance comparison between Myanmar cities using Vicenty and Haversine Formula

Cities	Vicenty Formula (km)	Haversine Formula (km)	Great Circle Distance of cities in Myanmar (km)
Yangon-Mandalay	572.295	575.047	575
NayPyiTaw-Mawlamyine	393.57	395.131	395
Bago-Pathein	196.114	196.019	196
Sittwe-Taunggyi	437.695	437.233	437
Dawei-Magway	756.627	759.471	759

As seen in the Table 1 above, Havevrsine Formula tracks fairly well to the Vicenty Formula and produces reasonably accurate results.

7 Conclusion

Nowadays, so many places are needed to go in our daily activities and it is necessary to know which task will perform at which location. Generally, users forgot which location they reach and the important job they have to do on that specific location. The proposed application helps them to remind all these kinds of activity to do on that specific location and makes their lives very easier.

Acknowledgments. Firstly, I would like to thankful to Dr. Aung Win, Rector, University of Technology (Yatanarpon Cyber City), for his supporting to develop this system successfully. Secondly, I would like to appreciate Dr. Soe Soe Khaing, Pro-Rector, Leader of Research Development Team for Domestic Pick and Place Robotic Arm Control System, University of Technology (Yatanarpon Cyber City), for her vision, chosen, giving valuable advices and guidance for preparation of this article. And then, I wish to express my deepest gratitude to my teacher Dr. Hninn Aye Thant, Professor, Department of Information Science and Technology, University of Technology (Yatanarpon Cyber City), for her advice. I am also grateful to Dr. Aye Mon Yi, Associate Professor, University of Technology (Yatanarpon Cyber City), for giving us valuable advices. Last but not least, many thanks are extended to all persons who directly and indirectly contributed towards the success of this paper.

References

1. Angadi, A.B., Shetti, N.V.: Friendly mobiminder-reminder based on user's location using GPS. IJETAE **3**(1), 269 (2013). ISSN: 2250-2459
2. Hedin, B., Norén, J.: Mobile location based learning reminders using GSM cell identification. IADIS **7**(1), 166–176
3. Ludford, J., et al.: Because I carry my cell phone anyway: functional location-based reminder applications. In: Proceedings of the SIGCHI conference on Human Factors in computing systems, Montreal, Canada, pp. 889–898 (2006)

4. Marmasse, N., Schmandt, C.: Location-Aware Information Delivery with Commotion, Lecture (2003)
5. Yeram, S., Patil, P., Gupta, D., Mule, K., Kulkarni, M.: Reminder Based on User's Location using Android Smart Phones [RBUOID] (An android application). IJSRCSEIT **2**(2), 119 (2017). ISSN: 2456-3307
6. Sohn, T., Li, K.A., Lee, G., Smith, I., Scott, J., Griswold, W.G.: Place-Its: a study of location-based reminders on mobile phones. In: Beigl, M., Intille, S., Rekimoto, J., Tokuda, H. (eds) UbiComp 2005: Ubiquitous Computing. Lecture Notes in Computer Science, vol. 3660. Springer, Heidelberg (2005). https://doi.org/10.1007/11551201_14
7. Zeimpekis, V., Giaglis, G., Lekakos, G.: A taxonomy of indoor and outdoor positioning techniques for mobile location services. SIGecom Exch. **3**(4), 19–27 (2002)
8. https://www.distancecalculator.net/country/myanmar
9. http://www.gpsvisualizer.com/calculators#distance_address
10. http://www.movabletype.co.uk/scripts/latlong-vincenty.html
11. https://whatis.techtarget.com/definition/GPS-navigation-system

Intelligent Systems

Front Caster Capable of Reducing Horizontal Forces on Step Climbing

Geunho Lee[1,2(✉)], Masaki Shiraishi[1], Hiroki Tamura[1], and Kikuhito Kawasue[1]

[1] Department of Environmental Robotics, University of Miyazaki, 1-1 Gakuen Kibanadai-nishi, Miyazaki 889-2192, Japan
{geunho,htamura,kawasue}@cc.miyazaki-u.ac.jp, futsal_fifa_of@yahoo.co.jp
[2] Ohnuma Robo & Design LLC, Miyazaki, Japan

Abstract. Obstacles such as ramps, steps and irregular floor surfaces are commonly encountered in homes, offices and other public spaces. These obstacles frequently limit the daily activities of people who use mobility aids. In this paper, a new offsetting mechanism for climbing a step while reducing the required horizontal climbing force and its step-climbing wheel prototype are proposed. Specifically, the proposed wheel allows to integrate into an existing personal assist devices. The proposed step-climbing wheel can help to reduce this limitation. The physical and mental burdens of caregivers and medical staff can also be reduced by making the users of the mechanism more self-sufficient. This paper provides details on this offsetting mechanism and its prototype. The mechanism is analyzed mathematically and its functionality is verified by experiments using a prototype.

Keywords: Personal mobile aid · Threshold and step
Step-climbing wheel · Axial translation

1 Introduction

Recently, various wheel-type assist devices such as wheelchairs [1] and walkers [2] have been developed and introduced. Basically, these assist devices are designed to be used under ideal conditions, namely flat floors with no dips and depressions. However, thresholds, uneven surfaces and ramps encountered in our daily routines hinder the use of such assist devices. As a result, their daily activities are limited. A wide variety of studies has been proposed to overcome these problems. For instance, in order to attach to typical wheelchairs, auxiliary wheel units have been designed [3]. On the other hand, replacement parts with augmented step climbing capabilities have been developed. Some of these replacement wheels combine multiple smaller wheels into a larger wheeled assembly [4–6]. Despite such a technical progress, the existing devices may expose weakness in high cost, bulky size, heavy weight and high complexity. Moreover, the existing devices are difficult to be integrated into low-cost self-propelled aids.

T. T. Zin and J. C.-W. Lin (Eds.): ICBDL 2018, AISC 744, pp. 233–239, 2019.
https://doi.org/10.1007/978-981-13-0869-7_26

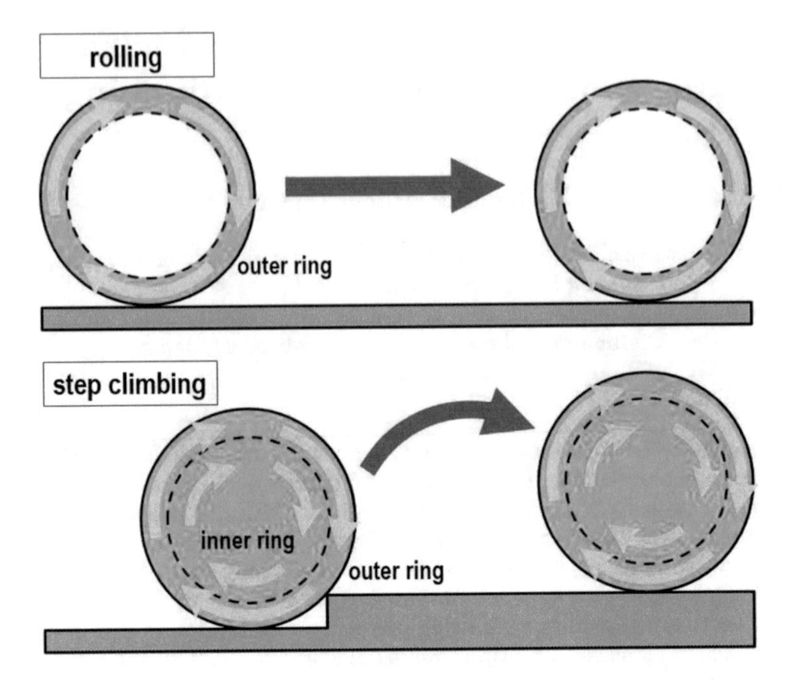

Fig. 1. Illustration of two different wheel motions: standard rolling (top) and step climbing (bottom)

In this paper, we introduce a new step-climbing wheel device, allowing to integrate into an existing personal aid. The wheel device is based on a new idea of increasing the rolling radius of a wheel, resulting in reducing the horizontal force. In detail, the change of the rolling radius is achieved by offsetting its rotating axis from the wheel centre vertically. In other words, the concept of a pseudo wheel diameter is to translate the axis at the center of the wheel to a position higher than the ground. Figure 1 illustrates different wheel motions such as standard rolling and step climbing according to the pseudo wheel diameter. In the standard rolling motion, the axle is centred at the wheel centre and undergoes a purely horizontal translation. In the step-climbing motion, the axle is offset from the wheel centre. Once the auxiliary centre is translated during step climbing, in the Fig. 1, the inner ring and the outer ring rotate together with respect to the centre. This paper describes the practical aspects of the use of the auxiliary centre in detail. This paper also provides details on this step-climbing wheel device and its control. The functionality of this device is analyzed mathematically and verified by experiments using a prototype. The technical contributions are, in particular, the realization of the proposed wheel unit for the step-climbing motion.

2 Design Concept of Step-Climbing Wheel

Three aspects are considered in the development of a step-climbing wheel mechanism. First, achieving a low cost design makes the proposed wheel mechanism accessible to a wider audience. Secondly, a compact design is required to make everyday use as convenient as possible. Thirdly, the ease of integration of the wheel mechanism will determine the extent to which it can be adopted for use with various existing devices.

As considering these aspect, the proposed wheel mechanism in this paper is capable of two distinct motions. The first motion is standard rolling, and the second is step-climbing as shown in Fig. 1. In the standard rolling motion, the axle is centred at the wheel centre and undergoes a purely horizontal translation. In the step climbing mode, the axle is offset from the wheel centre and can experience both horizontal and vertical translations.

3 Mechanical Configuration of Step-Climbing Wheel

The proposed wheel mechanism is designed for use in wheelchairs and hand trucks. In our work, the prototype is intended to be integrated into a commercial wheelchair. The wheelchair is equipped with the crank unit, the wheel unit and the geared motor unit as shown in Fig. 2. Two of the geared motor units drive the axle swivelling mechanism and the other drives the axle expansion mechanism. Specifically, the wheel diameter is 160 mm, the wheel width is 70 mm, the distance between two wheels is 530 mm and the height of the main hole from the ground and the auxiliary hole from the ground is 80 mm and 130 mm respectively.

First, the axle assembly called the crank unit is composed of two crank units and an expandable axle. The cranks swivel around pivots located on the wheel hubs at the midpoints between O and O' as shown in Fig. 2. The shaft consists of a screw and nut. By rotating the screw, the shaft can either be expanded or contracted, depending on the rotation direction. The axle is supported by the two crank units. By means of the crank units, it is possible to swivel the axle between O and O'.

Secondly, the wheel unit is composed of a bearing, a wheel hub consisting of two discs, a cone clutch part and an outer rim which is split into two rings. Here, ring-2 houses the outer cone of the clutch. The cone clutch consists of three guide bars, three springs and the inner cone. As shown in Fig. 2, the two axes about which the wheel can rotate are located O and O'. These two rotation axes are realized by two holes in the inner clutch cone. The main hole as shown at the top of Fig. 2, is located at O and is used when moving on flat surfaces. The auxiliary hole is located at O' and is used when climbing a step. The bearing allows for relative rotation between the rim and hub components. Two motions are made possible by use of the clutch mechanism. The rim normally rotates freely relative to the hub, but by engaging the clutch, it is possible to unify the motion of the rim and hub.

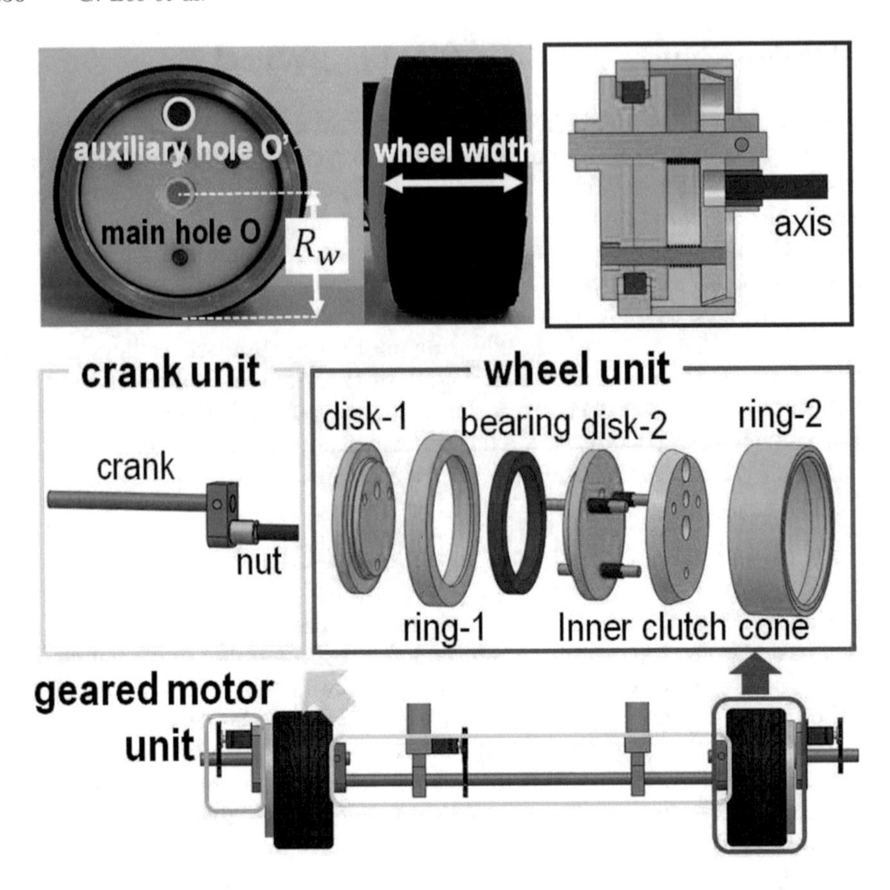

Fig. 2. Wheel prototype (top) and assembly details of the prototype (bottom).

Thirdly, the geared motor unit is composed of three servo-motors, their motor drivers, and gear assemblies. The motors are controlled by means of a micro controller and motor drivers. Two motors allow for bi-directional swiveling of the axle. The other motor is used for expanding and contracting the axle by means of the screw and nut mechanism.

The main hole in the clutch cone is a blind hole; whereas, the auxiliary hole is a through hole. When moving on a flat surface, the main hole is used. Upon expanding into the main hole, the shaft pushes against the bottom of the blind hole and disengages the clutch. As a result, only the rim rotates in this case. When switching between the main and auxiliary holes, the shaft is contracted and the springs around the guide rods engage the clutch. The shaft is then expanded through the auxiliary hole so that the clutch remains engaged. Consequently, the hub and rim rotate in unison during the step climbing procedure.

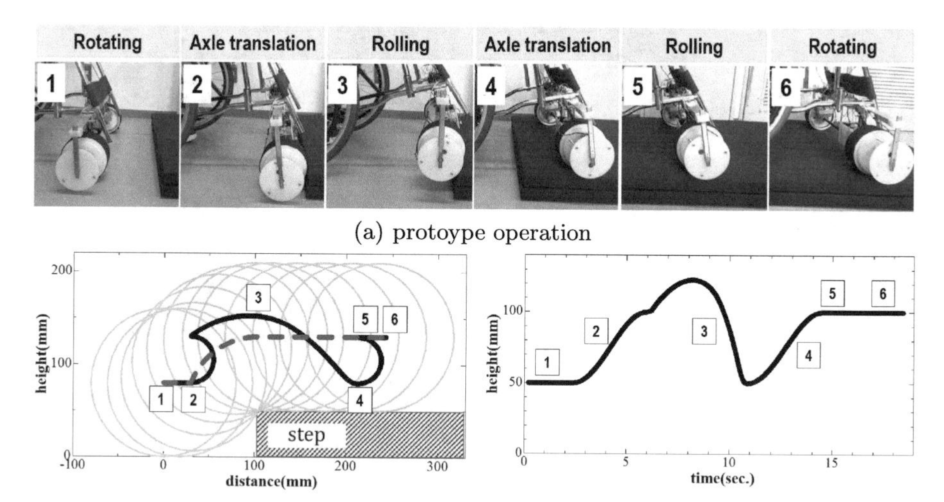

(a) protoype operation

(b) trajectories of axle and wheel's centre location

(c) variation of the axle hight during step climbing

Fig. 3. Experiment results for the step-climbing operations by using a commercial wheelchair

4 Evaluation Results and Discussion

In this section, the experimental evaluations of the proposed wheel mechanism are conducted to verify its effectiveness. Moreover, in order to show the utility of the wheel mechanism, the implemented prototype was installed on a wheelchair and extensive experiments were performed.

Figure 3 shows experiment results for the step-climbing mechanism after integrating it into a commercial wheelchair. In these evaluations, the trajectories of the axle during the step climbing procedure are examined and analyzed. The height of the axle is also investigated. First, Fig. 3(a) shows both continuous snapshots (from the snapshot numbered "1" to the snapshot numbered "6") under the control sequence. As we expected, the proposed control appeared to yield accurate motions according to each condition. Figures 3(b) and (c) display the variations of the axle position and the axle height during the step climbing sequence, respectively. In Fig. 3-(b), the red dotted line and the black bold line indicate axis trajectories of an existing device and the proposed wheel mechanism, respectively. Moreover, the gray lines indicate the proposed wheel contours at each time.

In Fig. 3(c), the black line indicates the variation of the axle height for the wheel mechanism. From these results, we confirmed that the proposed wheel mechanism could generate the step climbing motions based on the proposed offset control. Moreover, the axle height is constant while rolling along the flat surface up to $t = 3$[s]. The axle height increases by the axle offset operation when $3 \leq t < 8.5$[s]. The height of the axle due to the offset operation is maximum

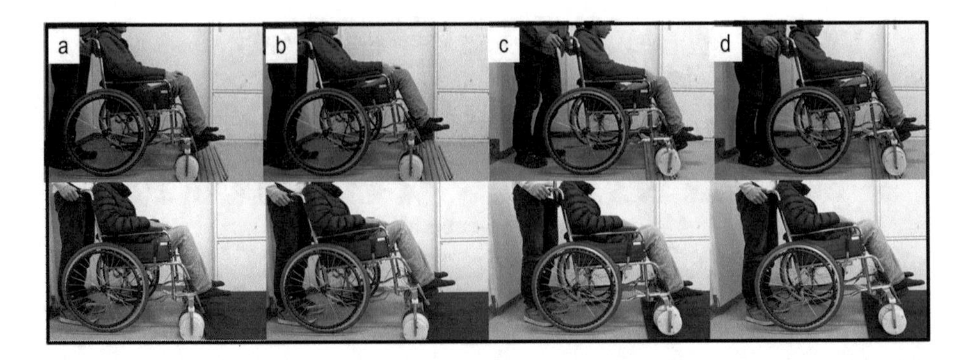

Fig. 4. Experimental results for step climbing by using the commercial wheelchair installed with the proposed wheel device

when $t = 8.5$[s]. For $8.5 \leq t \leq 11$[s], there is a large variation in the axle height due to rolling the wheel over the step. Due to the offset of the rotating axis of the wheel, the axle traverses the step smoothly. When $t = 11$[s], the axle is at its lowest point relative to the wheel centre in the step climbing sequence. Finally, the axle returns to its original position at O by reversing the offset motion for $11 < t \leq 14$[s].

In order to examine the feasibility of the proposed wheel device, experiments were set under the environmental conditions of daily life. Figure 4 shows snapshots for experiments performed according to the different conditions of steps such as height and material. These experiments included both the standard rolling and step-climbing motions. Specifically, Fig. 4 present the experimental scenes of step-climbing motions in a step height of 5 cm and 2 cm, respectively. As a result of these experiments, it was observed that the wheelchair installed with the proposed wheel could be climbed naturally in the steps without experiencing any troubles. We believe that our realization studies well under real world conditions, but the following issue remains to be addressed. In practical, the weight of the proposed wheel was about 5 kg . It would interesting to investigate several design performances such as lighter and more compact system by the choice of materials and off-the-shelf parts. Considering our future direction, a more reliable structure and more compact design will need to be incorporated into the next prototype.

5 Conclusions

A wheel mechanism and its prototype for the step-climbing motions were introduced. The wheel prototype was based on a new idea of increasing the rolling radius of a wheel when step climbing, resulting in reducing the horizontal force. In order to show the feasibility and effectiveness of the step-climbing wheel mechanism, we performed several indoor experiments and demonstrated the basic motion performances of the wheel prototype after the integration of a commercial wheelchair. From a practical standpoint, these experimental results were quite encouraging. From these results, we are able to confirm that the proposed wheel prototype could provide the users of self-propelled wheelchairs with easy maneuverability without requiring any horizontal forces when step climbing. Our next study directions is toward weight and size reduction with consideration of safety.

Acknowledgements. This work is partially supported by the Research Grant-A awarded by the Tateisi Science and Technology Foundation.

References

1. Carlson, T., Leeb, R., Chavarriaga, R., Millan, J.R.: The birth of the brain-controlled wheelchair. In: Proceedings of the IEEE/RSJ International Conference on Intelligent Robots and Systems, pp. 5444–5445 (2012)
2. Lee, G., Ohnuma, T., Chong, N.Y., Lee, S.-G.: Walking intent based movement control for JAIST active robotic walker. IEEE Trans. Syst. Man Cybern. Syst. **44**(5), 665–672 (2014)
3. Mori, Y. ,Katsumura, K., Nagase, K.: Development of a pair of step-climbing units for a manual wheelchair user. Trans. JSME 80 (820) (2010). (in Japanese). https://doi.org/10.1299/transjsme.2014dr0381
4. Quaglia, G., Franco, W., Oderio, R.: Wheelchair.q, a motorized wheelchair with stair climbing ability. Mech. Mach. Theory **46**(11), 1601–1609 (2011)
5. Razak, S.: Design and implementation electrical wheel chair for disable able to stairs climbing by using hydraulic jack. IOSR J. Electr. Electron. Eng. **7**(3), 82–92 (2013)
6. Sasaki, K., Eguchi, Y., Suzuki, K.: Step-climbing wheelchair with lever propelled rotary legs. In: Proceedings of the IEEE/RSJ International Conference Intelligent Robots and Systems, pp. 1190–1195 (2015)

Mobile Location Based Indexing for Range Searching

Thu Thu Zan(✉) and Sabai Phyu

University of Computer Studies, Yangon, Myanmar
{thuthuzan, sabaiphyu}@ucsy.edu.mm

Abstract. There had been numerous uses based on two dimensional indexing techniques in recent years. Such uses were generally intended to static data or no moving object data. Today indexing is essential for both static and dynamic data with advances in location services. It is not difficult to store and process for both of them but it need to be consistent for update between arbitrary and unpredictable number of moving object nodes and index structure. For this reason, this paper proposed moving object index structure based on presort range tree that will store moving locations conveniently. In this case, synthetic dataset is needed to generate accessing for several of moving location data. Therefore, this paper also proposed a procedure to generate a synthetic dataset that is the creation of dynamic two dimensional points. Besides, there is a comparison between different indexing techniques such as presort range tree and kd tree along with performance evaluation of tree construction and range searching over moving objects. Moreover, distance based range searching is added to compare between two of indexing.

Keywords: Location update policies · Location based service (LBS) Presort range tree · Distance-based range searching · 2D range query kd tree

1 Introduction

Everyone who is in IT field says "Today is the age of three things: Cloud Computing, Internet of Things, and Mobile." This word is true because there is no doubt that businesses can reap huge benefits from them [2].

Most of the location based services usually take their activities swiftly and accurately on getting locations. For this performing schedule, the basic location providers such as GPS to location management provider such as Google API are emerged along with the proliferation of mobile technologies. In fact, they support to get not only accurate and efficient position but also fast and right position depends on different location conditions. Different users have different desires and needful information so that all of these are sometimes built upon current location.

Since there are inconstant moving location objects, researchers think and look for the techniques of maintaining and handling over them. One of the solutions or techniques is balance structure indexing better than traditional indexing. In summary, this

T. T. Zin and J. C.-W. Lin (Eds.): ICBDL 2018, AISC 744, pp. 240–249, 2019.
https://doi.org/10.1007/978-981-13-0869-7_27

paper introduced presort range tree for dynamic attributes together with main contributions are as follows.

i. Presort Range tree structure is proposed for moving objects with the availability of dynamic circular range query.
ii. A procedure is proposed to generate virtual moving objects' synthetic dataset that is nearly the same in reality.

Besides, a model is added to deal with overall system and dynamic attributes in presort range tree are incorporated compactly. In fact, proposed tree outperforms when reducing the sorting time directly by database queries instead of using sorting in index structure. Furthermore, experimental results of using index trees (presort range tree and kd tree) and distance based range searching are inscribed lavishly in this paper.

2 Location Based Services

Location based services are services offered through mobile phones or location procurable devices' geographical location. Generally, a person who wants to inquire all of the information related to the routes or places before he goes. In this case, there are two types of location services for this: pull and push. In a pull type, the user has to actively request for information. In a push type of service, the user receives information from the service provider without requesting it at that instant [4].

2.1 Getting Mobile Location Framework

Most of the mobile phones are usually based on GPS to get their locations. In fact, different mobile phones have their own different features and functions to getting location. Some mobile phones have high power of cellular network features and they support accurate locations. Likewise, some mobiles are usually generating their locations based on network provider. Therefore, there is a tradeoff between different mobile phone features and functions; Google API is an optimal solution.

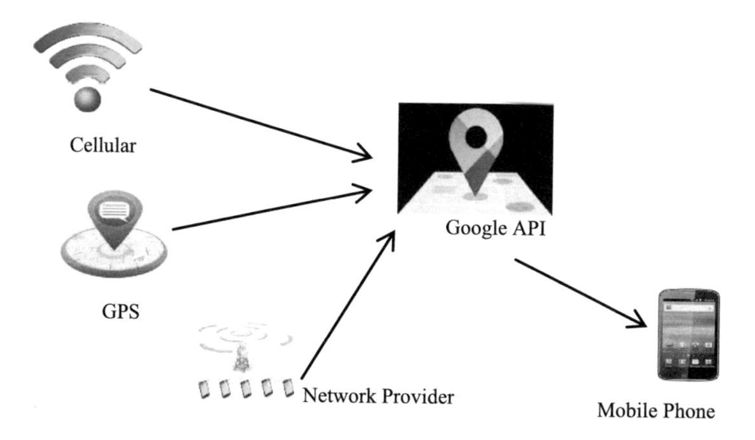

Fig. 1. Getting location for Mobile Application

This system has a framework that aids to get current location as fast as it can. This is shown in Fig. 1. It helps to provide a more powerful location framework than usual. This framework is intended to automatically handle location provider's support, accurate location, and update scheduling.

3 Location Update Policies

Various location update strategies are available in the mobile computing. They are divided into specific strategies like (1) Distance Based Location Update (2) Time Based Location Update (3) Movement Based Location Update (4) Profile Based Location Update and (5) Deviation Based Location Update [1].

3.1 Hybrid Update Algorithm

The distance-based update is a simple and easy location update scheme. In order to get its last and update location, each mobile host usually tracks the distance that its move. When the distance reaches its fix threshold, the mobile host gives out an update location.

But it is complicated because of the variation of cell sizes and the need to compute the distance a mobile has moved.

The time based location update scheme is also a simple strategy for location update. In this scheme, the mobile base station updates the location of user after reaching a particular time period or a timer. There is a limitation when the user is in stationary, unnecessary updates would be performed.

In this paper, hybrid location update algorithm is proposed based on time and distance so that it can significantly reduce location update overhead which improves the efficiency of mobility support mechanisms. The structure of mobile location update is shown in figure. The advantage of the proposed algorithm is that it reduces location update traffic, with a minimum increase in implementation complexity.

Algorithm: Hybrid Location Update
Input: Database of mobile locations contains the locations of registered mobile with time
Output: current registered mobile location

1. i=0; dis_threshold; time_threshold; time_scheduler;

2. Read the current location (L_{xi},L_{yi},t_i) and previous location$(L_{xi-1},L_{yi-1},t_{i-1})$ of registered mobile location

3. If the time_scheduler > time_threshold && $\sqrt{(L_{xi}\text{-}L_{xi}\text{-}1)^2 \; +(L_{yi}\text{-}L_{yi}\text{-}1)^2}$ > dis_threshold

   3.1. Update the database with current location $(L_{xi-1}, Lyi_{-1},t_{i-1})$ = current location (L_{xi},L_{yi},t_i),i+1.

   3.2. Total number of update=Total number of update+1;

4. Else current location(L_{xi},L_{yi},ti) = previous location $(L_{xi-1},L_{yi-1},t_{i-1})$, i=i+1;

4 Related Works

There are a number of papers that describe about moving objects' index tree structure and mobile update policies. Most papers are focus on using one index structure and one update policy. Some discuss combination of index trees called hybrid tree structure and comparison of using one index structure and it.

Kwon et al. introduced a novel R-tree based indexing technique which is called LUR-tree. This technique only updates its index structure when an object changes from its fix MBR (minimum bounding rectangle). If a new position of an object is in the MBR, it changes only the position of the object in the leaf node [5]. So, it remove unnecessary modification of the index tree while changing the positions because this technique updates the index structure only when an object changes from its fix MBR (minimum bounding rectangle.

Xia and Prabhakar proposed a novel indexing structure, namely the Q+R tree that is a hybrid tree structure which consists of both an R*tree and a QuadTree. In Q+R tree, quasi-static or no moving objects are stored in an R*tree and fast and frequent moving objects are stored in a Quadtree. By handling different types of moving objects separately, this index structure more accurately produces the reality and gives good performance. In their work, no assumption is made about the future positions of objects. There is no need for objects to move based on patterns and it has no limitations, as the maximum velocity [7].

Cheng et al. made a location update scheme in which update occurs when the movement threshold or the time threshold is reached. After doing an location update, the movement counters and the periodic location update timer reset to the current state. They used a convex function of the movement threshold. In fact, there is a value of the movement threshold that can reduce the signaling cost. They showed that the HMTBLU scheme has higher signaling cost than the MBLU scheme [3].

Casares-Giner and Garcia-Escalle proposed a location update scheme by combining two dynamic strategies, movement based and the distance based [6]. They showed that results obtained from these analytical model show that, with little memory requirements in the mobile terminal very good performances can be obtained. However, it required that after each movement, the mobile terminal has to search the identity of the new visited cell in a cache memory.

5 Proposed Approach

This paper is integrated by two major components: client side and server side. The overall system model is built and hybrid update algorithm that will aid to get last current location and reduce server update cost is proposed at the client side. Presort range tree procedure for moving objects is included in the server side.

5.1 System Model

A model, searchable model is built to incorporate dynamic attributes in presort range tree and query processing. This includes a server and a collection of registered mobile objects. Through a wireless communication network the registered mobile objects connected to database server. Where the current position of mobile object is stored and updated by a Google API at a fixed rate (e.g. 2 s). The database is managed by a DBMS which supports schedulers. In order to keep the location information up to date, these objects regularly send their updated positions to the server. Unnecessary updates wouldn't be performed at the server because Hybrid Update Algorithm is applied to the client side. A scheduler fires and updates the database when the update policies bound are reached. The server is used to range queries like "which mobiles are currently located within a disaster area?" To process such queries efficiently, the server maintains an index tree that, in addition to speeding up the query processing, is also able to absorb all of the incoming updates.

6 Presort Range Tree

A tree data structure is a useful tool for organizing data objects based on index. It is equally useful for organizing multiple mobile objects in terms of two dimensional relationships. There is an assumption for mobile locations that no two points have the same x- coordinate and also y-coordinate. To construct any tree structure, the first thing is preprocessing the data into the data structure. Then, queries and updates on the data structure are performed.

6.1 Proposed Presort Range Tree

The procedure of proposed presort range tree is the following;

Input: Lats=Array of two dimensional points sort on latitudes
Longs=Array of two dimensional points sort on longitudes

Procedure PRTree (Lats, Longs)

1. If Lats.length==1 then return new LeafNode(Lats[1]);
2. medium= [Lats.length/2];
3. Copy Lats[1….medium] to $Lats_L$ and Lats[medium+1….. Lats.length] to $Lats_R$;
4. for i=1 to Longs.length do
5. if Longs[i].x <= Lats[medium].x then append Longs[i] to $Longs_L$;
6. else append Longs[i] to $Longs_R$;
7. root= new Node((Lats[medium].x),One D Range(Y));
8. root.left= PRTree($Lats_L$, $Longs_L$);
9. root.right= PRTree($Lats_R$, $Longs_R$);
10. return root;

6.2 Circular Range Search

After preprocessing of tree construction is done, the structure allows searching circular range query for mobile objects. To determines whether registered mobiles are in service area or not so that this system has to **get bounding coordinates** with center and service distance: (centerLat, centerLong, bearing, distance).

bearingRadians = Radians(bearing);
lonRads = Radians(centerLong);
latRads = Radians(centerLat);
maxLatRads = asin((sin(latRads) * cos(distance/ 6371) + cos(latRads) sin(distance/ 6371) * cos(bearingRadians)));
maxLonRads = lonRads + atan2((sin(bearingRadians) * sin(distance/6371) cos(latRads)), (cos(distance/ 6371) − sin(latRads) * sin(maxLatRads)));

6.3 Example: Calculating Presort Range Tree with Center and Service Distance

Firstly, sort the mobile locations by latitudes and longitudes. Then the Presort Range Tree is built and shows as the following;

25.40319 98.11739
LEFT: 16.80958 96.12909
LEFT: 16.35099 96.44281
RIGHT: 16.77923 96.03917
RIGHT: 24.99183 96.53019
LEFT: 24.77906 96.3732
RIGHT: 25.38048 97.87883
RIGHT: 25.88635 98.12976
LEFT: 25.59866 98.37863
RIGHT: 25.82991 97.72671
RIGHT: 26.35797 96.71655
LEFT: 26.15312 98.27074
RIGHT: 26.69478 96.2094

The results of sample range search in centerLat, centerLng, distance: 26.693, 96.208, 100 km that are registered mobile locations to send notification as follows:

node (25.40319, 98.11739)
RIGHT: node (24.99183, 96.53019)
LEFT: node (24.77906, 96.3732)
RIGHT: node (25.38048, 97.87883)
RIGHT: node (25.88635, 98.12976)
LEFT: node (25.59866, 98.37863)
RIGHT: node (25.82991, 97.72671)
RIGHT: node (26.35797, 96.71655)
LEFT: node (26.15312, 98.27074)
RIGHT: node (26.69478, 96.2094)

7 Synthetic Mobile Dataset Generation

Synthetic datasets are needed to simulate mobile objects when they are unavailable for millions of mobile positioning data in reality. To generate synthetic datasets appropriately, moving objects are created dynamically by classifying their behaviors.

In fact, different moving objects have different velocities and movements. Therefore, location objects are generated by using random functions, which can be categorized by three different behaviors in such way that they seem realistic.

It is aimed to get realistic data for performance evaluation of presort range tree with indexing and storing dynamic location data.

It includes two main categories: (a) generate location which includes latitude and longitude between minlat, maxlat (−90, 90), minlon, maxlon (0,180) (b) update latitude and longitude consistency.

Procedure: Virtual Mobile Locations Generator

(1) random generate latitude and longitude within (maxlat minlat −90,90 and maxlon minlon 0,180)
  a. latitude = minLat + (Math.random() * ((maxLat-minLat) + 1));
  b. longitude = minLon + (Math.random() * (maxLon-minLon)+1);
(2) divide three types of latitude and longitude that are car, walk, and congested car based on distance and time for updating
(3) update latitude and longitude consistency
  a. Convert latitude and longitude to radian firstly
  b. updateLat = asin(sin(latitude) * cos(distance/ R) + cos(latitude) * sin(distance/R) * cos(BearingAngle));
  updateLon = longitude + atan2(sin(BearingAngle) * sin(distance/R) * cos(latitude), cos(distance/R) − sin(latitude) * sin(updateLat));
  Where,
  longitude, latitude = random generate latitude and longitude distance be 0.00832 km 0.01112 km 0.01020 km for each of car, walking and congested car, R = radius of earth =6378.1, bearingAngle = random generate angle between 0 to 360 and convert radian.

8 Simulation Results

The next experiment has been performed on a 2.60 GHz ASUS PC, with Intel (R) Core (TM) i7 CPU and 4 GB memory. For this experiment, this system uses the most popular testing framework in Java, JUnit. It is an open source testing framework which is used to write and run repeatable automated tests. The experiment was performed for computing query and execution time for range search, with number of data set points that are organized in two dimensions.

This Fig. 2 shows distance based range searching based on Mobile Region Check (MRC) procedure. For this experiment, this system takes center with latitude at 16.8548383 and longitude at 96.1349713. Dataset is started from 10000 to 500000 and

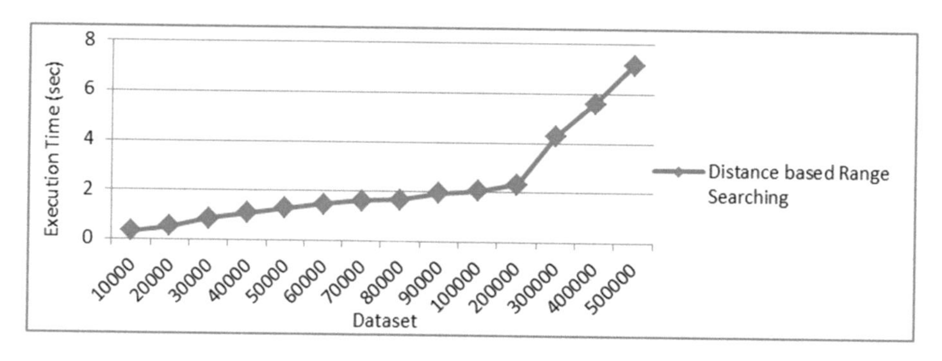

Fig. 2. Execution Time for Distance based Range Searching

there is a finding that the more dataset increase, the more execution time is needed by using distance based range searching.

In Fig. 3 shows two categories of kd Tree: tree construction time and range search time. There is no more change for range searching over numbers of moving mobile dataset. Tree construction takes time as long as numbers of dataset. However, the whole execution using kd tree (tree construction plus range search) consume time less than distance based range searching.

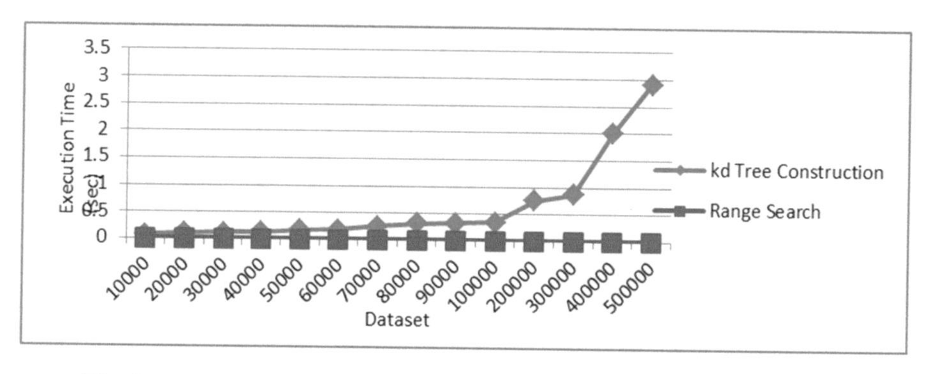

Fig. 3. Execution Time (Tree Construction) and (Range Search) using kd Tree

To construct presort range tree structure; the first thing is preprocessing the data into the data structure. In this process, presorting is done by each dimension and then it takes into the input parameter for building range tree. After taking preprocessing, queries and updates on the data structure are performed. The presorting time, tree construction time and range search time by different number of dataset is tested in Fig. 4.

In Fig. 5 discusses about the execution time required by three approaches. Results were quite different. Generally, query results of both index approaches are the same points. For balance tree structure, it may take nearly the same time for range search whatever the datasets increase dramatically. Besides, proposed presort range tree

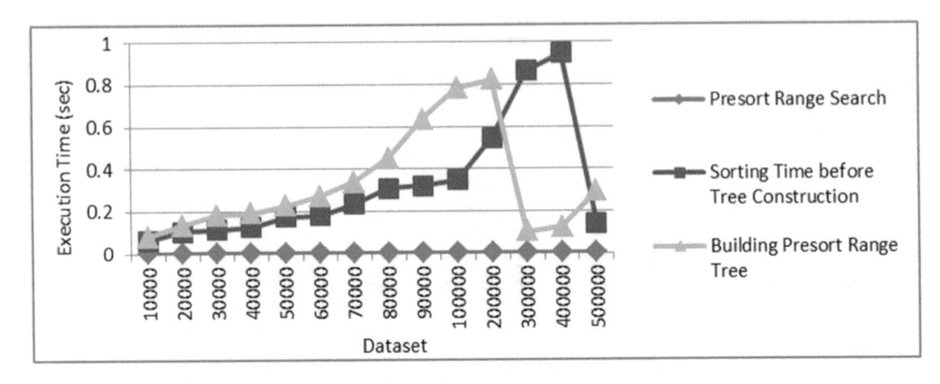

Fig. 4. Execution Time (sorting, building tree and range query) of Presort Range Tree

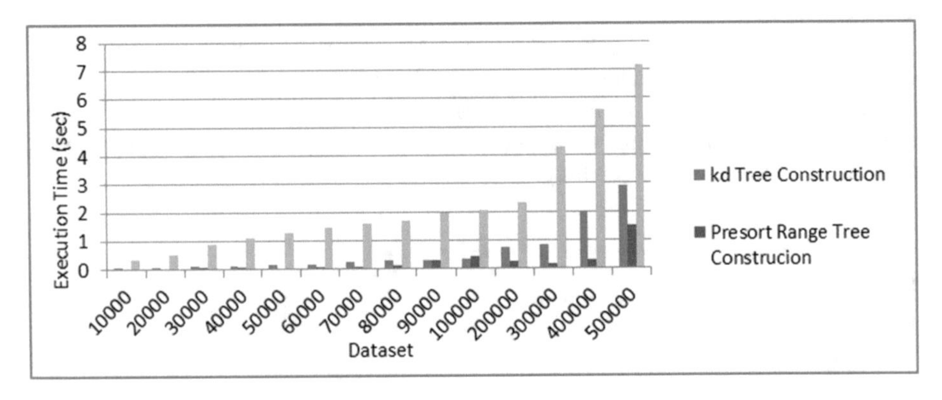

Fig. 5. Comparison of Total Execution Time of three techniques with range search

outperforms when reducing the sorting time by taking database queries such as order by latitude and longitude. As a result, better performance was achieved with larger number of data sets for range search including tree construction. The more volumes of data tests, the less number of seconds needs in presort range tree.

9 Conclusion and Future Works

In this paper, the main service task is handling mobile objects based on index tree structure. The system maintains the moving mobile locations and circular range query is available from the server. Therefore, the system is done for monitoring of mobile objects, to be able to efficiently locate and answer queries related to the position of these objects in desire time. The system will helps to be tradeoff frequency of update due to the locations of mobile objects and reduce server update cost. It also support range query with dynamic object locations.

For future works, the proposed Hybrid Update approach will be applied to other index structures (e.g. the quad tree, the K-D-B tree). Moreover, this proposed system can be used to storing other moving objects such as temperature, vehicle location and so on. The results obtained from the other index tree structure can be compared to this paper's results.

References

1. Kalpesh, A., Priyanka, S.: Various location update strategies in mobile computing. In: International Journal of Computer Applications® (IJCA) (0975–8887) Proceedings on National Conference on Emerging Trends in Information & Communication Technology (NCETICT 2013) (2013)
2. Rundle, M.H.: Future of technology whitepaper, UK (2015)
3. Cheng, P.F., Lei, X., Hu, R.Q.: Cost analysis of a hybrid-movement-based and time-based location update scheme in cellular networks. IEEE Trans. Veh. Technol. **64**(11) (2015)
4. Jensen, C.S., Lin, D., Ooi, B.C.: Query and update efficient B+-tree based indexing of moving objects. In: VLDB 2004 Proceedings of the Thirtieth international conference on Very large databases, vol. 30, pp. 768–779 (2004)
5. Kwon, D., Lee, S., Lee, S.: Indexing the current positions of moving objects using the lazy update R-tree. In: Third International Conference on Mobile Data Management. IEEE (2002)
6. Casares-Giner, V., Garcıa-Escalle, P.: An Hybrid Movement-Distance-Based Location Update strategy for Mobility Tracking, CICYT (Spain) for financial support under project number TIC2001-0956-C04-04 (2004)
7. Xia, Y., Prabhakar, S.: Q+Rtree: efficient indexing for moving object databases. In: Eighth International Conference on Database Systems for Advanced Applications (DASFAA 2003), 26–28 March 2003, Kyoto, Japan (2003)

Building Travel Speed Estimation Model for Yangon City from Public Transport Trajectory Data

Thura Kyaw(✉), Nyein Nyein Oo(✉), and Win Zaw(✉)

Department of Computer Engineering and Information Technology, Yangon Technological University, East Gyogone, Insein Township, Yangon, Myanmar
thurakyaw.ytu@gmail.com,
{nyeinnyeinoo,winzaw}@ytu.edu.mm

Abstract. Estimating travel speed is one of the most important steps to build an Intelligent Transportation System (ITS). In this paper, we propose a Travel Speed Estimation Model (TSEM) for Yangon City's Road Network by analyzing GPS data from buses using Machine Learning Techniques. The GPS data are collected from buses that pass through the most congested area of Yangon City. The paper designs the Travel Speed Estimation model in four steps. The first step is GPS data collection and Preprocessing by removing outlier points, reducing features by dimension reduction methods and selecting important features of raw GPS data to get a well-structured data set. The second step is road network analysis and map matching that extracts POI features vector from nearby road side places and segments the bus route by bus stop positions along the bus route by using KNN model. The next step is estimating travel speed of every road segment from the matched trajectory points. The final step is to calculate speed factors for road all segments and to store in a matrix that can be used in different urban transport applications.

Keywords: Travel Speed Estimation Model · GPS data · Speed factor matrix

1 Introduction

Estimating travel speed is an important issue for solving traffic congestion problems in all mega cities around the world. People are suffering from congested traffic on their ways to work and back to home. The numbers of vehicles get increases proportional to the explosive population of the cities. The existing road infrastructure becomes insufficient and it results the vehicles on the roads have to move in the stop-and-go manner. It strikes directly to the working force of a nation by consuming their working hours which effects the development of the country. Furthermore, the congested traffic causes increased fuel consumption and a vast amount of carbon dioxide emission. For these reasons, most of researchers get involved in the area of intelligent transportation system (ITS) research.

To estimate travel speed of a road network, it is the very first step to collect current locations and speed of the vehicles, the capacity of road network and the parameters

T. T. Zin and J. C.-W. Lin (Eds.): ICBDL 2018, AISC 744, pp. 250–257, 2019.
https://doi.org/10.1007/978-981-13-0869-7_28

that cause traffic congestion. There exist a number of technologies to collect traffic related data from on-road vehicles. Some researchers use road-side sensors, traffic ropes and traffic cameras which installed along the roads and at the junctions. But these techniques for data collection need a large amount of initial investment, regular maintenance of the devices and then need to perform complex translations of traffic related features from videos and road-side sensors. The next approach is to use GPS trajectory data collected from the cars, taxies and public transport vehicles and to find out traffic patterns using machine learning models. As the sensitivity of GPS sensors become more and more accurate and become cheaper than road-side sensor network installation due to advancements in electronic and communication technology, the second approach is more cost effective than the first approach.

In this paper, we design the travel speed estimation model based on GPS data collected form the public transportation system of Yangon, YBS route 21 which is running along the one of the most congested routes of Yangon City by using state of the art machine learning models. The key contributions are as follow.

(1) Extracting traffic delay related features from the road network data and place of interest (POI) data.
(2) Estimating average travel speed of each road segment along the running route of Bus No. 21 from the collected trajectory data and traffic delay related features.
(3) Storing the speed factor of road segments in a matrix by calculating the ratio of average travel speed to the speed limit of each road segment.

The paper is organized with the following sections. Section 2 describes related works. Section 3 depicts the proposed model of travel speed estimation. Sections 3.1 and 3.2 presents methods for extraction of traffic delay related features. Section 3.3 is about estimating average travel speed and building speed factor matrix of each road segments. Section 4 illustrates the experimental results and we conclude our work in Sect. 5.

2 Related Works

This section is about related researches on traffic congestion prediction, travel speed estimation, mining of travel patterns on urban road network. In [1] Xianyuan predicted citywide traffic volume of Beijing City using trajectory data. Their model is able to predict the travel speed of each highway, major and minor road of Beijing City by extracting special features from GPS trajectories based on traffic flow theory to provide speed-flow relationship. Network wide speed is predicted and then city wide traffic volume is predicted using machine learning technologies. They used traffic flow theories which lead to contain a number of assumptions in their model.

Swe Swe Aung and Thinn Thu Naing [2] proposed a model for traffic congestion prediction for Yangon City using Naïve Bayes Classifier on collected GPS data from ferry buses. The model was able to predict traffic congestion on a single junction due to they collected the trajectory data with android phones on their way to work and back to home.

Xiujuan Xu et al. [3] proposed an algorithm for Urban Traffic Detection by GPS data compression methods. Their algorithm was based on three keys features namely average travel speed, traffic density and traffic flow to predict road congestion. In [4] Yuqi Wang et al. proposed a framework to study the congestion correlation between road segments from multiple real world data. Their work includes three phases: extracting congestion information from GPS trajectories, extracting various features on each pair of road segments from road network and POI data and predicting congestion correlation based on the results of the first two phases.

Our work is to extract traffic delay features for POI and to estimate travel speed of every road segment along the bus route based on GPS data. After that we store traffic speed factors for all road segments in regular time window for traffic history data generation and further research.

3 Proposed System Model

Travel Speed Estimation Model proposed in our research is depicted in Fig. 1. Collected GPS data from buses are stored in a server, made data preprocessing to get clear data set. Map matching process is then performed to match the tracked buses onto the running route of bus route 21 and then average travel speed of each road segment is estimated. The color coded traffic map is used to display average travel speed of road network. The speed factor for each road segment is stored in a matrix for further reference.

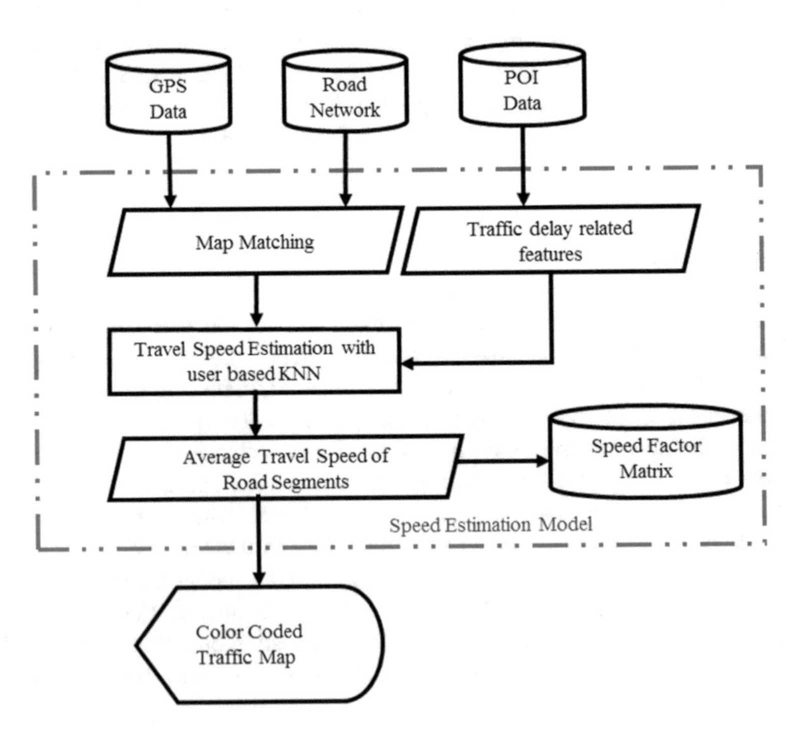

Fig. 1. Proposed system model for travel speed estimation

3.1 GPS Data Collection and Preprocessing

Data collection plays in an important role in every data analysis model. The integrity and clarity of collected data are essential properties to build a good data set. YBS bus no. 21 provides their bus tracking data for our research. They collect GPS data by installing Sino Track GPS trackers in the buses and trace the travel patterns, fuel consumption rate, mileage, engine on/off status and location of buses in every 20 s for 24 h a day.

GPS data by installing Sino Track GPS trackers in the buses and trace the travel patterns, fuel consumption rate, mileage, engine on/off status and location of buses in every 20 s for 24 h a day. We can access the live GPS data for every bus that services along the bus no. 21 route as well as history data for last three months.

The next step is to build a well-structured data set from the collected trajectories. The preprocessing of GPS data includes removing outlier points, dropping unwanted features by dimension reduction methods and selecting important features for Travel Speed Estimation Model (TSEM). Preprocessed trajectory can be expressed as:

$$T_i \in \{t_i, lat_i, lon_i, dir_i, v_i\} \tag{1}$$

where, T_i is the specific trajectory from a bus, t stands for time of the day in local time zone, position data (latitude, longitude) information is represented by (lat_i, lon_i), dir_i is the direction of the bus and v_i represents current speed of the bus. After preprocessing of raw GPS data, the data set is ready for analysis.

3.2 Road Network and Map Matching Process

The real map data for the whole city is too large to be used in Travel Speed Estimation Model. The map data can be divided into smaller data set. Suppose that each road R is a set of road segments e: $R \epsilon \{e1, e2, ..., en\}$ and each road segment consists of start and end points: $e \epsilon \{lat1, lon1, lat2, lon2\}$ where lat1, lon1 represents start point and *lat2, lon2* is end point of the road segment. The next important feature is Place of Interests (POI) within 50 m radius of road each segment e [1].

POI that appears moth frequently near road segment includes *Schools, Market places & Shops, Automobile services, Resident area, Shopping malls, Restaurants, Bank and ATM, Hotels and Entertainment services* within 50 m radius of road segment. We create a 9 dimensional vector where each column represents the occurrence of individual POI type as shown in Fig. 2.

Schools	Market places & Shops	Auto-mobile services	Hospital	Shopping malls	Restaurants	Bank and ATM	Hotels and Entertainment
2	5	1	1	1	4	2	2

Fig. 2. POI features vector for particular road segment

The next step is mapping trajectory to the map for average speed calculation and further processing. We use The map-matching algorithm proposed by Yuan et al. [5] is used to project bus GS points to the map. Their algorithm considers both the position context of GPS points and the road network topology. The interested bus route consists of 18 signalized junctions and 48 bus stops. We have to segment the bus route by the collected position (*lat, lon*) of every bus stop along the way. User based KNN model is used to find out the nearest position of bus to the bus stop's location.

3.3 Estimating Average Travel Seed

After matching GPS point p_i to each road segment, the average speed at particular point v_i, standard deviation dv and average travel peed v_{avg} of a bus on a particular road segment can be computed as:

$$v_i = \frac{Dist(pi, pi+1)}{(ti+1-ti)} \tag{2}$$

$$v_{avg} = \frac{\sum_i^n v_i}{n}, \; dv = \sqrt{\frac{\left(v_i - v_{avg}\right)^2}{n}} \tag{3}$$

where *Dist* is the Harversian distance that calculates the earth's surface curvature distance between the two GPS points *pi* and *pi + 1*, t is the time of the day in local time zone of point *p*. The map matching and speed computation are illustrated in Fig. 3. The estimated travel speed of a road segment can obtain directly if there is live time-position data from the bus is available.

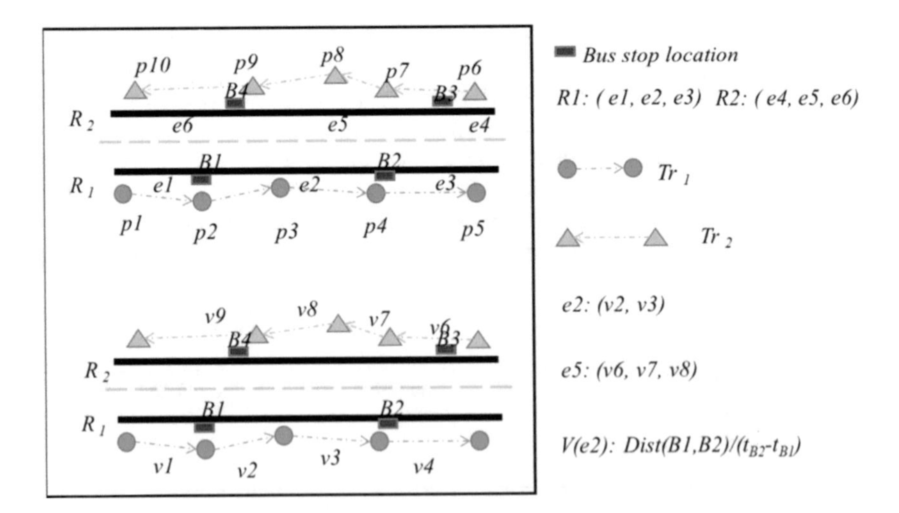

Fig. 3. Map matching and travel speed calculation

Building Seed Factor MatrixSpeed factor is the ratio of average travel speed of a bus to the speed limit of road segment. Due to limited number of running buses that can track in a particular time window (e.g. 8:00–8:10), there exist some road segments with no buses running on them. In order to solve this problem, we have to overlap the historical GPS trajectories to get the average travel speed of each road segment. The historical data such as the average speed, standard deviation, minimum speed and maximum speed of a road segment e are calculated at every 10 min time interval and recorded as statistical data.

POI feature vector of each road segment has significant consequence on traffic flow and traffic speed of bus route. Each POI category with its corresponding occurrence is used to define traffic delay factor that has effect on speed factor generation. The speed factor is calculated and updated in regular time interval from the statistical data, live data and traffic delay factors of each segment (Fig. 4).

Road Segment id	0000	0010	0020	.....	2350
1	1	0.98	0.78	...	0.98
2	1	0.95	0.72	...	0.93
3	1	0.90	0.80	...	0.95
...	...	...	...	...	...
n	1	0.85	0.74	....	0.97

Fig. 4. Speed factor matrix for each road segment

4 Experimental Results

This section depicts the experimental results of our proposed Travel Speed Estimation Model. The bus route 21 starts from the rural area, passes through the most congested area of Yangon City and ends at the city center as shown in Fig. 5. After matching the trajectory to the map the average speed v_{avg} standard deviation dv, maximum and minimum speed are calculated for each road segment by Eqs. (2) and (3). We extract the traffic delay feature from POI matrix of every road segment to generate traffic speed factor matrix.

Figure 6 shows the travel speed of a bus on a weekend (16.9.2017 Saturday, 12:30 P.M. to 1:30 P.M.) The graph shows the estimated travel speed, average speed, standard deviation, maximum speed and minimum speed of probe vehicle. From this graph we extract the statistical speed data of road segments and store them in the matrix. The slope of regression line declines gradually during the time interval 12:36 to 13:26. It can be seen clearly that there was serious traffic congestion on the route at 13:12 to 13:26 with almost no movement is identified by 0 km/h of estimated speed. After that the traffic delay features from nearby POIs are used to get speed factor matrix of road segments.

Fig. 5. Map matching results for collected GPS data to the bus route 21.

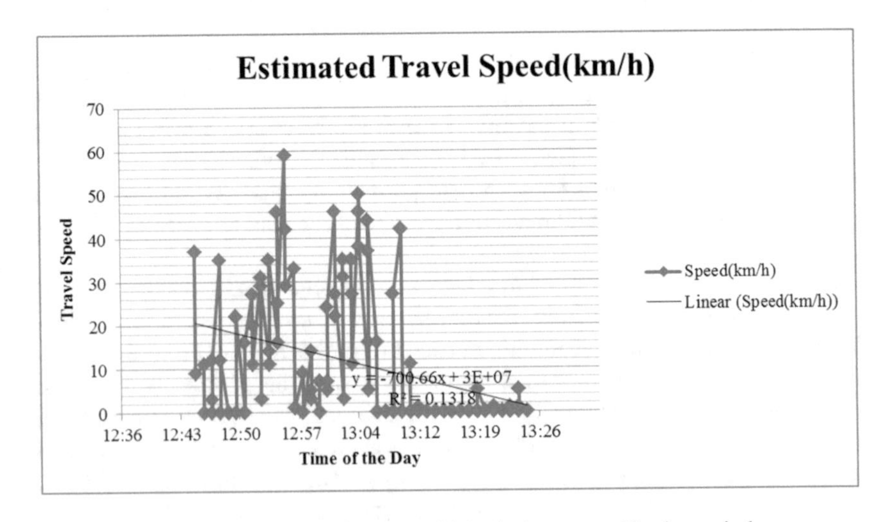

Fig. 6. Travel speed of probe vehicle during a specific time window

5 Conclusion

We developed a model to estimate travel speed of road network that uses machine learning techniques, linear and non-linear algebra. The sample results from the Travel Speed Estimation Model shows that average travel speed is highly correlated with day-of-the week and time-of-the-day. Traffic congestion information can be extracted from the graph by observing changes in travel speed and slope of the regression line. The mean error for this model is quite small with the value of R2 = 0.1318. The time-location data are collected from one of YBS bus routes. More accurate estimation and better results will come out if we could collect data from all of the bus routes running through Yangon City. Our next work will be traffic congestion prediction of Yangon City by using state of the art machine learning techniques in as an extension of current research.

Acknowledgments. This research was supported by Research Fund of Yangon Technological University, Yangon, Myanmar. We would like to extend our special thanks to all colleagues and friends who enthusiastically share their ideas and suggestions during research progress. Finally, we would like to thank YBS bus no. 21 for providing valuable data for our research.

References

1. Zhan, X., Zheng, Y., Yi, X., Ukkusuri, S.V.: Citywide traffic volume estimation using trajectory data. IEEE Trans. Knowl. Data Eng. **29**(2), 272–285 (2017)
2. Aung, S.S., Naing, T.T.: Naïve Bayes classifier based traffic detection system on cloud infrastructure. In: 6th International Conference on Intelligent Systems, Modelling and Simulation (2015)
3. Xu, X., Gao, X., Zhao, X., Xu, Z., Chang, H.: A novel algorithm for urban traffic congestion detection based on GPS data compression. IEEE (2016). 978-1-5090-2927-3/16/$31.00 ©2016
4. Wang, Y., Cao, J., Li, W., Gu, T.: Mining traffic congestion correlation between road segments on GPS trajectories. IEEE (2016). 978-1-5090-0898-8/16/$31.00 ©2016
5. Yuan, J., Zheng, Y., Zhang, C., Xie, X., Sun, G.Z.: An interactive-voting based map matching algorithm. In: Proceedings of IEEE International Conference on Mobile Data Management, pp. 43–52 (2010)
6. Liu, D., Kitamura, Y., Zeng, X., Araki, S., Kakizaki, K.: Analysis and visualization of traffic conditions of road network by route bus probe data. In: IEEE International Conference on Multimedia Big Data (2015)

Comparison Between Block-Encoding and Quadtree Compression Methods for Raster Maps

Phyo Phyo Wai(✉), Su Su Hlaing, Khin Lay Mon, Mie Mie Tin, and Mie Mie Khin

Myanmar Institute of Information Technology, (MIIT), Mandalay, Myanmar
Phyophyowai84@gmail.com, susu07.su@gmail.com, khin_lay_mon@miit.edu.mm, miemietin1983@gmail.com, miemie.khin9@gmail.com

Abstract. The compression methods are essential for big data, such as maps, images and large data, because the storage spaces are more required as long as data are being overloading. In this paper, quadtree compression method and block-encoding compression method are compared by their lossless compression results based on the raster maps. Quadtree is a method which is very mature and popular in segmentation and compression area. Block-encoding method is a 2D version of run-length encoding method.

Keywords: Quadtree · Block-Encoding · Lossless compression

1 Introduction

The maps are useful to locate the area where the place is. The original maps sizes are usually extremely large. The storage spaces of an android device are very small if it is comparing with computer storage spaces. With the android devices are very popular, maps that uses in android devices are also needed to compress for storage space. Raster maps are grid cells based structure of the pixel. One pixel can be representing a place where is the importance area. Therefore, lossless compression is very important for maps. When some pixels are lost at the compress time, the information on the map will also be missed information. Therefore, lossless compressions of raster maps are essential and very effective. When compressing the maps, many lossless compression techniques can be used. Quadtree compression technique is very popular, and quadtree has very effective segmentation result. Block-encoding technique is very similar to quadtree data structure in working. But block-encoding can work on the sizes of the images such as width and height of the image are not equal to length while the quadtree can only work on the square images.

T. T. Zin and J. C.-W. Lin (Eds.): ICBDL 2018, AISC 744, pp. 258–263, 2019.
https://doi.org/10.1007/978-981-13-0869-7_29

2 Related Works

Mohammed and Abou-Chadi [4] investigated block truncation coding for image compression. In their paper, the original block truncation coding (BTC) and Absolute Moment block truncation coding (AMBTC) algorithms were based on dividing the image into non overlapping blocks and used a two-level quantize. The two techniques were applied to different grey level test images which each contains 512 $\times$ 512 pixels with 8 bits per pixel. Li and Lei [5] presented a novel block-based segmentation and adaptive coding (BSAC) algorithm for visually lossless compression of scanned documents such as photographic images, text and graphic images. Klajnšek, Rupnik and Špelic [6] developed a novel algorithm which is lossless compression for volumetric data. Their algorithm was based on their previously presented algorithm, which used quadtree encoding of slices of data for discovering the consistency and matches between sequential slices. Li and Li [7] described that large spatial databases management were usually used multi-level spatial index techniques. In that paper, a hybrid structure of spatial index was presented based on multi-grid and QR-tree. Cho, Grimpe and Blue Lan [8] discussed that the quadtree compression technique was the most common compression method applied to raster data. Quadtree coding stored the information by subdividing a square region into quadrants, each of which may be further subdivided in squares until the contents of the cells have the same values. Zhao, Lu, Wang, and Yao [9] described that Mobile GIS had developed into a popular and important research direction of GIS. They were implemented the browse query of mobile GIS on the mobile terminal applications, using Mobile Widget and Mobile Maps Widget as Technology platform. Yu, Zhang, Li, and Huang [10] were witnessing the exploding development of Mobile Geoinformation Services (MGS), which was a complex system engineering involving many different technological fields or disciplines.

3 True Color and Color Maps

True color is a radiometric resolution which exceeding the human eyes color resolving power. Computer graphics apply to radiometric resolutions of at least 24 bits equivalent to 2^{24} (16.8 Million) colors and hi-color is a radiometric resolution of up to 16 bits (65536 colors) [1, 2].

Color maps are also represented to as color palettes or color look-up tables. They use an indirect or pseudo-color representation which assigns an index value instead of actual color values to each pixel. These index values represent addresses used for looking-up the actual color values in the previously established color table, which allow a significant reduction in the amount of data when applied to images with relatively few colors, because the index values can be kept much smaller than actual color values such as complete red, green and blue (RGB) triplets. By typically, index values are stored as 4-bit or 8-bit integer values in contrast to the color map elements, which are usually stored as 24 bits (3x8 bits per color). Color tables are not suitable for

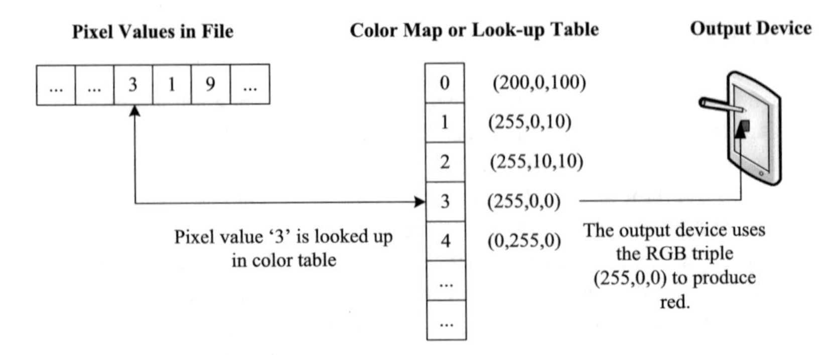

Fig. 1. Example of using a color map to specify a color

storing images which contain more than 256 colors, such as photographs. Color maps are very suited and often used for cartographic raster data with their limited number of colors frequently corresponding to individual color layers [3]. In this paper, 24-bits true colors are used for lossless compression of maps (Fig. 1).

4 Quadtree Compression Method

In our quadtree method, firstly, when we input the rater map, the quadtree method will divide the map into segments. After doing the segmentation, we will get the segmentation regions and calculate the color pixel for each segment. And then, we will get the pixels with color values. At the same time, when we get the segmentation regions, we will do spatial indexing creation, and then we will get pixel neighborhood relationship.

5 Block-Encoding Method

In block encoding method, it consists of two parts, namely flag encoding and block data encoding. Flag encoding records the location and color information of the block targets, and block data encoding records get block data and their boundary information.

6 Overall Performance

To evaluate the performance of this system, it computes the compression ratios for the images in the dataset by using following equation.

$$Compression\ Ratio = \frac{\text{size of the output stream}}{\text{size of the input stream}} \tag{1}$$

In the following, it calculates the compression ratio of the one image from Map Set #1 by choosing randomly. Image name is T1_1 image, and its original size is 1024 ×

1024 pixel size bitmap file. Original file size is 3 MB. The output file size, after compressing, is 1.3 MB. When calculating the compression ratio, the compression ratio output is 0.4. And, it is calculating by using this equation $100 \times$ (1- compression ratio) for a reasonable measure of compression performance. When inserting the compression ratio into the equation, $100 \times (1 - 0.4) = 60$ it get the output result is 60. The result value of 60 means that the output stream occupies 40% of its original size or that the compression has resulted in savings of 60%. In this system, time is not effective as much as run on the computer. Because this system is run on the android device that has the OS version is v4.4.2 (KitKat), Chipset is Qualcomm MSM8226 Snapdragon 400, CPU is Quad-core 1.2 GHz Cortex-A7, GPU is Adreno 305, internal memory storage is 8 GB and RAM is 1.5 GB.

7 Experimental Results

In this system, for maps, 1024×1024 map sizes, 49 maps for MapSet#1, 5 maps for MapSet#2 and 4 Maps for MapSet#4 are experimented. Their some detail results and maps are shown in below (Figs. 2, 3) and Table 1.

Name and Size Of Original Map	Compression Approaches	Analysis Results
 T1_02.bmp Original File Size = 3073 KB Image Size = 1024x1024	Quadtree	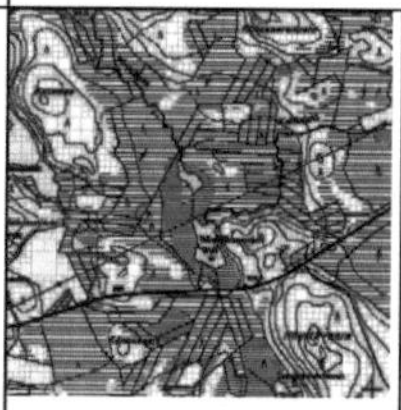 CompressionTime= 21.456100096 Image Size = 1024,1024 Minimum Block Size = 1 QuadTree Node = 294919 QuadTree Level = 10 DecompressionTime= 4.001779425 Compress File Size = 3457 KB
	Block-Encoding	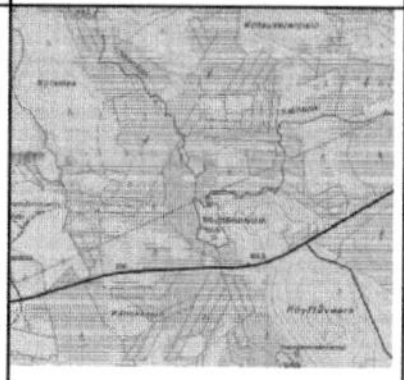 CompressionTime= 11.17007432 Image Size = 1024,1024 Block Count = 16384 DecompressionTime= 2.380869582 Compress File Size = 1863 KB

Fig. 2. Compression results

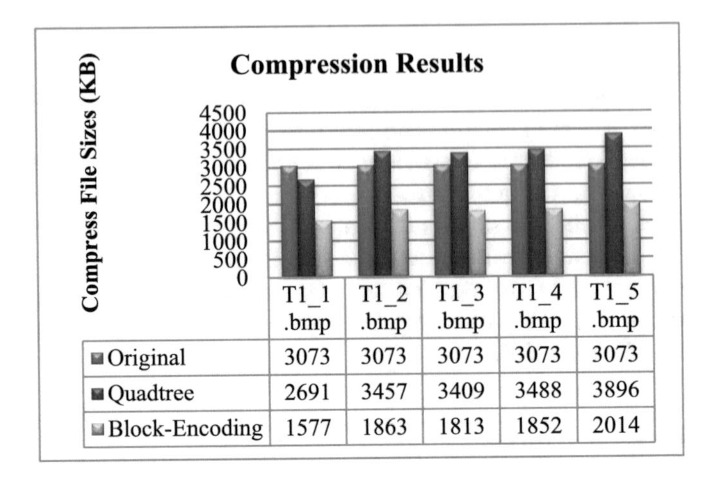

Fig. 3. Compression results chart

Table 1. Compression ratio results

File name and size	Quadtree	Block-Encoding
T1_1.bmp, 3073 KB	0.875691507	0.513179304
T1_2.bmp, 3073 KB	1.124959323	0.606247966
T1_3.bmp, 3073 KB	1.109339408	0.589977221
T1_4.bmp, 3073 KB	1.135047185	0.602668402
T1_5.bmp, 3073 KB	1.267816466	0.655385617
Total	56.13016596	30.20240807
Average	1.145513591	0.616375675

8 Conclusion

By conclusion, although quadtree is mature for the segmentation of the images and effective for black and white images, it is not effective for true color maps lossless compression. Sometimes, it can be large compression file size more than original file size of the map. Block-encoding method can compress the file nearly 2:1. Compression time is also effective. So, block-encoding method is more effective than quadtree method for this paper.

References

1. Munay, J.D., VanRyper, W.: Encyclopedia of Graphics File Formats. O'Reilly & Associates Inc., Sebastopol (1994)
2. Nebiker, S.: Spatial Raster Data Management for Geo-Information Systems - A Database Perspective. Dissertation ETH No. 12374 (1997)

3. Palette Homepage. http://www.wikipedia/Palette (computing) - Wikipedia, the free encyclopedia.htm
4. Mohammed, D., Abou-Chadi, F.: Image compression using block truncation coding. Cyber J. Multi. J. Sci. Technol., J. Sel. Areas Telecommun. (JSAT) (2011). February Edition
5. Li, X., Lei, S.: Block-based Segmentation and Adaptive Coding for Visually Lossless Compression of Scanned Documents. IEEE (2001). 0-7803-6725-1/01/200 01
6. Klajnšek, G., Rupnik, B., Špelic, D.: An improved quadtree-based algorithm for lossless compression of volumetric datasets. In: 6th WSEAS International Conference on Computational Intelligence, Man-Machine Systems and Cybernetics, Tenerife, Spain, 14–16 December 2007, pp. 264–270 (2007)
7. Li, G., Li, L.: A hybrid structure of spatial index based on multi-grid and QR-tree. In: Proceedings of the Third International Symposium on Computer Science and Computational Technology (ISCSCT 2010), Jiaozuo, P. R. China, 14–15 August 2010, pp. 447–450, (2010). ISBN 978-952-5726-10-7
8. Cho, C., Grimpe, E., Lan, Y.-C.B.: Method and apparatus for blockwise compression and decompression for digital video stream decoding. In: International Application Published Under The Patent Cooperation Treaty (PCT), World Intellectual Property Organization, International Bureau, EP1994762 A1, 26 November 2008 (2008)
9. Zhao, X., Lu, H., Wang, H., Yao, P.: Research and implementation of mobile GIS based on mobile widget technology. A. J. Eng. Technol. Res. **11**(9) (2011)
10. Yu, M., Zhang, J., Li, Q., Huang, J.: Mobile Geoinformation Services—Concept, Reality & Problems. AsiaGIS2003 (2003)

Video Monitoring System and Applications

A Study on Estrus Detection of Cattle Combining Video Image and Sensor Information

Tetsuya Hirata[1(✉)], Thi Thi Zin[1(✉)], Ikuo Kobayashi[2(✉)], and Hiromitsu Hama[3(✉)]

[1] Faculty of Engineering, University of Miyazaki, Miyazaki, Japan
hl13037@student.miyazaki-u.ac.jp,
thithi@cc.miyazaki-u.ac.jp
[2] Field Science Center, Faculty of Agriculture, University of Miyazaki, Miyazaki, Japan
ikuokob@yahoo.co.jp
[3] Osaka City University, Osaka, Japan
hama@ado.osaka-cu.ac.jp

Abstract. In Japan, the detection rate of estrus behavior of cattle has declined from 70% to 55% in about 20 years. Causes include the burden of the monitoring system due to the aging of livestock farmers and oversight of detection of estrus behavior by multiple rearing. Because the time period during which estrus behavior appears conspicuously is nearly the same at day and night, it is necessary to monitor on a 24-h system. In the method proposed in this paper, region extraction of black cattle is performed by combining frame difference and MHI (Motion History Image), then feature detection of count formula is performed using the characteristic and features of the riding behaviors. In addition, as a consideration of the model experiment, a method of detecting the riding behavior by combining the vanishing point of the camera and the height from the foot of the cattle was proposed. The effectiveness of both methods were confirmed through experimental results.

Keywords: Estrus detection · Frame difference · Vanishing line

1 Introduction

Presently, the detection rate of estrus behavior of castle has been decreasing year by year from about 70% to 55% from about 20 years ago to the present. Causes include the burden of monitoring system due to aging of livestock farmers and oversight of detection of estrus behavior by multiple rearing. Furthermore estrus happens at the nearly same proportion for day and night, so we need to monitor on a 24-h system [1]. In this research, we aim to reduce the burden on livestock farmers by automatic detection of bovine estrus behavior using image processing technology, and to improve the artificial insemination rate accordingly.

T. T. Zin and J. C.-W. Lin (Eds.): ICBDL 2018, AISC 744, pp. 267–273, 2019.
https://doi.org/10.1007/978-981-13-0869-7_30

2 Propose Method

Behavior that appears most notably in estrus behavior of a cattle is mounting and standing action. Mounting is the state that only the cattle performing the riding behavior estruses and the cattle being ridden doesn't estrus. In addition, the standing is the state that the cattle allowing a ride also estrus. Here we propose the method to detect these two states. Figure 1 shows the flowchart of the proposed method.

2.1 Region Detection Using Frame Difference

First of all, the differences for every 3 frames were used to confirm the change of movement. This is because there is a big difference between normal behavior and estrus behavior as a feature of frame difference. In this research, we combined the frame difference with MHI (Motion History Image) and extracted the object area using the energy intensity for each coordinate [2]. Now, let's express the difference in the current frame as FD_t, the exercise intensity in the current frame as E_t, the energy update rate as α, and the energy as C_E, then they can be expressed by the following equation:

$$E_t(x, y) = \delta_t(x, y) + E_{t-1}(x, y) \times \alpha \tag{1}$$

$$\delta_t(x, y) = \begin{cases} C_E & \text{if } FD_t(x, y) \in FD_{t-1}(x, y) \\ 0 & \text{otherwise} \end{cases} \tag{2}$$

At this time, the energy update rate α and the energy function C_E were set as 0.48 and 80, respectively.

Two original frames and the results of difference are shown in Fig. 2.

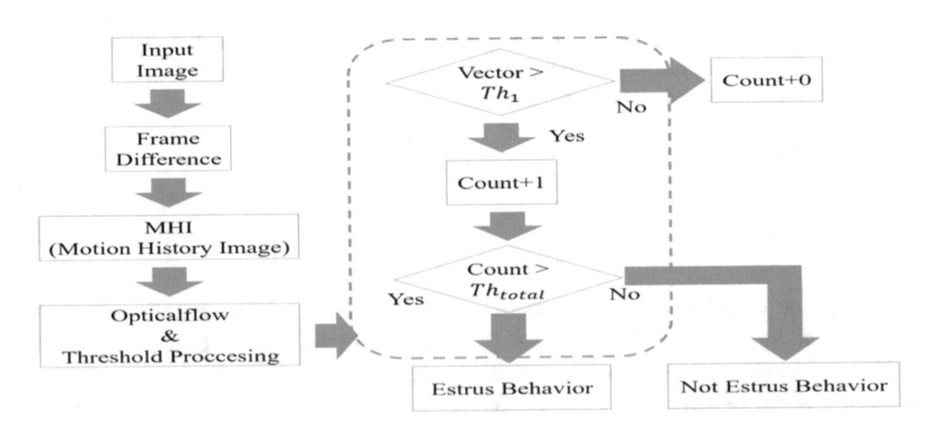

Fig. 1. Flowchart of the proposed method to detect mounting and standing.

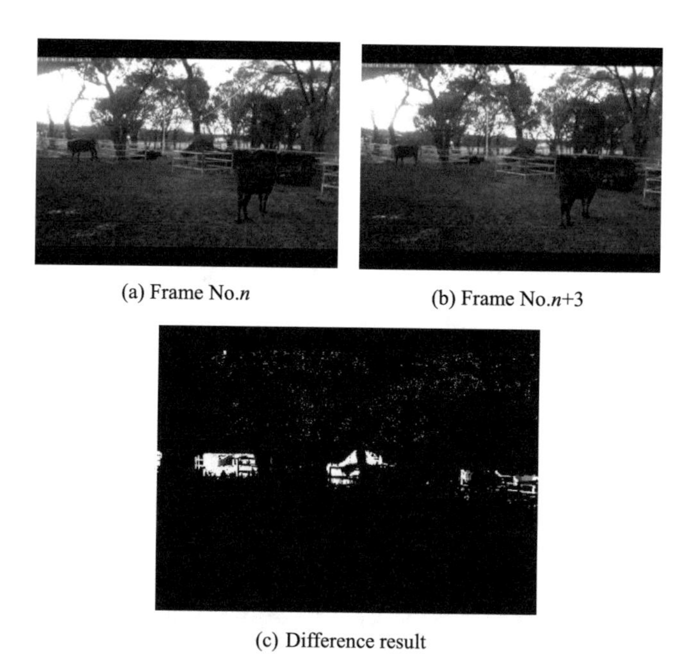

(a) Frame No.n (b) Frame No.n+3

(c) Difference result

Fig. 2. Example of frame difference processing.

2.2 Detection of Estrus Behavior Using Optical Flow

As a detection method, we detected the riding behavior using optical flow. Optical flow is a vector representation of the motion of an object in a digital image and is a method used for detecting moving objects and motion analysis. We focused on the fact that the movement of the object obliquely upwards during the riding behavior, and we treated so as to remain only when the angle of the vector was 0° to 180°. An example of before and after processing is shown in Fig. 3.

2.3 Evaluation Method Using Optical Flow

The vector in the upward direction exceeding the threshold value decided in advance is set as the criterion for starting time of the riding behavior and the system counts. After that, when the vector in the horizontal direction continues for above a certain level, the system continues to count and judges whether the behavior is estrus or non-estrus according to the count. The magnitude of the vector used here for detection is expressed by the size of a block of 40×40.

2.4 Estrus Detection Using Vanishing Point of Camera

As another method, we tried a method of detecting the estrus of a cattle using the vanishing point of the camera. The vanishing point and the vanishing line are defined as follows. When straight lines lining up in 3D space are extended in the depth direction, they reach the vanishing point at infinity. Straight lines parallel in 3D space have the feature that they have the same vanishing point. A vanishing line can also be considered as a set of vanishing points. When the optical axis of a camera is parallel to the ground, the vanishing point of any point at the same height as the lens center of the camera is on the vanishing line. Focusing on this feature, we conducted a model in which the lens center of the camera is set at the same height of the back of a cattle. The certain threshold is set at the height of the vanishing line, further set it slightly higher to give tolerance. If the back of the cattle exceeds the threshold, it is assumed that a riding behavior is occurring. By setting the threshold, it becomes possible to adjust along the height of each individual cattle's back. The aspect of simulation experiment is shown in Fig. 4.

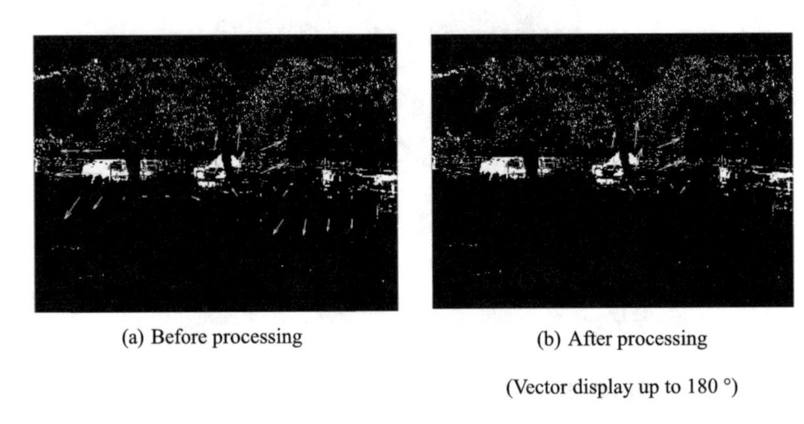

(a) Before processing

(b) After processing

(Vector display up to 180 °)

Fig. 3. Features using optical flow.

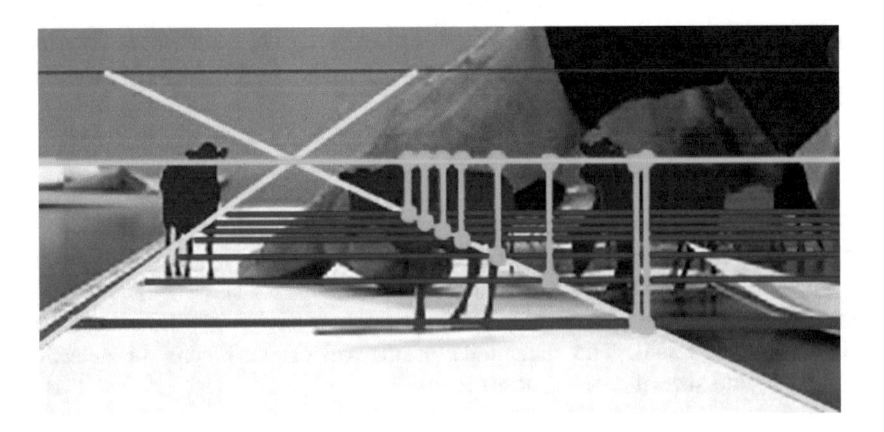

Fig. 4. Simulation experiment using vanishing line.

The red line is the central horizontal line on an image of the camera. The two yellow lines which are parallel lines in the depth direction on 3D space are straight lines on 2D image, and they are used to find the vanishing point. The point where the yellow lines intersect is the vanishing point. The blue lines are the coordinates of several cattle's feet positions. The green lines are lines of height from the feet of teach cattle to the back. The light blue line is an approximate line obtained from the coordinates of the captured cattle's back. This is regarded as the vanishing line, and detection of a riding behavior is performed.

3 Experimental Result

3.1 Detection Result Using Frame Difference

The experiment was carried out at the ranch owned by University of Miyazaki. The training data and the number of test data used to evaluate the proposed method are summarized in Table 1. In the experiment, images of every three frames were used, counting those satisfying the threshold condition was added, and when the addition exceeded a certain count, it was regarded as an estrus. The experimental results and the final detection rate are summarized in Table 2 and in Table 3, respectively.

Table 1. Number of data used (Video).

Training data	Test data
10	5

Table 2. Types of estrus behaviors and detection results.

Test No	Types of estrus	Result
1	Standing	○
2	Mounting	○
3	Standing	△(○)
4	Mounting	○
5	Mounting	○

Table 3. Final detection result of estrus behavior.

Number of estrus	Detection count	Detection rate (%)
5	5	100

Although estrus behaviors could be detected by our system, estrus behaviors occurred in the vicinity of the camera in the detection result in Test No 3 in Table 2. Since the result of the frame difference appears large, the movement of the vector appears to be large and it can be detected, but in our system, the standing and the operation of the mounting occur at the same time in other places. These two images are shown in Fig. 5. From Fig. 5, it is ambiguous whether it reacts to either one or one of them, but this time it is treated as being able to detect it.

3.2 Detection Result Using Vanishing Line of Camera

The result of detection of the riding behavior using the vanishing line of the camera is discussed in this section. The number of data used and the detection results are summarized in Tables 4 and 5. In this way we were able to detect correctly.

Table 4. Number of data used.

Training data	Test data
5	10

Table 5. An example of detection result.

Test No	Types of estrus	Result
1	Not Estrus	○
2	Not Estrus	○

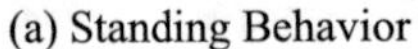

(a) Standing Behavior

(b) Mounting Behavior

Fig. 5. Estrus behavior in Test 3.

4 Conclusion

The proposed method in this study combined frame difference and MHI, focused on the direction of vector by optical flow, and detected the estrus behavior. The effectiveness of this method has been confirmed through the experimental results with the high detection rate. However, the number of test data was small, and it is thought that it is necessary to acquire a lot of data in estrus behavior and determine the threshold value. In this method, since estrus was detected while leaving noise in region extraction, it was affected by noise. From the results of the frame difference, there is a problem that camera noise appears as noise. Since the magnitude of noise itself is not large, it can be considered that more accurate detection can be performed. Also, since there is a difference in the total score depending on the distance, it can be considered that more accurate detection can be performed by setting a threshold according to the distance.

Finally, by combining sensors, it is thought that it is possible to estimate the distance from the image and to perform more useful estrus detection.

In the method using the vanishing line, the detection accuracy was 100% because of the simulation experiment using models. For practical use in the actual environment, there are some problems left to be solved. The camera must be set so that it does not move outdoors. Also, it is necessary to set marks to obtain the vanishing line of the camera. If these problems are cleared and the coordinate of the back of the cattle is known, it is considered that more useful estrus detection system can be made.

Acknowledgment. This work was supported in part by SCOPE: Strategic Information and Communications R&D Promotion Program (Grant No. 172310006) and JSPS KAKENHI Grant Number 17K08066.

References

1. Livestock improvement work group conception study results. http://liaj.or.jp/giken/gijutsubu/seieki/jyutai.htm. Accessed 10 Jan 2017
2. Tsai, D.-M., Huang, C.-Y.: A motion and image analysis method for automatic detection of estrus and mating behavior in cattle. Comput. Electron. Agric. **104**(2014), 25–31 (2014)
3. Gu, J., Wang, Z., Gao, R., Wu, H.: Cow behavior recognition based on image analysis and activities. Int. J. Agric. Biol. Eng. **10**(3), 165–174 (2014)

Behavior Analysis for Nursing Home Monitoring System

Pann Thinzar Seint and Thi Thi Zin(✉)

Graduate School of Engineering, University of Miyazaki, Miyazaki, Japan
th17054@student.miyazaki-u.ac.jp,
thithi@cc.miyazaki-u.ac.jp

Abstract. In this paper, we describe the nursing home monitoring system based on computer vision. This system is aimed for an effective and automatic take care of aged persons to monitor appropriate medication intake. Skin region detection for mouth and hand tracking and color label detection for water and medication bottles tracking are mainly performed for initialization. To differentiate the hand and face region, we use the regional properties of head with online learning. Tracking is done by the minimum Eigen values detection. The overlapping area ratios of desired object to body parts are simply used as feature vectors and Pattern Recognition Neural Network is proposed for the decision of simplest action. This paper presents the 7 types of simple actions recognition for medication intake and our experimental results give the promising results.

Keywords: Medication intake monitoring · Mouth and hand tracking
Eigen values · Feature vectors · Pattern recognition neural network

1 Introduction

Nowadays, the population of elderly people are increasing worldwide [1], among them the number who spend their lives with "disabilities" grow faster than caregivers. To overcome these problems, smart monitoring becomes popular as "extensive caring unit", integrating with the information technology. Technology can significantly improve the life's quality of persons having physical and mental deficiencies. The field of the video surveillance attracted a lot of interests but it was little directed toward controlling medication intake [2].

The recognition of medication intake is based on the concept of scenarios. The transitions between simple states are strongly needed to define for various types of scenarios. By analyzing the state duration statistics, the complete or incomplete status can be checked. To realize the system model, the correctness of feature vectors is in important role. Hands, mouth and bottles detection are needed to be robust. Generally, the system consists of color segmentation, object representation, feature extraction and action classification. The interaction between "the mouth and hands", "four color-labeled bottles and hands" and "mouth and water bottle" are monitored to recognize for each simplest action. Vision based system with detecting and tracking areas are challenging because of the dramatic variation of hand and head appearance, illumination and environment. In this paper, we propose the simple basic actions of

T. T. Zin and J. C.-W. Lin (Eds.): ICBDL 2018, AISC 744, pp. 274–283, 2019.
https://doi.org/10.1007/978-981-13-0869-7_31

medication intake recognition using neural network by the contribution of tracking method.

2 Related Works

The video surveillance system are attracted to many fields of application areas. The monitoring system of medication intake becomes popular in recent years. In [3], the tracking of fingers are used to handle the bottle. But fingers are not visible every frame. And, template matching is used to detect and track the face and the mouth is localized with lip color. In [4], the author firstly modeled the head with an ellipse and the local search determines the ellipse which matches with the head. The author proposed to track the face contour based on particle filter [5].

To differentiate the face from the hand, the author checked the eyes using Ada-boost and hand region is detected based on contours using Sobel operator [1]. To know handling the bottle, distance between the centroids of two hand and bottles are calculated and they recognize 97% of the complete event for medication intake by using Petri net in [2]. Multi-resolution HOG features are used to detect several hand postures in [6]. These methods require offline training for some kinds of appearance and also require representative training dataset and storage space. In this paper, we don't use the earlier training for face and hand detection. Regional properties of face area are only used for face detection by online learning and then hand can be easily detected by differentiating face from skin area.

3 Proposed System

In this paper, color information is mainly used and then followed by appearance information. Supervised learning with neural network is introduced to classify for each simplest action and the proposed system is shown in Fig. 1. The proposed method is started when three skin areas are detected.

3.1 Color Segmentation

(i) *Skin Region Segmentation*

It is the pre-processing step in the video sequence. YC_bC_r color space is used to perform skin region segmentation. Two different chrominance values C_b and C_r with heuristic threshold are chosen by subtracting luminance from blue and red components. Due to separation of illumination and chrominance components, this color space is effective for skin color modeling [7].

The skin color based on C_b and C_r values can provide good coverage of different human races. The threshold chosen by $[C_r1, C_r2]$ and $[C_b1, C_b2]$ is classified as skin color tone if the input values $\{C_b, C_r\}$ fall within the threshold. We get the noisy image after skin color segmentation. Morphological operations and median filtering are utilized for refining the skin regions extracted from the skin pixel segmentation for the various irregularities and noisy area. After binarization of the image, the face and hand region blobs are obtained as shown in Fig. 2(a).

(ii) *Color-Labeled Segmentation*
In this case, RGB color space is used to detect the color bottles. Red, Blue and Green color for medication bottles and Yellow for water bottle are extracted by the color thresholding. For detecting the color bottle, the RGB image is firstly converted into gray scale image and then differentiated with desired color channel image. And the predefined color threshold value is applied to resulted image. Finally, all of the desired object are represented by the Bounding Box (BB) information as shown in Fig. 2(b).

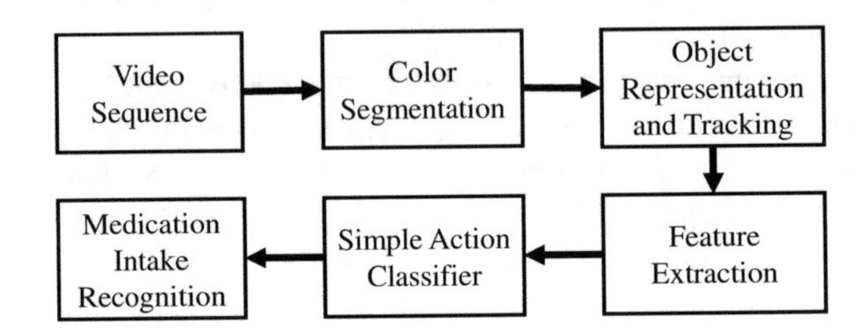

Fig. 1. Overview of proposed system

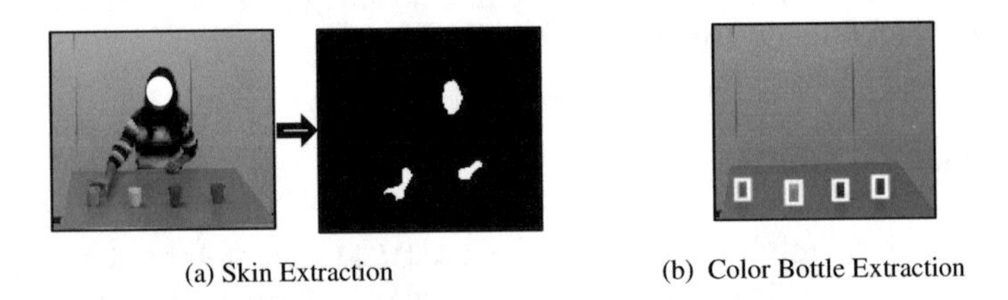

(a) Skin Extraction (b) Color Bottle Extraction

Fig. 2. Object representation by bounding box

3.2 Mouth Detection and Tracking

(i) *Differentiation of Head and Hand Region*
For the mouth detection, we propose the online learning method to save time and memory consumption. We have tested with Viola Jones' face and mouth detector. It is very good at frontal face detection. But, when the head is changed with dramatic variations, that detector cannot work. In our proposed method, Head region is firstly extracted according to regional properties (centroids, major axis length and minor axis length) which are in the range between possible values. It is the very simplest way to extract the head region from the skin area in this experiments as shown in Fig. 3(b). We also easily get the hand region by differentiating the head at the skin region as shown in Fig. 3(c). And then, the tracking process is needed to consider for taking medicine activity which is very straightforward. It is important to track the mouth in partial occlusion condition.

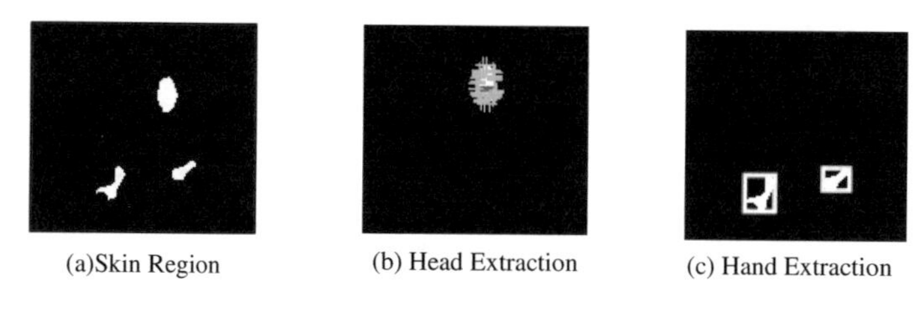

(a)Skin Region (b) Head Extraction (c) Hand Extraction

Fig. 3. Differentiation of head and hand region

(ii) *Mouth Tracking*
We extract the skin area for every video sequences. After extracting the head region, we can detect the mouth region. For the context of medication intake scenario, the detection of head region was done using geometric constraints such as the ratio of major to minor axis length of skin region in previous research [2]. When the image becomes larger in the case of closing hands to the camera system, these constraints may lead to wrong detection. In our work, this situation could be avoided by additional centroid information of head. For taking medicine and drinking water actions, mouth can be occluded by hand or bottles. The detection cannot be performed well in such kind of situation. So, Eigen values are computed for mouth detection and tracking.
Eigen Geometry gives two different kinds of geometric transformation and the rotation can be described by angle and axis (quaternion). The quaternion is a complex number system and can be represented in the following.

$$q = a + bi + cj + dk \tag{1}$$

where a, b, c, d are real numbers and i, j, k are the fundamental quaternion units and there are only three degrees of freedom which represent the rotation [8]. Geometric transformation contributes the transformation matrix which is obtained by mapping the greatest number of point pairs between before occlusion and occluded images. In this paper, we use the 'Similarity Matrix' and its Eigen values can be described by $\{1, se^{i\theta}, -se^{-i\theta}\}$ as a rotation and isotropic scaling [9]. In this way, we perform the mouth tracking as shown in Fig. 4. The orientation tracking by Eigen values is shown in Fig. 5. Our proposed method gives the promising results for face orientation. Eigen value detector gives the orientated Bounding Box because of moving Eigen values.

(iii) *Bounding Box Localization*
In this experiment, we use the rectangle box information for all tracing object. For face tracking with Eigen values, we get the orientated bounding box information. The orientated box has the rotation of θ degree and is described by 2D axis information with 4 corner points (8 coordinates) as shown in Fig. 6(b). Generally, the rectangle box has 4 coordinates (x, y, height, width) as shown in Fig. 6(a). Therefore, normalized rectangle box is converted to get the same dimensions of all bounding box information. These rectangle locations are

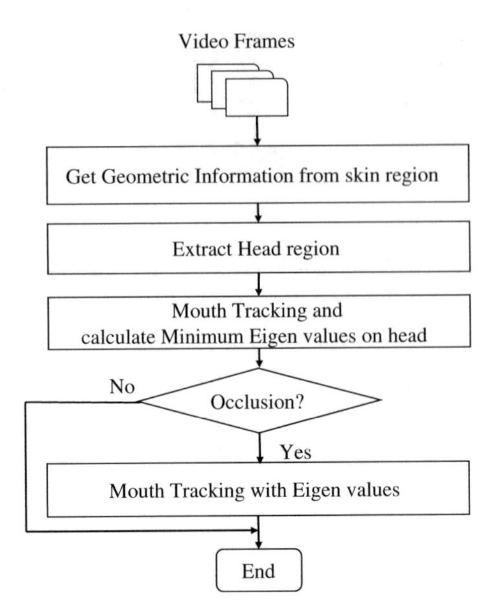

Fig. 4. Flowchart for mouth tracking

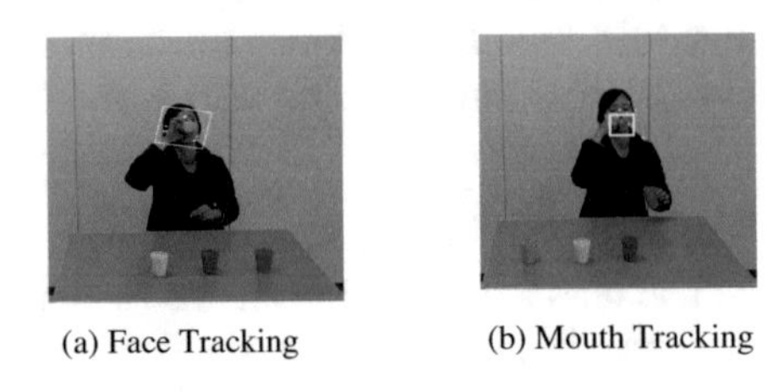

(a) Face Tracking (b) Mouth Tracking

Fig. 5. Tracking with Eigen values

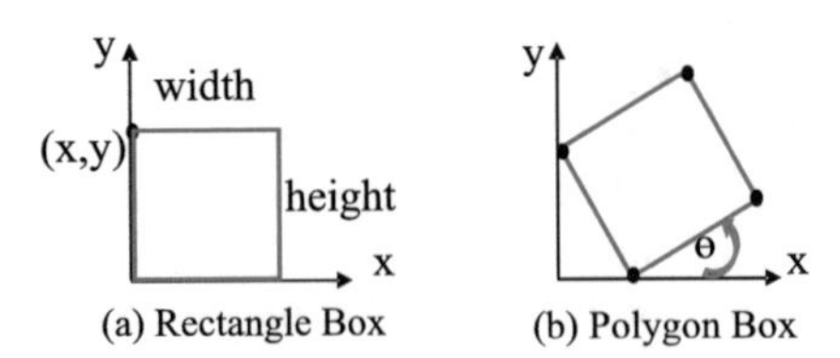

(a) Rectangle Box (b) Polygon Box

Fig. 6. Bounding box rotation

tracked for every video sequence. Such kinds of information are then used to calculate the feature vectors for the simple actions classification.

3.3 Feature Extraction

The overlapped-area ratios of Bounding Box (BB) are mainly used as the feature vectors for every video sequence. Overlap Ratio can be calculated by the area ratio of intersection between BB_A and BB_B, divided by the area of the union of the two [10].

$$OverlapRatio = \frac{BB_A \cap BB_B}{BB_A \cup BB_B} \tag{2}$$

Here, 11 feature vectors are introduced to the training section for the interaction between skin region and color objects. 8 feature vectors for "hands to four objects", 2 vectors for "hands to mouth", 1 for "water bottle to mouth" can be obtained per frame. These features are input patterns to the neural network.

3.4 Decision Method for Simple Action

Pattern Recognition Neural Network (PRNN) is used to classify for each simple action. It is a supervised learning technology and the network is trained by Scale Conjugate Gradient Training function. The network makes the rules and decision making. And, a reasonable answer is provided for all possible inputs which take into account their statistical variation. After learning the patterns of behavior, it will give what type of actions performed.

In this system, a single hidden layer with 20 neurons worked well. A training pattern is an input vector p with $(p_1, p_2,\ldots, p_{11})$ whose desired output vector t *with* $(t_1, t_2,\ldots, t_7)$. Figure 7 shows the network architecture of our experiment.

4 Experimental Result

The PRNN is used to train for 7 types of simple actions as shown in Fig. 8. These 7 short terms actions (A1, A2, A3, A4, A5, A6 and A7) are described as follows.

Action 1: Normal
Action 2: Pick Green Medicine bottle
Action 3: Pick Red Medicine bottle
Action 4: Pick Blue Medicine bottle

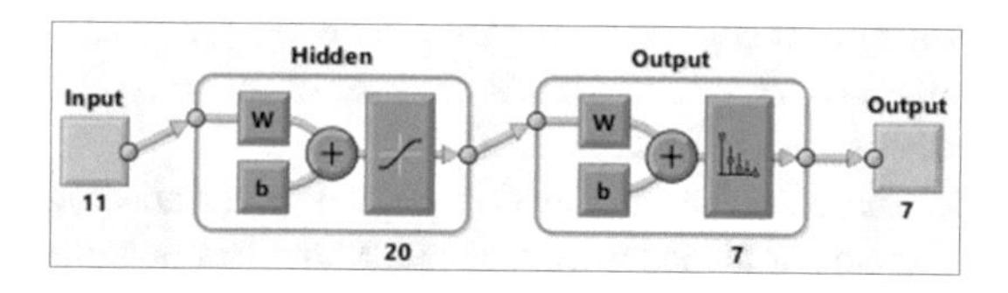

Fig. 7. Neural network architecture

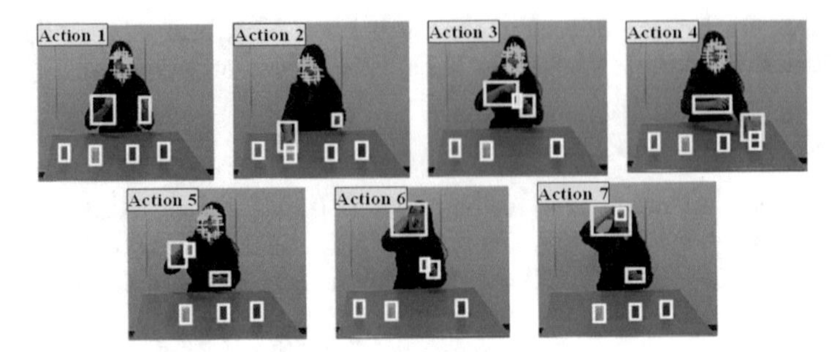

Fig. 8. Simple actions of "hand to object" and "hand to mouth" interaction

Action 5: Pick Yellow Water bottle
Action 6: Pick and Take Medicine bottle
Action 7: Drink Water

We used the hand region extracted from skin extraction. Therefore, there may be incorrect intersection between hand region and objects when the person wears short-sleeve shirt condition as shown in Fig. 9. In this condition, there is wrong recognition with red and blue bottle because of the hand area including arm region although the color detection is correct.

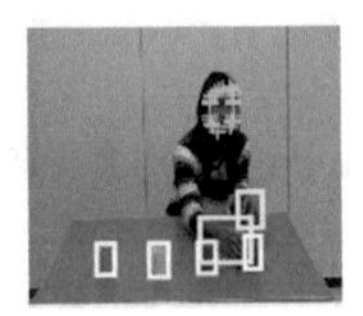

Fig. 9. Wrong- recognition of medication bottle picking

Simple action recognition of medication intake scenario was tested with 4 video sequences by different persons. Video sequence 1 and 2 are with long-sleeve shirt and sequence 3 and 4 are with short-sleeve shirt. SONY 4K FDR-AX100 video camera is used in our experiment.

Figure 10(a) shows the action recognition results with long-sleeve shirt. There has been wrong results in some number of frames with "pick up blue" to "take medicine" and "pick up red" to "normal" in frame number 26 and 52. Before taking the medicine, there are miss-classification with "still picking up medicine" to "take medicine" in frame number 58, 59, 63, 75 and 89. Figure 10(b) gives the results with short-sleeve shirt. The system wrongly recognize with "pick blue" to "red" for initial pick up and put down in frame number 63, 64, 78 and 79 because of the arm region. There has been another similar errors as shown in Fig. 10(a).

By summing up, the color bottle and skin detection are absolutely correct. Mouth tracking is almost correct for all video sequences. Some errors happened before taking medicine. And, when one hand was hidden behind the object, there was some mistakes.

But it does not badly influence the complete medication intake scenario. The proposed system gives the overall accuracy of 90% for simple action recognition. The following Table 1 shows the actions recognition results of testing video sequences.

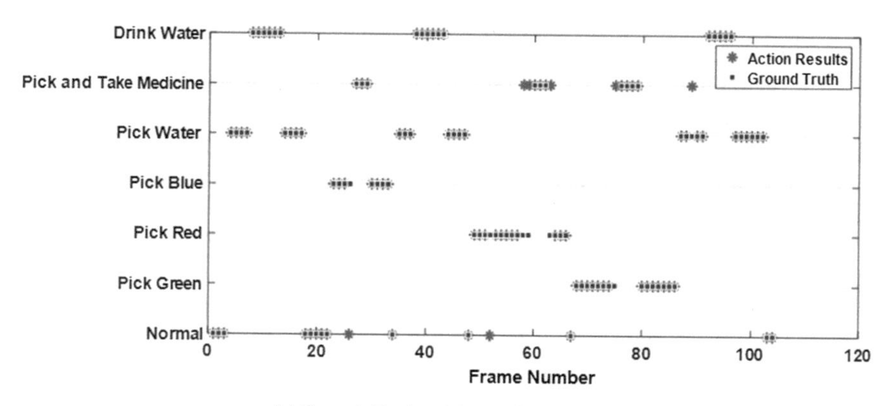

(a) Ground- Truth and Action Results of Video Sequence 1

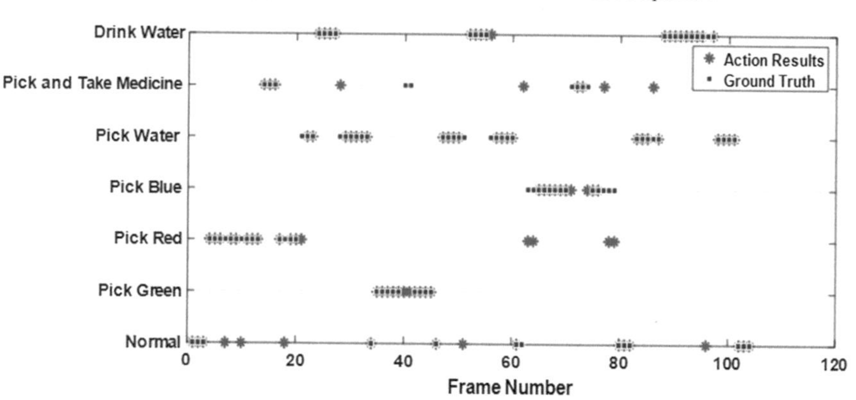

(b) Ground- Truth and Action Results of Video Sequence 3

Fig. 10. Recognition results of medication intake scenario

Table 1. Correct recognition rate of simple actions

Video sequences	Correct rate
Sequence 1	93.27%
Sequence 2	94.54%
Sequence 3	83.65%
Sequence 4	88.76%

5 Conclusion and Future Works

The nursing home monitoring system is performed to monitor the actual act of medication intake of aged person. On the given constraints of medication environment, we proposed the simple short term actions recognition. The important restriction is the collaboration of user who doesn't wear the clothes having high intensity value of color bottles. This paper presented the differentiation and tracking of face and hand region from skin region and color bottles for the context of medication intake. There is no specific pre-training for such kind of tracking. PRNN classifier was simply designed to realize the system model. Our result gives the good performance and erroneous results sometimes caused by initial picking up bottle in short- sleeve shirt and other hidden situations.

In our future work, we will do to observe using two-camera system. Also, hand region detection needs to be more robust for accurate features. To improve the accuracy of simple action recognition, we will add more video training data sources because different sequences of style and action occur in different person and they can also have the different speed for medication intake. By analyzing the statistics of each sequence of simple actions with Markov chain model using deep learning, "complete event of medication intake or not" will be checked.

Acknowledgment. This work is partially supported by the Grant of Telecommunication Advanced Foundation.

References

1. Huynh, H.H., Meunier, J., Sequeria, J., Daniel, M.: Real time detection, tracking and recognition of medication intake. World Academy of Science, Engineering and Technology, pp. 280–287 (2009)
2. Ammouri, S., Bilodeau, G.-A.: Face and hands detection and tracking applied to the monitoring of medication intake. In: IEEE Canadian Conference on Computer and Robot Vision, pp. 147–154 (2008)
3. Batz, D., Batz, M., Lobo, N.D.V., Shah, M.: A computer vision system for monitoring medication intake. In: 2nd Canadian Conference on Computer and Robot Vision, pp. 362–369 (2005)
4. Valin, M., Meunier, J., Starnaud, A., Rousseau, J.: Video surveillance of medication intake. In: 28th IEEE Annual International Conference on Engineering in Medicine and Biology Society, pp. 6396–6399 (2006)
5. Rui, Y., Chen, Y.: Better proposal distribution: object tracking using unscented particle filter. In: Proceedings of IEEE Computer Society Conference on Computer Vision and Pattern Recognition (CVPR), vol. 2, pp. 786–7934 (2001)
6. Zho, Y., Song, Z., Wu, X.: Hand detection using multi-resolution HOG features. In: IEEE International Conference on Robotics and Biometrics (ROBIO), pp. 1715–1720 (2012)
7. Maheswari, R.: Korah: enhanced skin tone detection using heuristic thresholding. Biomed. Res. **28**, 4147–4153 (2017)
8. Quaternion. https://en.wikipedia.org/wiki/Quaternion. Accessed 11 Feb 2018

9. Hartley, R., Ziserman, A.: Multiple View Geometry in Computer Vision, 2nd edn, pp. 37–64. Cambridge University Press, New York (2003)
10. Bboxoverlapratio. https://uk.mathworks.com/help/vision. Accessed 12 Jan 2018
11. Kriesel, D.: A brief introduction to neural networks. Scalable and Generalized Neural Information Processing Engine, pp. 80–90 (2005)

A Study on Detection of Abnormal Behavior by a Surveillance Camera Image

Hiroaki Tsushita[1] and Thi Thi Zin[2(✉)]

[1] Graduate School of Engineering, University of Miyazaki, Miyazaki, Japan
hcl1046@student.miyazaki-u.ac.jp
[2] Faculty of Engineering, University of Miyazaki, Miyazaki, Japan
thithi@cc.miyazaki-u.ac.jp

Abstract. At present, an enormous amount of accidents and terrorisms has been occurred all over the world not only Japan. Due to the spread of security cameras, the number of occurrences of theft and robbery incidents has been decreasing more and more. Nonetheless, the arrest rate has not improved so much and improvement and rising of the arrest rate are required.

The objective of this paper is detection of snatching that involves an event between two persons, and we made an effort to detect snatching in various kinds of situations by using some video scenarios. This video scenarios include the scene of snatching with a bicycle and the scene of non-snatching with normal pedestrian passing. Our proposed methods consist of several steps: background subtraction, pedestrian tracking, feature extraction, and snatch theft detection. We focused on the feature extraction process in details and used weighted decision fusion system based on these parameter, area feature, motion feature, and appearance feature in the paper [1]. We attempted to detect the snatching event from diverse features.

Keywords: Snatching · Feature extraction · Weighted decision fusion

1 Introduction

1.1 Background

It is said that the total population of Japan will be reduced to 86.74 million in 2060. Furthermore the elderly population which is over 65 years old is estimated to increase to approximately 40% [2]. Snatching is a criminal act that occurs frequency comparing with other serious criminals, and it is characterized by that socially vulnerable people such as the elderly and women are easily targeted.

In this paper, we aimed to develop a surveillance camera system that automatically detects the occurrence of snatches on the street. Surveillance cameras have been wide spreading, and active detection of incidents is also required for cost reduction of security problems.

T. T. Zin and J. C.-W. Lin (Eds.): ICBDL 2018, AISC 744, pp. 284–291, 2019.
https://doi.org/10.1007/978-981-13-0869-7_32

1.2 Objectives

In this paper, we made the system of detecting the snatch theft. The main purpose is to detect the theft such as snatching in various situations. Moreover we would like to detect the events of snatching with bicycle and the events of snatching in various environments in real time from the occurrence situations of snatching recently [3].

We believe that development of this research is necessary because it leads to cost reduction of security problems and prevention of accident incidents.

2 Proposed Method

In this section, we proposed an effective method for detecting snatch thieves. The overall system flowchart is shown in Fig. 1. Our proposed system consists five steps to detect snatching event. Moreover we introduced the point system to detect snatching event efficiently by using three feature parameters, area, motion, and appearance. We need to find out its characteristics, identify it as abnormal behavior and recognize it. These features are used in this study to detect whether snatching event has occurred and to distinguish between normal pedestrian and suspicious person.

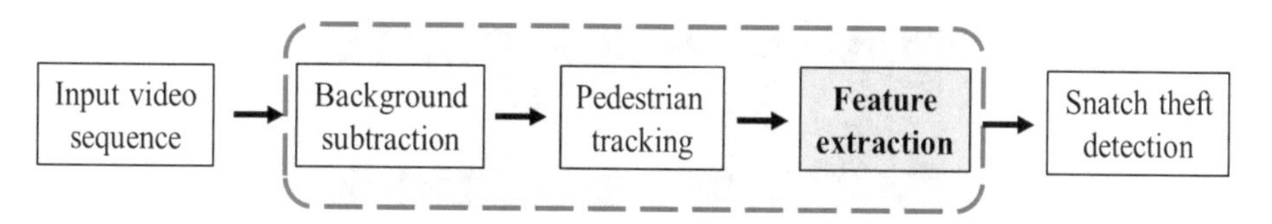

Fig. 1. The overall flowchart to detect snatch theft

2.1 Background Subtraction

Foreground image is made by difference between background and input image. It is not accurately extracted if we calculate the amount of white pixel without removing of the shadow of person silhouette [4]. That's why we cut the silhouette about 10% from the bottom of bounding box (BB). The process result is shown in Fig. 2.

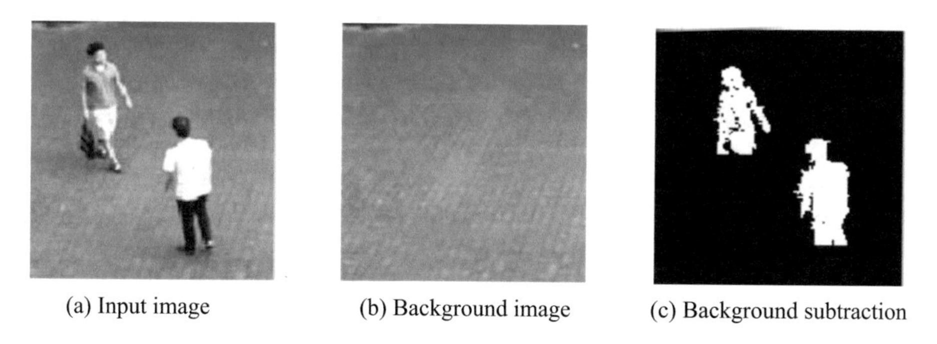

Fig. 2. The process of background subtraction and cutting the shadow

2.2 Feature Extraction

This step is main process in our proposed system to detect the snatching. As mentioned in some of the previous chapters, it is possible to judge abnormal behavior by extracting a person from the input image and considering the characteristic of the person [5].

We compared with the intersection events before and after, and the change amount compared with before and after, is used as feature value (FV). We describe these features in details from following.

Spatial Feature

This feature value is most weighted in our point system. This feature is used to gain the information in which direction the two persons in the video frame are moving. Also we focused only on frames after intersection has occurred in this area feature [6]. We divided the area where a person moves in eight areas (Fig. 3). Then if the moving directions of the two persons after the intersection are the same, it is identified that there is a high possibility that snatching has occurred in the spatial feature.

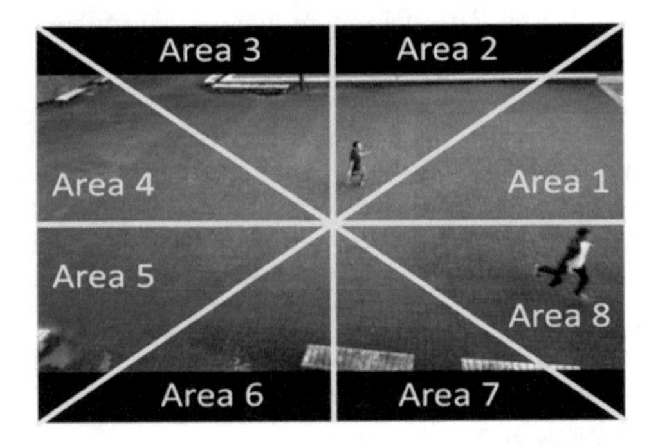

Fig. 3. The frame is divided into eight areas

Motion Feature

The feature of the movement used this time is not just the moving velocity but the moving acceleration of a person. In the case of a person who is just walking in the frame, no significant change in the moving acceleration can be seen because of walking at a constant speed in the frame. On the other hand, after a suspect rubs a bag, its feature value of acceleration is estimated to take high value because a suspect runs away. The difference of acceleration between walking and running person is shown in Fig. 4. Moreover we focused on the frames before and after intersection in this feature. The average values are taken using the feature values every five frames before and after intersection. The equation for calculating the feature value, FV_M of the moving acceleration for every five frames is shown in below.

(a) walking pattern

(b) running pattern

Fig. 4. The difference in acceleration between walking and running person

Appearance Feature

The reason for adopting the feature of appearance originally is that the silhouette of the bag increases and decreases in that case the snatching event happens. The histogram of before and after snatching event is shown in Fig. 5. Even with appearance feature, we focused on the frames before and after intersection as well as the feature of motion. Also we take the average value of each before and after intersection by using the feature value every five frames. The equation for calculating the feature value, FV_A of the appearance for every five frames is shown in below.

$$FV_A = \frac{total\ number\ of\ white\ pixel\ in\ bounding\ box}{height\ of\ bounding\ box} \tag{2}$$

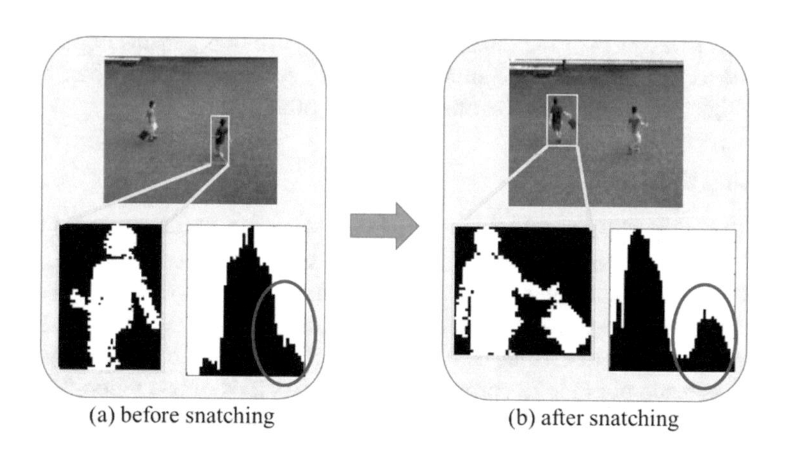

(a) before snatching (b) after snatching

Fig. 5. Comparing with the histogram in a victim case

2.3 Weighted Point in Our System

As a method of determining the specific gravity, it was decided considering the occurrence situation and various environments snatching event happens in diverse scenes [7]. Our proposed system has 10 points in total, and we classified into three outputs “Snatching”, “Potential Snatching”, “Non Snatching”. The flowchart of classification is shown in Fig. 6.

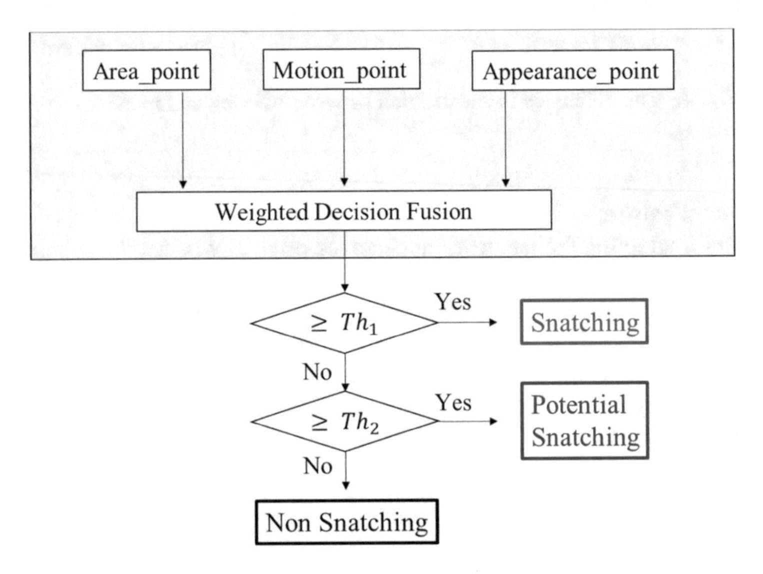

Fig. 6. The flowchart of classification into each output

In area feature, we supposed to add 5 points if the moving direction of two persons is the same area. On the other hand, we add 4 points if the each moving direction is the adjacent area. In motion feature, 3 points are added if the feature value of the motion of either person is larger than the threshold value (Th_M) in two persons appearing in the video sequence. In appearance feature, 2 points is set to be added when the feature value of either person is more than the threshold value (Th_A).

2.4 Setting the Threshold

We used 4 training videos to set the threshold in this thesis. The contents of the events that occur within the training videos are as follows: snatching with bicycle, snatching without bicycle, non-snatching (crossing each other), and non-snatching (passing the bag). The state of the events that occur in the training videos is shown in Fig. 7. By including all these different kinds of scenes in the training videos and setting threshold values, it was made possible to correspond to various kinds of test video. Each feature value in each of training videos is shown in Table 1. The following Eqs. (3) and (4), were used to calculate the thresholds of the feature of motion and appearance as Th_M and Th_A.

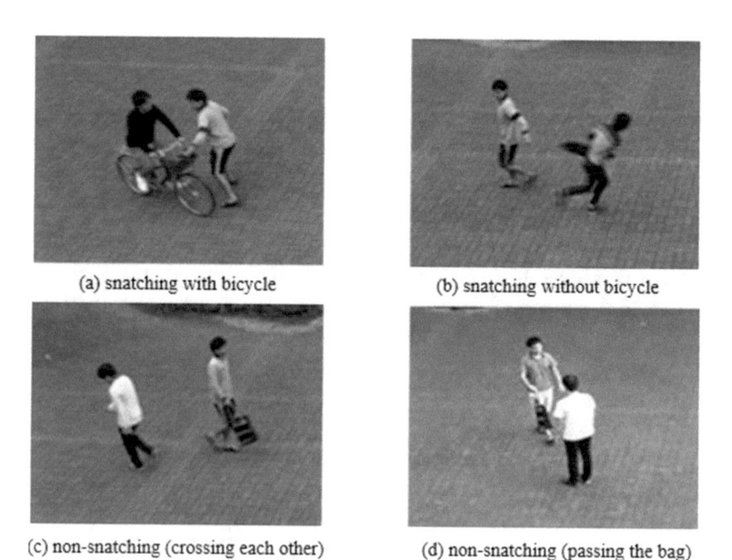

(a) snatching with bicycle (b) snatching without bicycle

(c) non-snatching (crossing each other) (d) non-snatching (passing the bag)

Fig. 7. The state of the events in 4 training videos

Table 1. Each feature value in 4 training videos

Data	Person	Area	Motion	Appearance
Training 1	Suspect	4	0.051	16.9
	Victim (with bicycle)	4	0.111	-17.9
Training 2	Victim	1	0.034	-12.1
	Suspect	1	0.164	26.8
Training 3	Pedestrian A	1	0.034	12.2
	Pedestrian B	5	0.054	-14.5
Training 4	Pedestrian A	5	0.064	-0.4
	Pedestrian B	1	-0.020	-1.1

□: Data with snatching □ : Data without snatching

$$Th_M = \frac{\sum_{i=1}^{8} FV_{Motion(i)}}{8} = 0.061 \tag{3}$$

$$Th_A = \frac{\left|\sum_{i=1}^{8} FV_{Appearance(i)}\right|}{8} = 12.8 \tag{4}$$

3 Experimental Results

In this experiment, the total number of videos used is 19 videos. They consist 4 videos for training to determine the thresholds and 15 videos for test whether snatching is detected correctly or not. The test data contained 9 video data with snatching events, and 6 video data without snatching events. The video data with snatching includes scenes using bicycles and scenes where snatching was occurred between humans. On the other hand, the video data without snatching includes scenes where two persons pass each other and scenes of handover of bags. Thus we examined the snatching detection in various situations, and gained correct classification result in this paper. The result is shown in Table 2.

Table 2. The classification result in total test video data

	Classification Result			Total
	Snatching	Potential Snatching	Non Snatching	
Video with snatching	8	1	0	9
Video without snatching	0	0	6	6

4 Conclusion

We have proposed and tried the methods to effectively detect snatching event in the modern life where the surveillance cameras are spreading more and more. As the method of introducing the point system, we conducted research to help solve the snatching incidents that occur in various environments and situations. Detection of snatching was attempted by using mainly three kinds of feature parameters. Also taking into consideration the features before and after intersection of two persons, we focused on the amount of change in each feature parameters.

In recent snatching cases, there are many different types of snatching cases such as using cars and motorcycles. It is also predicted that snatching event will occur not only during the day time but also in the evening and night time. Thus solving the theft case is a demand in the future, it becomes a challenging research area.

References

1. Penmetsa, S., Minhuj, F., Singh, A.: Autonomous UAV for suspicious action detection using pictorial human pose estimation and classification. Electron. Lett. Comput. Vis. Image Anal. **13**(1), 18–32 (2014)
2. Development of a High-Performance Next Generation Intellectual Security Camera System for Automatic Detection of Crime Situation. https://kaken.nii.ac.jp/ja/grant/KAKENHI-PROJECT-26350454/. Accessed 25 Dec 2017

3. Ibrahim, N., Mokri, S.S., Siong, L.Y., Mustafa, M.M., Hussain, A.: Snatch theft detection using low level features. In: Proceedings of the World Congress on Engineering, London, UK, vol. 2, pp. 862–866, July 2010
4. Tsushita, H., Zin, T.T.: An effective method for detecting snatch thieves in video surveillance. In: International Conference on Artificial Life and Robotics (ICAROB 2017), Miyazaki, Japan, pp. 303–306, January 2017
5. Chang, Y.-H., Lin, P.-C., Jeng, L.-D.: Automatic motion trajectory analysis for dual human interaction using video sequences. Int. J. Comput. Electr. Autom. Control Inf. Eng. **9**, 1294–1301 (2015). World Academy of Science, Engineering and Technology
6. van Huis, J.R., Bouma, H., Baan, J., Burghouts, G.J.: Track-based event recognition in a realistic crowded environment. In: Proceedings of SPIE, vol. 9253, pp. 92530E-2–92530E-7. Copyright Society of Photo-Optical Instrumentation Engineers (SPIE), The Netherlands, September 2014
7. Yang, H., Du, Q., Ma, B.: Weighted decision fusion for supervised and unsupervised hyperspectral image classification. In: Proceedings of IEEE International Geoscience and Remote Sensing Symposium (IGARSS), Honolulu, USA, pp. 875–879, December 2010

A Study on Detection of Suspicious Persons for Intelligent Monitoring System

Tatsuya Ishikawa(✉) and Thi Thi Zin(✉)

Graduate School of Engineering, University of Miyazaki, Miyazaki, Japan
hl12001@student.miyazaki-u.ac.jp,
thithi@cc.miyazaki-u.ac.jp

Abstract. Currently, surveillance cameras are proliferating for prevention of crimes worldwide and early detection of emergency situations, and they play a very important role in the field of crime prevention and verification against various crimes. With regard to crime recognition and crime arrest, the number of surveillance cameras has been on the rise since it became widespread, and this has also led to crime prevention. However, in most cases, it will be coped after the occurrence of crime, and as for ongoing surveillance for 24 h, the burden on the surveillance side is heavy and there are cases where suspicious people are overlooked. In this paper, by focusing on the action of "Loitering" performed by a criminal using various characteristics of a person, it is possible to automatically determine whether the target person is a "Normal Pedestrian" or "Suspicious Pedestrian". We will develop algorithms to make it recognizable and confirm the usefulness in terms of crime prevention and crime verification. It is also expected that establishing detection technology will contribute to crime reduction as a deterrent against crime.

Keywords: Crime prevention · Loitering · Surveillance system
Suspicious person

1 Introduction

Currently, terrorist attacks and suspicious cases are taking place all over the world, and in turn, monitoring is carried out 24 h a day in places where security and crime prevention measures are strengthened [1]. With the spread of surveillance cameras, the number of criminal suspects successfully apprehended has been increasing. Although there is a monitoring method by UAV (Unmanned Aerial Vehicle), surveillance cameras are more prevalent [2]. These important offenses do not occur suddenly, and are in many cases planned crimes. In the planning stage, in order to perform preliminary inspection and confirm a crime plan on the target crime scene, the perpetrators often go back and forth the aforementioned place, which is a suspicious behavior from an objective point of view [4].

Note: Reference [3] should be mentioned before Reference [4].

Various crimes have been carried out from suspicious behavior, and a typical case of this is the "Boston marathon bombing terror case". This incident occurred in Boston, Massachusetts, USA, where a bombing occurred at around 14:45 (local time) on April

T. T. Zin and J. C.-W. Lin (Eds.): ICBDL 2018, AISC 744, pp. 292–301, 2019.
https://doi.org/10.1007/978-981-13-0869-7_33

15, 2013 (3:45 PM Japan time), which was during the 117th Boston marathon competition. According to the investigation by the US government at the time of the occurrence of this case, the person thought to be a suspect was acting unnaturally compared to the general person. Also, the time of explosion is the time period when the most runners reached the goal in the previous competition, hinting that the criminal had been planning from before the day of the event [5].

If most terrorist attacks and events have this planning phase and it is possible to catch the perpetrator in this planning stage, it is possible to prevent it beforehand, but there are some problems regarding prevention of serious crime. Firstly, there is a limitless pattern of human behavior, and it is difficult to recognize which behavior is suspicious behavior among them. Secondly, it is desirable to arrest a target as a suspicious individual before a crime occurs, but in the case of surveillance with the current 24-h system, it will be a huge burden to the monitoring side. Moreover, if it is not visually recognized, it will be difficult to determine whether the person is suspicious or not. The merit of monitoring with a human's eyes is recognition which cannot be done with machines will be possible, but as a disadvantage there is the possibility of overlooking details and false recognition due to the physical load on the monitoring side. Third, there are cases where depending on places of venue, actions taken by suspicious individuals could mean serious crimes in the making.

These perceptions are largely dependent on whether the target site is an important place for crime or not [6]. For example, when a person takes a suspicious action in a specific place such as station home or in front of a bank, he/she recognizes it as a person of caution. However, taking the same action in places that are nothing around does not necessitate special attention. As described above, it is necessary to recognize the suspicious person detection by adding factors such as "place", "action", and "time". In response to the above problem, if it is possible to recognize automatically whether the target is a suspicious individual by processing videos from video camera using image processing, it is thought that prevention and deterrence of crime and prevention of the burden on the monitoring side are possible. In this paper, we propose a method for analyzing whether human beings in a camera are Normal Pedestrians or Suspicious Pedestrians.

This paper is organized as the follows: In Sect. 2, we described about some related works and loitering event detection using computer vision and image processing technology. Section 3 explain about the detail of the proposed system and the experimental results are presented in Sect. 4. Finally, the conclusion and some future works are described in Sect. 5.

2 Related Works

In this session, we will explain about some related works concerning with the suspicious person detection using computer vision and image processing technology. In paper [8], the author proposed a robust knowledge discovery tool able to extract queue abnormalities by focusing on the detecting of people who are loitering around the queue and people going against the flow of the queue or undertaking a suspicious path. They first detect and track the moving individuals from a stereo depth map and then automatically learn queue zone by taking into consider the trajectory of the detected mobile by applying soft computing-based algorithm. Then the abnormalities behaviors are detected using the statistical properties on zone occupancy and transition between zones.

In paper [9], the authors proposed a complete expert system by focusing on the real-time detection of potentially suspicious behaviors in shopping malls. Firstly, they performed image segmentation to locate the foreground objects in scene. Then, tracking is performed by using two-steps method in which the detection-to-track is done by Kalman filter with optimization of the Linear Sum Assignment Problem (LSAP). They also handle the occlusion by using the Support Vector Machine (SVM) based on the features of Global Color Histogram (GCH), Linear Binary Pattern (LBP) and Histogram of Oriented Gradient (HOG). Finally, the resultant trajectories of people are used for analyzing the human behavior and identifying the potential shopping mall alarm situations, especially for the places such as shop entry or exit and suspicious behaviors such as loitering and unattended cash desk situations. The experiments are performed on the CAVIAR dataset and the tracking method achieve the accuracy of 85% in occlusion situations.

In paper [5], the authors proposed a loitering detection method by applying an associating pedestrian tracker in public areas. In the proposed method, the authors analyze the spatio-temporal features to detected the loitering people and generate alarm if loitering people are detected. The object identification is performed by applying the mean square error based on the histogram of oriented gradients feature. The tracking is done by using Euclidean distance based on HSI color model and the color and shape difference for each consistent labeling tracking. The performance evaluation is performed on the PETS2007 dataset and got the 75.45% averaged recall rate and 87.12% averaged precision rate.

3 Proposed System

The system flow of a general video surveillance system is shown in Fig. 1. When detecting a suspicious individual or a crime, a person is extracted by performing "background subtraction" from "input image". Thereafter, "feature extraction" and "tracking" are performed, and "behavior recognition" of the target person is performed to determine the target's action. Although this system is a system for finding offenders, there are several kinds of crimes in a single word even though it is a crime. Here, it is classified roughly into the following three groups. There are 3 kinds of suspicious event detections: "caused by human-object interaction", "one human being", and "multiple human beings" [7].

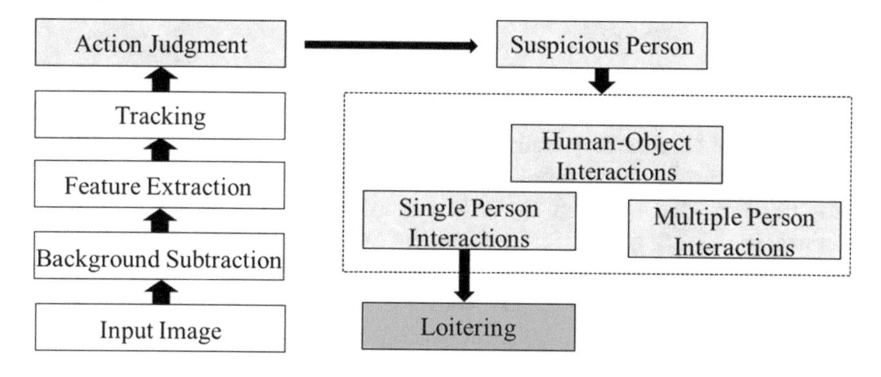

Fig. 1. General video surveillance system

In this proposed system, we will focus on the detection of "Loitering" person who is consistently travel around the site for a long time. This action is an action that frequently occur at the planning stage before causing the crime and also difficult to recognize at first glance. Therefore, the detection of this behavior is a big step towards prevention of crime. Since this research is composed as part of the overall video camera system research, in this paper we will perform processes beginning from "post-tracking" to "behavior recognition" to determine whether the target person is a susof the target personof the target personpicious pedestrian or not. The proposed system is composed of four main analyzing processes for the detection of suspicious person. The first one is grid-based analyzing which use the position information. The other are direction-based, distance and acceleration based analyzing processes which can generate the important score for the detection of suspicious person. Figure 2 shows the system flow of the proposed system.

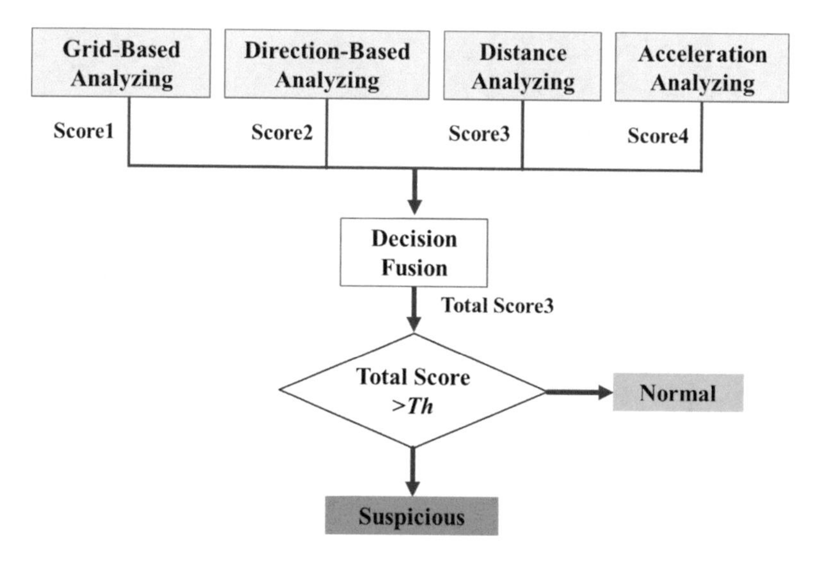

Fig. 2. System flow of the proposed system

3.1 Grid-Based Analyzing

Loitering person is a person who keep walking, stopping and go around the spot for a long time for some purpose. Especially in city streets, most of the pedestrians never stay in that place because they walk to a destination. However, suspicious individuals tend to walk about in the vicinity in order to ascertain the people and surroundings around the site. This becomes as an important factor for removing the ordinary pedestrians. In order to get that information, we used the grid-based analyzing method. Firstly, we divide the incoming video frame into 25 blocks by vertically and horizontally dividing the image with 5 divisions with respect to the original size as shown in Fig. 3. Then, find the block number in which the coordinates of the feet of the target person is located and count the frequency of the same block number for every frame. In order to distinguish the suspicious pedestrian and normal pedestrian who stop and wait for some reason, we defined the threshold filter. When the frequency of the count of the block increases, the filter is applied in eight peripheral directions of the block as shown in Fig. 4.

1	2	3	4	5
6	7	8	9	10
11	12	13	14	15
16	17	18	19	20
21	22	23	24	25

Fig. 3. Example of block classification

1	2	3	4	5
6	7	8	9	10
11	12	13	14	15
16	17	18	19	20
21	22	23	24	25

1	2	3	4	5
6	7	8	9	10
11	12	13	14	15
16	17	18	19	20
21	22	23	24	25

Fig. 4. Example of the filter

Then, scoring is performing over the grid-based analyzing results using the following rules:

- Block numbers whose target block count is less than 30 frames are 0 points per block number.
- For block numbers whose target block count is 30 frames or more and less than 50, 1 block number per block number.
- For block numbers whose target block count is 50 or more, 3 blocks per block number.

Furthermore, when a block number whose target block count is 50 or more is detected, the following score addition is also performed.

- When the target block count is one or more block numbers, count the block numbers in the eight directions around the target block number and calculate 1 point for each block number whose count is 10 frames or more.
- When there are plural block numbers whose target block count is 50 or more Count the block numbers in the eight directions around the block number with the largest count among the target block numbers and calculate the block number 1 One point for each.

3.2 Direction-Based Analyzing

An ordinary pedestrian can make multiple turns in the process of heading to the destination and the direction change can vary depending on the destination and shape of the road. But the normal pedestrians proceed in one direction, so that the frequency of changing the direction is also under limited number. However, in the case of "Loitering", the target person travels back and forth within the sense, so that the number of turning the direction tends to increase more than a normal pedestrian. In order to detect the direction changing, we calculate the angle using Eq. (3.1) based on the coordinates of the target person in each frame.

$$\theta = arctan\frac{y_t - y_{t-1}}{x_t - x_{t-1}} \tag{3.1}$$

We divide the angle of moving direction into two categories: 0 to 180° and 0 to −180° in order to define the positive direction and negative direction respectively. After that, each direction is further divided into four divisions. If the direction of the target person changes from the positive direction to the negative direction or from the negative direction to the positive direction, we count as one turning of direction. After counting the number of turns of the target person, we may determine if target is a normal or suspicious pedestrian using the threshold value. Figure 5 shows the classification of angle of the direction change.

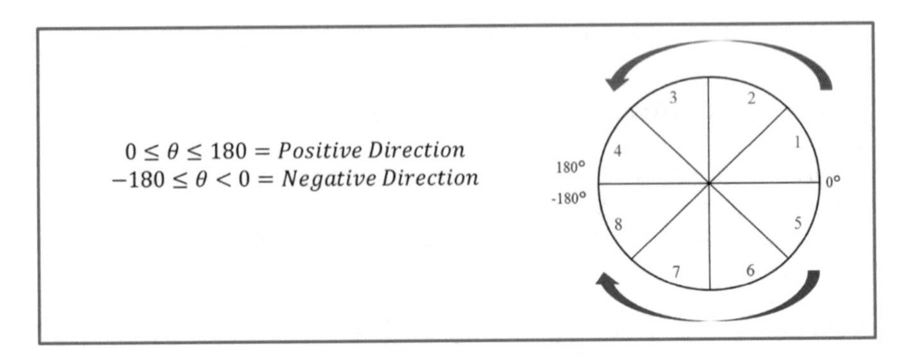

Fig. 5. Angle classification of direction change

The scoring rules for the direction-based analyzing method are as the following:

- When the direction change number of the target person is less than 1, 1 point.
- When the direction change number of the target person is 1 or more and 3 or less, 2 points.
- When the direction change number of the target person is 4 or more, 3 points.

3.3 Distance-Based Analyzing

At the time of the normal pedestrian is passing through the scene of surveillance camera, the distance between the start location and the location of the target after every 10 s is also useful information for detecting the suspicious pedestrian because the distance value of the loitering person can be greater than the normal pedestrian. Even though this distance value can depend on the traveling direction of the target person, for the scene of the public road, this rule is true because all pedestrian are moving along the road. By using this rule, it is possible to analyze whether the target person is walking continuously or not as shown in Fig. 6.

Fig. 6. Example of pedestrian movement

We calculate two kinds of moving distances. The first one is the position $s = (p_1, q_1)$ when the target person entered the frame and the position where it went out of the frame or the position at the last l frame of the video $l = (p_l, q_l)$ (Distance 1). The second distance is the distance between the coordinates $n = (p_n, q_n)$ of the target person every 10 s. We calculate the distance 1 by using the Eq. (3.2) and the distance 2 by using the Eq. (3.3).

$$D_1(p, q) = \sqrt{(p_n - p_{n-1})^2 + (q_n - q_{n-1})^2} \tag{3.2}$$

$$D_2(p, q) = \sqrt{(p_l - p_1)^2 + (q_l - q_1)^2} \tag{3.3}$$

After calculating the distance, the results are analyze using the following rules:

- When the distance value of the target person for every 10 s is less than the threshold value 0 point per value.
- When the distance value of the target person every 10 s is equal to or larger than the threshold value 1 point per value.

3.4 Acceleration-Based Analyzing

When comparing ordinary pedestrians and suspicious individuals, there are differences in terms of distance traveled and direction change characteristics, but there are differences in acceleration as well. A normal pedestrian walks with an acceleration that does not change so much and heads towards a specific destination. However, in the case of "Loitering", the behavior of "stopping pedestrians" is more apparent than that of ordinary pedestrians because they travel back and forth through the place without passing by immediately. Therefore, by repeatedly calculating the acceleration, by comparing the number of times the target person's acceleration has become extremely low with the data of the ordinary pedestrian, it is possible to analyze whether the target person is an ordinary pedestrian or not it can.

3.5 Decision Fusion

After calculating the score for each analyzing method, we perform the decision fusion process by adding the scores of each method. Then, we apply some threshold value in order to make the final decision according to the following rules:

- If the total score of the target person is less than 10, we classify it as "Normal" person.
- If the total score of the target person is 10 or more, we classify as "Loitering" person.

4 Experimental Results

In order to perform the experiment, 6 videos including both normal pedestrians and loitering pedestrian have been taken. Among them, one video is used as the training data for threshold determination, data is taken from one "suspicious pedestrian (Loitering Person)" and five "ordinary pedestrians (Normal)". After determining the threshold, the remaining five were used as testing data whether the person in the video is either "Suspicious Pedestrian" or "Normal Pedestrian". Examples of video images actually used are shown in Fig. 7(i) and (ii). Experimental results are shown in the Table 1.

(i)

(ii)

Fig. 7. Examples of video images

Table 1. Overall experimental result

	Conventional method [8]	Proposed method
Loitering (Accuracy)	100%	100%
Normal (Accuracy)	91.0%	100%
Number of subjects (Loitering)	1	4
Number of subjects (Normal)	9	27

5 Conclusion

In this paper, we conducted behavior detection of the target person focusing on the act of "suspicious pedestrian" in the crime scene, in particular "wandering". For the behavior detection, it is recognized whether or not the target person is a "suspicious pedestrian" by performing four analyzes "block division analysis", "direction change analysis", "movement distance analysis", and "acceleration analysis", and finally compared the results of the experiment with the results of the conventional method. Compared to the conventional method, we were able to improve the number of people, such as recognition of two "Loitering" behaviors, recognition of 18 "Normal" behaviors. In the future, if it is possible to increase the flexibility of multi-adaptability of this algorithm, we can judge the "suspiciousness" of the target person in any situation, various cost reductions, early detection of crime, prevention of prevention. We believe that it can contribute not only to important crimes such as terrorism but also to the deterrence of familiar minor offense.

References

1. Zin, T.T., Tin, P., Hama, H., Toriu, T.: A triplet Markov chain model for loitering behavior detection. ICIC Express Lett. Part B Appl. **6**(3), 613–618 (2015)
2. Chau, D.P., Bremond, F., Thonnat, M.: Object tracking in videos: approaches and issues. In: The International Workshop "Rencontres UNS-UD", Danang, Vietnam, arXiv:1304.5212 [cs.CV], 18 April 2013 (2013)
3. Zin, T.T., Tin, P., Toriu, T., Hama, H.: A Markov random walk model for loitering people detection. In: Proceedings of 2010 Sixth International Conference on Intelligent Information Hiding and Multimedia Signal Processing, Darmstadt, Germany, pp. 680–683 (2010)
4. Wikipedia, Boston marathon bombing terrorism incident, (Updated on 22nd December 2017) (2017). (https://ja.wikipediadia.org/wiki/%E3%83%9C%E3%82%B9%E3%83%88%E3%83%B3%E3%83%9E%E3%83%A9%E3%82%BD%E3%83%B3%E7%88%86%E5%BC%BE%E3%83%86%E3%83%AD%E4%BA%8B%E4%BB%B6)
5. Nam, Y.: Loitering detection using an associating pedestrian tracker in crowded scenes. Multimed. Tools Appl. **74**(9), 2939–2961 (2015)
6. Xiao, J., Peng, H., Zhang, Y., Tu, C., Li, Q.: Fast image enhancement based on color space fusion. Color Res. Appl. **41**(1), 22–31 (2016)
7. Ishikawa, T.: A study of detection of suspicious person using image processing techniques. Miyazaki University Faculty of Engineering Department of Electrical System Engineering Graduation thesis, February 2016
8. Patino, L., Ferryman, J., Beleznai, C.: Abnormal behaviour detection on queue analysis from stereo cameras. In: 12th IEEE International Conference on Advanced Video and Signal Based Surveillance (AVSS), pp. 1–6, August 2015
9. Arroyo, R., Yebes, J.J., Bergasa, L.M., Daza, I.G. Almazán, J.: Expert video-surveillance system for real-time detection of suspicious behaviors in shopping malls. Expert Syst. Appl. **42**(21), 7991–8005 (2015)

A Study on Violence Behavior Detection System Between Two Persons

Atsuki Kawano[1(✉)] and Thi Thi Zin[2(✉)]

[1] Graduate School of Engineering, University of Miyazaki, Miyazaki, Japan
hl14010@student.miyazaki-u.ac.jp
[2] Faculty of Engineering, University of Miyazaki, Miyazaki, Japan
thithi@cc.miyazaki-u.ac.jp

Abstract. Lately, surveillance cameras have been widely used for security concerns to monitor human behavior analysis by using image processing technologies. In order to take into accounts for human rights, costs effectiveness, accuracy of performance the systems so that an automatic—human behavior analytic system shall be developed. –In particular, this paper focused on the action of two person violence and detecting two person fighting each other will be considered. Some experimental results are presented to confirm the proposed method by using ICPR 2010 Contest on Semantic Description of Human Activities (SDHA 2010) dataset.

Keywords: Action detection · Morphology processing · Violence detection
Security camera

1 Introduction

1.1 Background

In late years the violence in the station yard particularly the violence for the station employee has been brought into question. There have been 800 cases in only the violence for the station employee in West Japan Railway in 2014. When it is midnight time zone, the assailant side often drinks, and approximately 70% is a wicket and a home. After the correspondence by the security camera setting was carried out, too, but a case happened by the recording of the camera; is coped, and it is difficult to prevent a case. In addition, there is a problem of fatigue of a problem and the watchdog of personnel expenses, the oversight by the negligence when a human being watches it using a security camera. Therefore, it thinks that it can be for one method to solve problems such as reduction or the oversight of the staff of the watchman to perform the real-time detection. In addition, even if a watchman cannot rush, I am careful by light or a sound and rouse it, and it thinks that it is helpful for decrease of the violence to put the poster of a purpose detecting violence automatically in real time [1–3].

T. T. Zin and J. C.-W. Lin (Eds.): ICBDL 2018, AISC 744, pp. 302–311, 2019.
https://doi.org/10.1007/978-981-13-0869-7_34

2 Proposed Method

Recently, deep learning is popular, but the processing for an animation and a large quantity of images is unsuitable because processing to perform the deep learning for a still image is the center, and a great deal of training data are necessary. Therefore, we suggest technique to classify actions in without using deep learning this time. In order to handle the animation obtained from the security camera every frame, we extract quantity of characteristic from the shape of a provided silhouette image and classify actions. We consider three kinds of Hugging, Handshaking, Pointing, six kinds in total as three kinds of Punching, Pushing, Kicking, non-violence as violence.

2.1 Background Subtraction

First we compare two images of the input image which a background image and the person whom a person does not enter, and the background difference share is technique to make the image which has been taken only the part different between the images by performing a difference share, and appeared. Figure 1(a) shows an input image Fig. 1(b) shows a background image. We used a color image, gray image, three kinds of images of the edge image this time to perform a difference share more exactly.

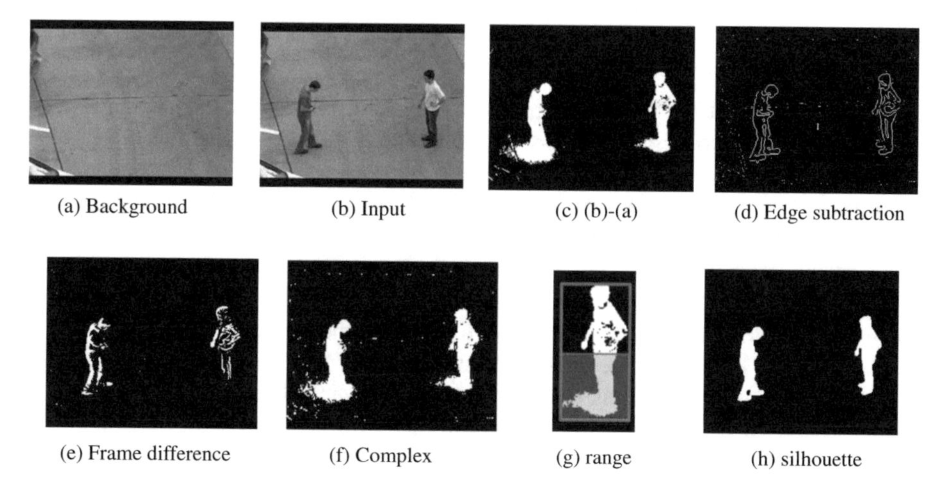

(a) Background (b) Input (c) (b)-(a) (d) Edge subtraction

(e) Frame difference (f) Complex (g) range (h) silhouette

Fig. 1. The way to generate a silhouette image

With the background differences and Color image minute, three dimensions Euclid distance of this RGB color space for the pixel of the same position of a background image and the input image of the color image are performed to make the new image which is to be substituted the value for. This result is expressed with gray image, and a difference is white as a small pixel, and a difference becomes black as a big pixel as shown in Fig. 1(c).

The kind of the image is different from the background differences and Gray image share in a method same as a background differences and color image share. This makes the threshold equally, too and performs a background difference share.

An edge image is the image which picked up the part that a color greatly changes. The extracting method of the edge included various methods, but extracted an edge using the zero-crossing method this time. With the zero-crossing method, a value is technique to detect 0 (zero) points from the result of a second Laplacian filter differentiating it for the value of a pixel next to each other. When the second differential expression to use for a Laplacian filter assumes the value of the pixel of x line eyes f(x),

$$\mathrm{f}^{\wedge\prime\prime} = (\mathrm{f}(\mathrm{x}+1) - 2\mathrm{f}(\mathrm{x})\mathrm{f}(\mathrm{x}-1)) \tag{1}$$

Image of the edge is shown in Fig. 1(d).

Then, we perform the difference share between the frames. The difference share between the frames is technique to take the differences between input image and input image minute before 1 frame. I can in this way check the quantity that moved between 1 frames. I show the result for the difference between the frames in Fig. 1(e). We put three kinds of background difference shares and difference shares between the frames that we made together and make one piece of image until now. The complex image described in Fig. 1(f).

2.2 Noise and Shadow Processing

The image of Fig. 1(f) includes many noises. Firstly, we confirm the volume of all noises by labeling processing, and size removes a small noise than the threshold. When a person may look small to you in the input image and establishes the threshold to delete a big noise, a person may disappear. Therefore, you must make the threshold of the degree not to disappear even if a person looks small to you. As a result, we delete it using aspect ratio because a big noise is left. Mostly BB becomes lengthwise, but removes only the label which is against the silhouette of the person type clearly when 2 people come into contact to have possibilities to become oblong.

And, next is Shadow processing. The image of Fig. 1(g) contains many noises. To overcome this problem, labeling processing is used to calculate the volume of all noises and removing noises that are smaller than threshold value. When setting the threshold value to remove big noise and the person may look small in the input image, the person may disappear because it is considered as noise. Therefore, it is important to set up the appropriate threshold value for getting the actual foreground object. The shadow is removed by using aspect ratio because shadow noise is located in the left side of the BB. And removing only the label which is not the actual silhouette of the object, while the two people come into contact.

2.3 Morphology Processing

As for the image which removed a shadow, there is a hole in the silhouette of the person, and a running part is seen. Therefore, we tie a missing part and handle morphology to cover it. The morphology processing [4, 5] is a generic name of the shape-based image processing operation in a process to handle an image. Firstly, we performed Majority processing to tie a missing part. With the Majority processing, we make 3*3 filter and, in eight pixels of circumferences of the leading pixel, substitute the value of a certain pixel for a leading pixel more than five pixels. I work to cover it next. By the stopgap processing, we handle close to bury a part surrounded by white pixels in. By the stopgap processing, we used close processing to bury a part surrounded by white pixels in. By the above-mentioned processing, a clean silhouette image can be obtained. The silhouette image is shown in Fig. 1(h). In addition, this silhouette image is used to extract quantity of characteristic using this image.

2.4 Estimate a Background

To perform a background difference share, it is necessary for the background of the input image to agree with a background image by all means. Because we cannot cope with a change of the light quantity and the change of the shadow to an input image for fixing a background image, it is necessary to estimate a background image and update it.

We prepare for number of update sections frame to estimate a background and find the median of each pixel. By estimating the background image without the person we can in this way make it because we can substitute the value that appeared a lot in update section for a pixel.

However, a person is recognized with a background when we stay at the same position with frames more than half between the update sections. By substituting a pixel level of the background image for the position corresponding to the white pixel of the synthetic image of Fig. 1(h) of the input image to solve this. Even if a person remains with the majority frame of the update section by using this technique, only a part remaining in can refer to a background before one of the update and can estimate a background with characteristic quantity extraction, and versatility becomes higher.

3 Extract Quantity of Feature

3.1 The Area of the Silhouette Image

We calculate the size of the label of the silhouette image to count how many people enter the frame. It cannot grasp the number of people well when we make the threshold the fixed number when it is near when a person is far from a camera. Therefore at first divide the area of the person by the longitudinal length of BB and am normalized. Then decide the threshold using the maximum of the y coordinate of 2 people next. We understand that y coordinates increase when based on the left of the image so that a person is near to a camera. If far-off at the greatest area for the one in the value of a normalized area when it is the nearest being about 48 pixels in the case of this time S

$$S = -a(y - b)^{\wedge}2 + c \tag{2}$$

We can express this formula (a, b and c is constant). The flexibility of the distance conditions from a camera to a person increases by deciding the threshold dynamically.

There is the number of white pixels in BB in the case beyond S two. In other words we can think that we are coming into contact.

3.2 Make a Waist Image

By extracting the position of the waist to know the position of 2 people we surround both in BB and pull out the position of the approximate waist and assume it an interest domain. There is the interest domain of this time in the range of Fig. 2(a). However, the position of the waist of the each 2 people is delicious and is not detected because only one BB is made when 2 people came into contact. The state that 2 people came in contact with is shown in Fig. 2(b). Therefore it is necessary to put out a contact part between 2 people so that BB two make it when 2 person intervals touched it. When 2 people come into contact, we calculate the grand total of the longitudinal white pixel and can put out the connection between people of 2 by making all the values of the line that a value is small 0 (black). Figure 2(c) shows the grand total of the longitudinal white pixel. It is removed all other than a top and bottom part of the center of gravity of the body by this processing, and the number of the labels may increase. Therefore we can turn off a useless label by the size of the label extracting only two maximum things. I show the image which I made in Fig. 2(d) and call it a human trunk image. We make BB for this human trunk image and quite take out the height thought to be the waist as an interest domain. We can demand the position of each waist of 2 people by demanding the center of gravity. The image which extracted only the waist is shown in Fig. 2(e) and call it a waist image.

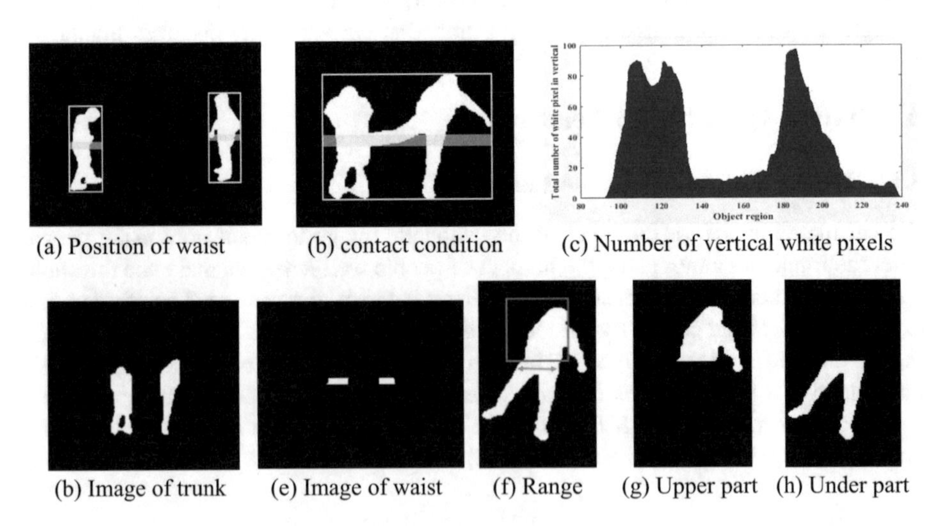

(a) Position of waist (b) contact condition (c) Number of vertical white pixels

(b) Image of trunk (e) Image of waist (f) Range (g) Upper part (h) Under part

Fig. 2. Extract quantity of feature

3.3 Make a Lower Part of the Body Image, An Upper Body Image

We make BB only for lower parts of the body. It is right, and half of the bottoms of the silhouette image cannot calculate it only by making BB when we put up a case and a foot lowering the hand. For a silhouette image, we remove an upper part than a waist image by slightly bigger width a little than the BB width of the waist image to solve this point. The range that removed is shown in Fig. 2(f). The foot can in this way delete the upper body when we raise a foot without disappearing. However, we cannot turn off the handles which protrude aside when it is this method. Therefore we calculate the size of the label again and can solve the size of the label in the silhouette image by leaving only the biggest thing. The resulting image is described in Fig. 2(g) and call it a lower part of the body image.

We can make the image only for the upper body by taking the differences between silhouette image and lower part of the body image minute next. Figure 2(h) shows the image only for the upper body and call it an upper body image.

3.4 Detect the Position of the Head

It is delicious and cannot detect the position of the head when we consider several percent simply as a position of the heads from BB of the silhouette image as a position of the heads when I give a case and the hand which 2 people touched. Therefore we calculate a center of gravity using a human trunk image as an interest domain in 10% from BB.

3.5 Silhouette Similarity Between 2 People

The nonviolent action often resembles it in the action between 2 people. Therefore when a silhouette came in contact, I divide it to the left person with the right person and calculate a resemblance degree of the form of the silhouette of the person of right and left. We do it with the center between 2 people of the center of gravity of the head which I calculated in 5.4 as a division border of the people of right and left.

We calculate the length of both, the dispersion of the grand total of each lateral white pixel and assume the value that is multiplied by a similar degree. Because we use the product of the dispersion, the value that is almost 0 is provided so that the action of both is similar.

4 Detect an Action

Using quantity of characteristic that is extracted in the above the flow chart of the action detection is described in Fig. 3.

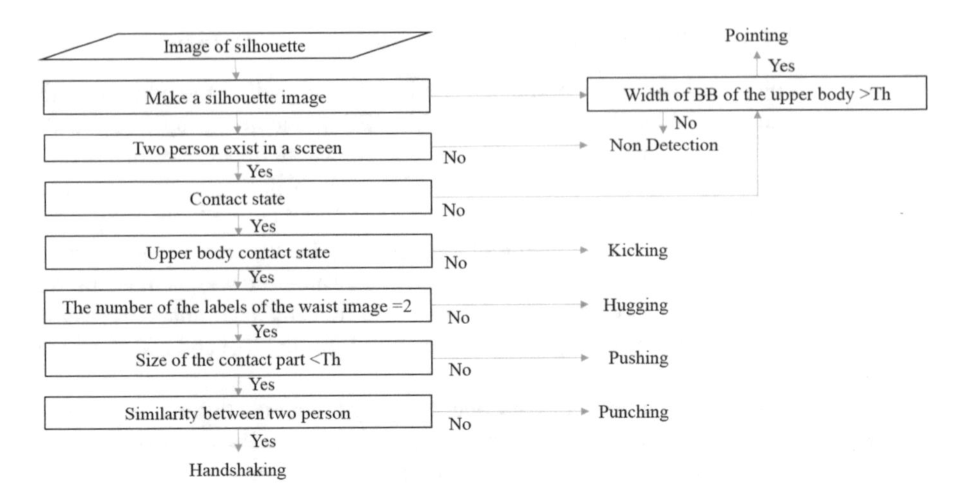

Fig. 3. Flow chart

4.1 Pointing Detection

Pointing does not have the contact of 2 people and has the characteristic that the BB width of the upper body image increases. Therefore, regardless of the number of people in the frame, the BB width of the upper body image defines it as Pointing for a number frame when it is more than the threshold.

4.2 Kicking Detection

When the number of the labels of the lower part of the body image becomes 1 in the state that 2 people came in contact with, Kicking defines it.

4.3 Punching and Pushing Detection

Punching and Pushing calculated the length of the longitudinal arm which it touched and classified it. Pushing comes to have a big total of the length of the longitudinal white pixel to use both hands. On the contrary, in the case of Punching, it becomes small because it is one hand. When it is less than Pushing, the threshold when the length of the longitudinal white pixel is more than the threshold, using this characteristic, we define it as Punching.

4.4 Hugging Detection

There is characteristic that it has a big contact area of the person of 2 in Hugging. Therefore, we can confirm a person of 2 thing needing it in a screen, and the number of the labels of the waist image defines time when it was to 1 as Hugging.

4.5 Handshaking Detection

Handshaking is time when the silhouette of the person of in the state that 2 people come in contact with and right and left resembles it. Therefore, the resemblance degree that we calculated in 5.5 extends over the number frame and defines it as Handshaking when it is less than the threshold.

Note No 5.5 in this paper.

5 Experiment Environment and Evaluation Method

Using "2010 ICPR Contest on Semantic Description of Human Activities" which are available to the public as for the data set that used this time [6]. Videos from Video1 to Video10 are included in this data set. Some of six kinds of actions that they classified this time are random order, and these are included, and a photography place, a person model, clothes, brightness are different each. We used Video1 and Video2 as training data this time and assumed Video10 test data from Video3.

6 Result and Consideration

6.1 Result of Experiment

Firstly we acted with each frame and, about the calculation of the result, classified it. Because a startup and a definition of the action end are difficult, we define the end as one action after the silhouette of both began to come in contact this time and detect an action with each frame. Because the detection may be difficult, I check a result to in front of 4 frames of the input image and, only with 1 frame, detect the action that there was the most. A detected result is right in half or more of the action frame; if can detect it, do it with detection success. O means detection success, × means detection fails, and – meanst the behavior pattern is not included in data set (Table 1).

Table 1. Result

Video	1	2	3	4	5	6	7	8	9	10
Handshaking	O	O	O	O	O	O	O	O	O	×
Hugging	O	O	O	O	O	O	O	O	O	O
Pointing	O	O	O	O	O	O	O	×	O	O
Pushing	O	O	O	O	–	O	O	×	O	O
Punching	O	O	–	O	–	–	O	×	O	O
Kicking	O	O	×	–	O	–	O	O	O	O
Accuracy	100	100	80	100	50	100	100	100	100	83

6.2 Considerations

The problem that is the biggest as a cause of the behavior patterns which was not able to detect is that non-to confirm in a frame person of 2 moves. Particularly, there are many such frames on the video which had low rate of detection. Because we cannot extract a silhouette well when a silhouette image comes in contact with the human being except the object, it is due to not able to extract quantity of characteristic. Figure 4(a) shows an input image and a silhouette image when the person except 2 people was reflected. In addition, the processing of the shadow was not performed without head neighborhood and the silhouette of the right person being fogged, and others being recognized with a lower part of the body in this example.

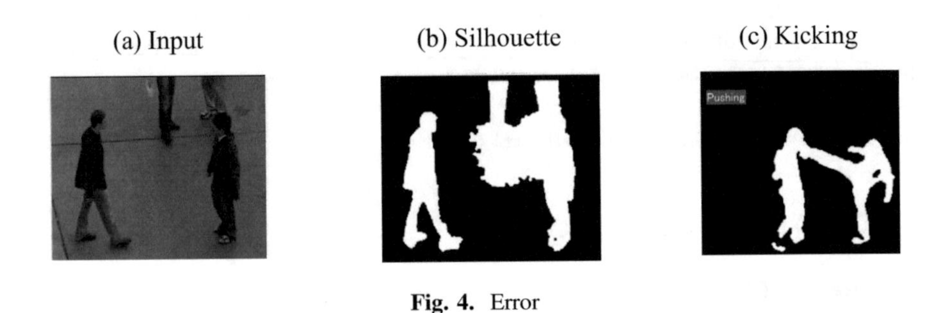

Fig. 4. Error

Because way of kicking was different from other data set, in Kicking of Video3, we were not able to detect it. I was as contact of the lower part of the body by a definition of Kicking, but Punching was detected because of gaving a foot highly in Video3 without the contact of the lower part of the body being detected. A state of Kicking of Video3 is shown in Fig. 4(b).

7 Conclusion

We used image processing by detection of the violence between 2 people in this article. The detection rate when only 2 people began frame was high and knew that this technique was effective as a violence detection method of this case. When there is violence because this technique classifies human actions with 4 frames, it is shortest and performs warning by a sound or the light on the site within one second after contact and can tell a watchman. However, a detection rate when others were reflected greatly decreased. Because various situation including the violence except the person of 2 interval is assumed when non-target person comes out in the real environment, you must improve it.

About the problems that a silhouette is piled up, we photograph it from the higher place and can decrease the possibility by using plural cameras.

It is necessary to examine methods to decide the threshold to use for a method of the processing when the label of the silhouette has been connected with others for the extraction of a more correct silhouette as the future prospects, shadow processing dynamically.

References

1. Yun, K., Honorio, J., Chattopadhyay, D., Tamara, L., Samaras, D.: Two-person interaction detection using body-pose feature and multiple instance learning. In: IEEE Computer Society Conference on Computer Vision and Pattern Recognition Workshops (CVPRW 2012), 16–21 June 2012, pp. 28–35 (2012)
2. Shu, Z., Yun, K., Samaras, D.: Action detection with improved dense trajectories and sliding window. In: Agapito, L., Bronstein M., Rother, C. (eds.) Computer Vision - ECCV 2014 Workshops. ECCV 2014. Lecture Notes in Computer Science, Springer, vol. 8925, pp. 541–551 (2014)
3. Wang, H., Schmidt, C.: Action recognition with improved trajectories. In: Proceeding of IEEE International Conference on Computer Vision (ICCV), Sydney, Australia, pp. 3551–3558, December 2013
4. MathWorks, R2017a. (https://jp.mathworks.com/help/images/morphological-dilation-and-erosion.html). Accessed 30 Jan 2018
5. Chang, Y., Lin, P., Jeng, L.: Automatic motion trajectory analysis for dual human interaction using video sequences. World Acad. Sci. Eng. Technol. Int. J. Comput. Electr. Autom. Control Inf. Eng. **9**(5), 1336–1343 (2015)
6. ICPR 2010 Contest on Semantic Description of Human Activities, Interaction Challenge Sample. (http://cvrc.ece.utexas.edu/SDHA2010/Human_Interaction.html). Accessed 30 Jan 2018

Image and Multimedia Processing

Object Detection and Recognition System for Pick and Place Robot

Aung Kaung Sat(✉) and Thuzar Tint(✉)

Department of Information Science, University of Technology (Yatanarpon Cyber City), Pyin Oo Lwin, Myanmar
aungkaungsat93@gmail.com, thuzartint1984@gmail.com

Abstract. Object recognition system plays a vital role in controlling the robotic arm for applications such as picking and placing of objects. This paper is directed towards the development of the image processing algorithm which is the main process of pick and place robotic arm control system. In this paper, soft drink can objects such as "Shark", "Burn", "Sprite" and "100 Plus" are recognized. When the user specifies a soft drink can object, the system tries to recognize the object automatically. In the system, the target object region and the motion of the object are firstly detected using Template Matching (Normalized Cross Correlation) based on YC_bC_r color space. The detected image is segmented into five parts horizontally to extract color features. In feature extraction step, mean color and Hue values are extracted from each segmented image. And then, Adaptive Neural Fuzzy Inference System (ANFIS) is employed to recognize the target object based on the color features. After recognizing the user specified object, the robotic arm pick and place it in the target region. Experimental results show that the proposed method is efficiently able to identify and recognize soft drink can objects.

Keywords: Template matching · Mean color · Hue values · ANFIS

1 Introduction

Vision based control of the robotic system is the use of the vision sensors as a feedback information to the operation of the robot. Vision based configuration is employed with web camera. With the use of robotic arms, object recognition is the vital for navigating and grasping tasks. The vision system is used to obtain and analyse images to detect objects and compute their location information.

This paper is related to the focus of automatic extraction and analysis of useful information from image sequences taken by using video cameras. To perform automatic analysis tasks such as motion detection, segmentation, features extraction and classification of moving objects on the image sequences, the different kind of computerized video surveillance systems that can automatically perform the above analysis tasks without human effort.

Many researchers and computer scientists are still trying to make their new effective algorithms which can be used in motion detection, feature extraction and other image analysis tasks to build the perfect computerized video surveillance systems. In [1],

T. T. Zin and J. C.-W. Lin (Eds.): ICBDL 2018, AISC 744, pp. 315–323, 2019.
https://doi.org/10.1007/978-981-13-0869-7_35

there are many template matching techniques such as naïve template matching and image correlation matching. The problem that occurs in Naïve Template Matching is in computing the similarity measure of the aligned pattern image. And there are many correlation methods for template matching. Normalized cross correlation is the improved model of the traditional cross correlation methodology. There are many challenges in template matching process such as illumination and clutter background issues. As a result of these challenges, this method detects the wrong objects. Therefore, feature extraction process is used to solve this problem.

In [9], conventional mathematical tools are not well appropriate for solving imprecise systems. By contrast, a fuzzy inference system corresponds to set of fuzzy IF-THEN rules that have learning capabilities to approximate non-linear functions without using accurate analyses. So, the fuzzy inference system is considered to be a universal approximator. Takagi and Sugeno former discovered this fuzzy identification system methodically. The fuzzy inference system can apply various real-world applications in controlling, prediction and implication.

This paper is organized as follow; second section gives overview of system design followed by the third section that describes color space transformation, template matching. Fourth section gives about the feature extraction and fifth section describes the classification processes. The sixth section shows the experimental results based on various testing.

2 System Design Overview

The system is intended to detect and recognize the specified soft drink can object. Figure 1 illustrates the overall procedure of the object recognition system. In the system, there are two main parts such as training and testing. And there are three components in the system such as preprocessing stage, feature extraction stage and recognition stage.

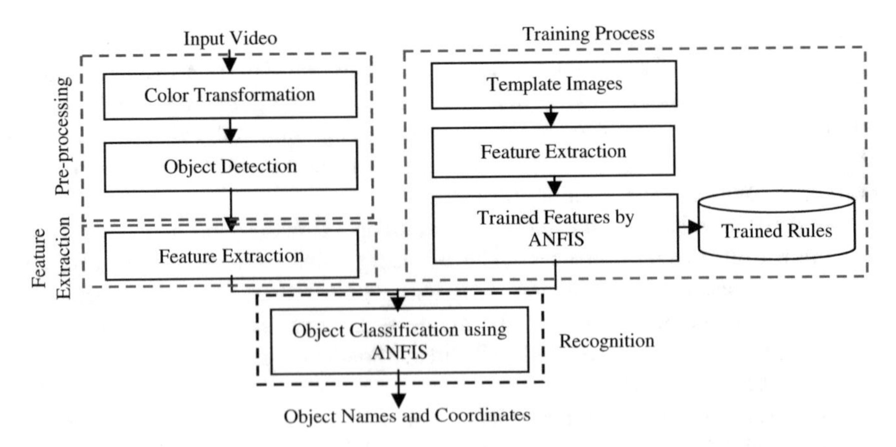

Fig. 1. The overall procedure of the object recognition system

In the pre-processing stage, color transformation process and object detection process are used. In the feature extraction stage, the detected object is segmented into five parts horizontally and then mean and Hue color features are extracted from these five regions. In the recognition stage, ANFIS is employed based on training features and target object features to recognize the specified objects. Finally, the system gives the coordinates (x and y) of the recognized objects.

3 Preprocessing Stage

In the object recognition system, pre-processing stage consists of two main processes. They are color transformation and object detection.

3.1 Color Transformation

In the system, the incoming frame from video camera is RGB color space. The system converted the RGB color frame into YC_bC_r color frame. The YC_bC_r color may be computed from RGB using different forms of transformations. Y is luminance component and C_b and C_r are the blue difference and red difference chroma components.

The transform uses an RGB input value with each component in the range of [0 – 255] and transform it into Y, C_b and C_r, in the ranges [0.0, 255.0], [–128.0, 127.0], and [–128.0, 127.0], respectively. The Y-component is level-shifted down by 128, so that it also falls into the [–128.0, 127.0] range [2].

The conversion equations from RGB to YC_bC_r color transformation is given below.

$$\begin{bmatrix} Y \\ C_b \\ C_r \end{bmatrix} = \begin{bmatrix} 0.299 & -0.168935 & 0.499813 \\ 0.587 & -0.331665 & -0.418531 \\ 0.114 & 0.50059 & -0.081282 \end{bmatrix} \begin{bmatrix} R \\ G \\ B \end{bmatrix} \tag{1}$$

The resultant image of YC_bC_r color space which is displayed in Fig. 2(b) yield reasonable good results. The original and YC_bC_r color transformation of input image is shown in Fig. 2(a) and (b).

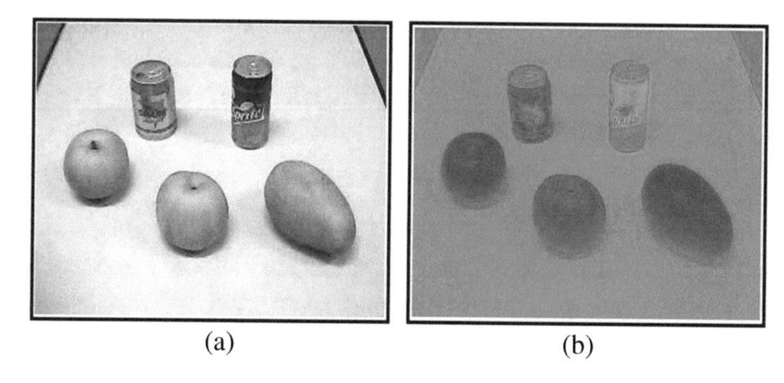

(a) (b)

Fig. 2. (a) Input image (b) YC_bC_r color image

3.2 Object Detection

After applying YC_bC_r color transformation process, template matching approach is applied to detect the specified object region. If a template describing a specific object is available, object detection becomes a process of matching features between the template and the image sequences. Object detection with an exact match is generally computationally expensive and the quality of matching depends on the details and the degree of precision provided by the object template. There are two types of template matching, fixed and deformable template matching. When object shapes do not change with respect to the viewing angle of the camera, fixed templates are useful [3]. In this paper, twenty-eight fixed templates (7*4 = 28 templates) for each soft drink cans (Shark, Burn, Sprite and 100Plus) are used for template matching.

For finding matching points in different images, Normalized Cross Correlation is a standard method. The NCC method is a simple template matching technique that defines the location of a desired form. The template image is matched to all possible positions in a source image by shifting pixel-by-pixel and computes a location that directs how the template best matches the image in that position [4, 5].

Normalized cross correlation uses the following general procedure.

1. Calculate cross-correlation in the spatial or the frequency domain, depending on size of images.
2. Calculate local sums by precomputing running sums.
3. Use local sums to normalize the cross-correlation to get correlation coefficients.

4 Feature Extraction Stage

After detecting the specified object, features extraction stage will be performed. In the feature extraction stage, there are two parts. They are segmentation process and color feature extraction process.

4.1 Segmentation Process

Before extracting the color features, the detected object which is resulted from template matching is segmented into five parts horizontally to extract color features from each segmented image. In this paper, a new image segmentation technique is proposed to improve the accuracy of the recognition process. The detected object or template image and the resultant image of segmentation process which contain the five sub-parts are presented in Fig. 3(a) and (b).

4.2 Color Feature Extraction

After segmenting the detected object, color features are extracted from each segmented region. In RGB color image, a color is composed of three color channels such Red, Green and Blue. For the entire image, there are 3 metrics and each one channel represents color features [7].

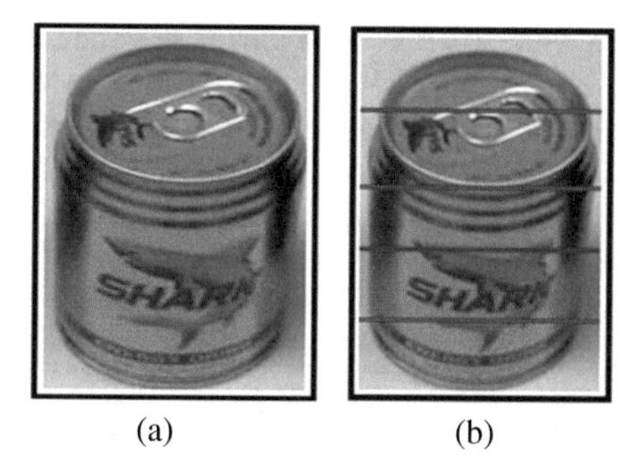

(a) (b)

Fig. 3. (a) Soft drink template (shark) (b) The resultant image of segmented image which contains five sub-parts

The color distribution is the same as a color composition in the probability theory. A histogram can be seen as the discrete probability distribution. The color histogram that represents the joint probability of the intensities of the image is the most well-known color features for feature extraction. From the probability theory, a probability distribution can be considered by its moments. Thus, if a probability distribution is taken by the color distribution of an image, moments can be used to distinguish the color distribution. The first order (mean), the second (standard deviation), and the third order (skewness) color moments have been shown to be proficient in representing color distributions of the images [8]. In this paper, among the three color moments, first order (mean or color averaging) is used to extract the features from the images. If the value of the ith color channel at the jth image pixel is p_{ij}, then the first order (mean) color moment is as follows.

$$E_i = \frac{1}{N}\sum\nolimits_{i=1}^{N} P_{ij} \quad (2)$$

For color image, color moments are very compact representation features compared with other color features since 15 numerical values (3 values for each segment) are used to represent the color content of each image channel. The average color images are shown in Fig. 4(a) and (b).

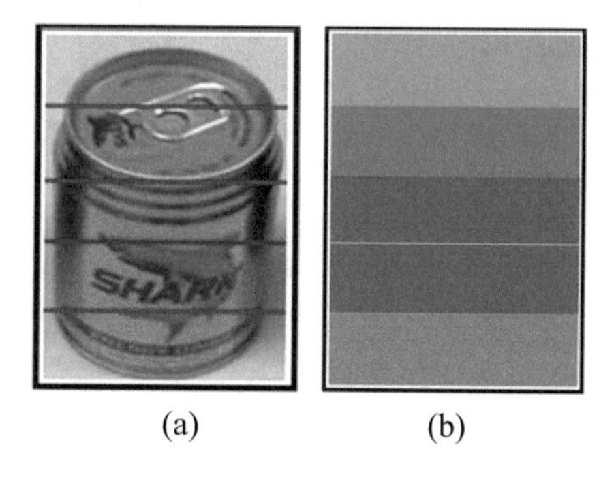

(a) (b)

Fig. 4. (a) The segmented image which contains five sub-parts (b) Mean color values of segmented image (Shark)

These average color values for each segment are transformed into HSV color values. The HSV (Hue, Saturation, Value) color space is a simple transform from the RGB color space, in which all the existing image formats are presented. The HSV color space is a popular choice for manipulating color. So, Hue values which are the representative color values are extracted from each segmented image. Therefore, five hue values are ready to trained and recognized for object classification. The HSV color images for five segmented images are shown in Fig. 5(a) and (b).

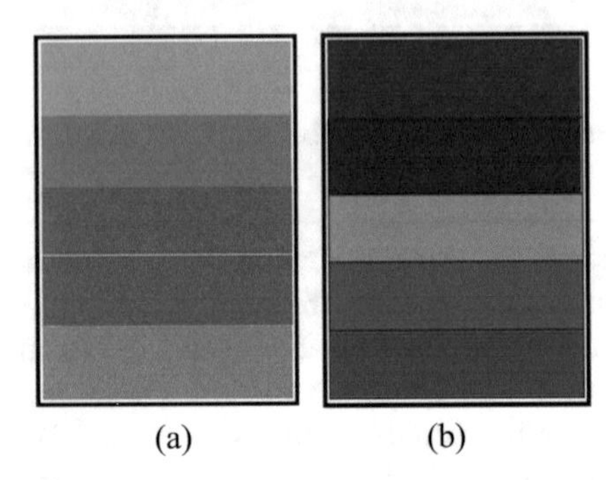

Fig. 5. (a) Mean color values of segmented images (Shark) (b) HSV color values of segmented image (Shark)

5 Recognition Stage

After extracting the color features (five Hue values) from segmented image, the next step is recognition. In this step, the extracted color features are recognized with ANFIS (Adaptive Neural Fuzzy Inference System). Adaptive neuro-fuzzy inference system (ANFIS) is a kind of feed-forward adaptive neural network that is based on Takagi-Sugeno fuzzy inference system. Since it integrates both neural networks and fuzzy logic principles, it has potential to capture the benefits of both in a single framework. Its inference system corresponds to a set of fuzzy IF-THEN rules. So, ANFIS is considered to be universal approximator. For this reason, ANFIS will be much more reliable than any other methods and techniques. Using the ability of ANFIS to learn from training data, it is possible to create ANFIS with limited mathematical representation of the system [9].

In this ANFIS architecture, there are five input values which are Hue color values for each segmented image. And the output is the classified object name (Shark, Burn, Sprite and 100Plus). In this model, three triangular membership functions are used. So, ANFIS model gives 243 trained rules ($3^5 = 243$) for classification.

After the recognition step, the specified object name and coordinates are produced. According to the results, the robotic arm pick and place the user specified object into the target region.

6 Experimental Results

In this section, the proposed system is engaged and evaluated to prove the effectiveness and strength of the proposed method. The proposed system is tested with various objects such as bottles, fruits and soft drink cans.

In this paper, two testing will be described in details. In the testing, the first is for "Shark" object and the second is for "Sprite" object. When the user command "Shark" via the voice, the system will find out the "Shark" object. If the system found this "Shark" object, the information dialog box which contains object ID, object name, Number of objects, X-Y coordinates and Position will be displayed. According to the information dialog box, the robotic arm will pick and place the commanded object into the target region. The results of the system for finding "Shark" object and "Sprite" object are shown in Figs. 6 and 7.

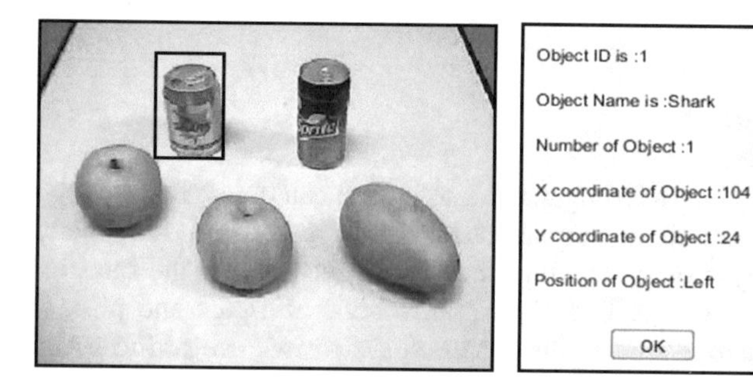

Fig. 6. "Shark" object detection and recognition

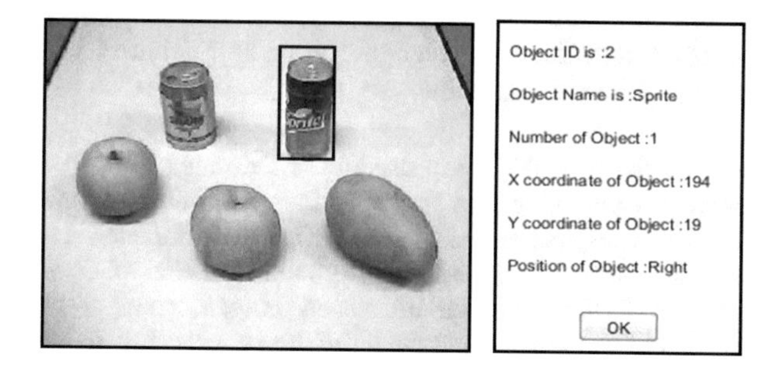

Fig. 7. "Sprite" object detection and recognition

For performance measure for object detection and recognition, the following equations are used:

- Average Process Time (sec) = Total Process Time for each object/Total Number of Images
- Accuracy (%) = ((Total Number of Images – Total Number of Error Images)/Total Number of Images)/100

Total Number of Error Images contain total number of false acceptance images and false rejection images. The experimental result of proposed system is as shown in Table 1.

Table 1. Experimental result of proposed system

Object names	No of frames	No of false acceptance images	No of false rejection images	Average process time	Accuracy (%)
Shark	400	–	2	2.66 s	80%
Sprite	400	–	1	1.48 s	90%
Burn	400	1	1	2.85 s	80%
100 Plus	400	–	1	2.62 s	90%
Total Average Process Time and Accuracy				2.4 s	85%

7 Conclusion

The paper is proposed to recognize the soft drink can object for picking and placing of robotic arm. In the system, template matching approach is used for object detection and ANFIS is employed based on color features for recognizing the specified object. After recognizing the user specified object, the robotic arm pick and place it in the target region. The system evaluated the proposed system with regard to accuracy and computation time. The experimental result shows that the proposed object detection and recognition system for soft drink can object gets 85% accuracy and average process time is 2.4 s for four hundred frames. Though the proposed approach can detect the desired objects in real time, illumination is sensitive to detect the objects when the light is bright or dim. Thus, detection and recognition of soft drink can object of various conditions will be the future work in this area.

Acknowledgments. Firstly, I would like to thankful to Dr. Aung Win, Rector, University of Technology (Yatanarpon Cyber City), for his supporting to develop this system successfully. Secondly, I would like to appreciate Dr. Soe Soe Khaing, Pro-Rector, Leader of Research Development Team for Domestic Pick and Place Robotic Arm Control System, University of Technology (Yatanarpon Cyber City), for her vision, chosen, giving valuable advices and guidance for preparation of this article. And then, I wish to express my deepest gratitude to my teacher Dr. Hninn Aye Thant, Professor, Department of Information Science and Technology, University of Technology (Yatanarpon Cyber City), for her advice. I am also grateful to Dr. Thuzar Tint, Lecturer, University of Technology (Yatanarpon Cyber City), for giving us valuable advices. Last but not least, many thanks are extended to all persons who directly and indirectly contributed towards the success of this paper.

References

1. Swaroop, P., Sharma, N.: An overview of various template matching methodologies in image processing. Int. J. Comput. Appl. (0975–8887) **153**(10), 8–14 (2016)
2. Prentic Hall.: Digital Image Processing: Pratical Application of Image Processing Techniques.
3. Nath, R.K., Deb, S.K.: On road vehicle/ object detection and tracking using template. Indian J. Comput. Sci. Eng. **1**(2), 98–107 (2010)
4. Briechle, K., Hanebeck, U.D.: Template matching using fast normalized cross correlation, Institute of Automatic Control Engineering, Germany
5. Singh, S., Ganotra, D.: Modifications in Normalized Cross Correlation Expression for Template Matching Applications, GGS Indraprastha University, Delhi, India
6. Alattab, A.A., Kareem, S.A.: Efficient method of visual feature extraction for facial image detection and retrieval. In: Proceedings of 2012 Fourth International Conference on Computational Intelligence, Modelling and Simulation (CIMSiM), pp. 220–225 (2012)
7. Venkata Ramana Chary, R., Rajya Lakshmi, D., Sunitha, K.V.N.: Feature extraction methods for color image similarity. Adv. Comput. Int. J. (ACIJ) **3**(2), 2 (2012)
8. Stricker, M., Orengo, M.: Similarity of color images. In: Proceedings of SPIE Storage and Retrieval for Image and Video Databases III, vol. 2420, pp. 381–392, February 1995
9. Roger Jang, J.-S.: ANFIS: adaptive network-based fuzzy inference system. IEEE Trans. Syst. Man Cybern. **23**(3), 665–685 (1993)

Myanmar Rice Grain Classification Using Image Processing Techniques

Mie Mie Tin(✉), Khin Lay Mon, Ei Phyu Win, and Su Su Hlaing

Myanmar Institute of Information Technology, Mandalay, Myanmar
{mie_mie_tin, khin_lay_mon, ei_phyu_win, su_su_hlaing}@miit.edu.mm

Abstract. Modern technologies are being used in agriculture such as quality control and classification of grains that are very important for more productive and sustainable production. Classification of the similar small rice grains can be also made with the help of image processing techniques. This paper studies different characteristics of Myanmar rice grains and their varieties. The classification of various varieties of rice grains is made by using image processing techniques and algorithms. Five types of rice grains in Myanmar such as Paw San Hmwe, Lone Thwe Hmwe, Ayeyarmin, Kauk-Nyinn-Thwe and Kauk-Nyinn-Pu are considered for present study in classifying the rice seeds and quality. Firstly, each grain image is preprocessed to enhance the grain image and then segmented by using the edge detection methods such as threshold method. Five morphological features are extracted from each grain image. This system emphasizes on the development a computer vision-based system that is combined with proper heuristic algorithms for automatic classification of Myanmar's rice grain samples. This research is very significant in Myanmar because Myanmar is great producer of different qualities of rice grains and therefore the study and basic implementation would greatly help the researchers, agriculturist and other stakeholders of agricultural growth.

Keywords: Image processing · Enhancement · Segmentation · Classification Rice grain · Myanmar

1 Introduction

Rice is the main staple food in Myanmar and is grown on over 8 million ha, or more than half of its arable land. Myanmar is the world's sixth-largest rice-producing country. Therefore, in our country, it is very important to improve and become automated these agricultures by using the advanced and new technology [8].

The analysis of grain quality and types can be rapidly assessed by visual inspection of experts [8]. However, as the different shapes and appearances of rice samples, specialists may have difficulties to manually identify and classify the various rice grains. Therefore, it is still a challenging task to select and find a particular type of rice among varieties of rice grains [7].

In the present work, a digital image processing has been devised in order to investigate different types of characteristics to identify the rice varieties. In this system,

T. T. Zin and J. C.-W. Lin (Eds.): ICBDL 2018, AISC 744, pp. 324–332, 2019.
https://doi.org/10.1007/978-981-13-0869-7_36

images of five different varieties of rice samples are captured by using Canon PowerShot SX60 HS camera. Each grain image is segmented by using the edge detection methods. After image segmentation, the primary five features are created based on some shape, size and color features which are the quality indices to distinguish rice among bulk of rice samples. To have high classification accuracy, it is necessary to select the best classifier. This study emphasizes to develop a computer vision-based system combined with appropriate meta heuristic algorithms for automatic recognition and classification of bulk grain samples.

2 Related Work

Zhao-yan et al. (2005) suggested an image analysis based research work to identify six types of Zhejiang rice grain. This system was based on seven colors and fourteen morphological features to classify and analysis the rice seeds. In this system, 240 kernels were used as the training data set and 60 kernels as the testing data set in neural network. The identification accuracies were between 74% and 95% for six varieties of rice grain. The robustness was however missing in this research [3].

Ozan AKI et al. (2015) studied the classification of rice grain by using image processing and machine learning techniques. In this study, four types of rice grain in Turkish were considered for classifying. Each grain image was segmented and six attributes were extracted from each grain image related to its shape geometry. Weka application was used for evaluation of several machine learning algorithms. For real time quality assessment of rice grain, nearest neighbor with generalization algorithm achieved the classification accuracy 90.5% [4].

Kaur et al. (2015) proposed a method that extracted seven geometric features of individual rice grain from digital images. The varieties of rice grains were classified into three different classes. Calibration factor was calculated to make a method without depending of camera position. In this method, it was tested on five varieties of grain. The proposed method compared with the experimental analysis that was used by digital vernier caliper and the error rate of measuring the geometric features of rice grains was found between −1.39% and 1.40% [10].

Birla et al. (2015) presented an efficient method for the quality analysis of Indian basmati Oryza sativa L rice grain by using machine vision and digital image processing. Machine vision proved as an automatic tool for reliable quality assessment of seeds instead of the analysis of human inspectors. This proposed algorithm used the perfect threshold, object detection and Object classification to calculate the number of chalky rice and broken rice with the improved accuracy [6].

Silva Ribeiro (2016) proposed to use methods of data analysis of shape, color and texture extracted from digital images for grain classification. From the results obtained it was demonstrated that the use of patterns of morphology, color and texture extracted from images using the digital imaging processing techniques are effective for grain classification. The LBP texture pattern proved the most efficient information among the three, and with it alone was possible to reach a 94% hit rate. Combining addition to the pattern shape of LBP information with FCC and color with HSV was possible to improve the success rate to 96% [11].

Based on extensive literature survey of classification of different types of grains, both national and international level; few major challenges were reported as limitations in existing research contributions, few of them are:

- In most of the works, the implementation is not robust;
- There is less work for color images of rice grains; and
- Most importantly, limited research was found on Myanmar rice grains.

The above limitation, especially last one has been major factor for taking up the present research work as classification of Myanmar rice grains.

3 Myanmar Rice Grain

Actually, there are varieties of rice grains in Myanmar and in fact, Myanmar exports huge quantity of rice to other neighboring counties. Though, few prominent rice grains have been considered for this research work so that generalization of this work could be applied to develop a framework of image processing tools or techniques for classification of rice grains. Five varieties of Myanmar rice grains such as Paw San Hmwe, Lone Thwe Hmwe, Ayeyarmin, Kauk-Nyinn-Thwe and Kauk-Nyinn-Pu were used for classification in this study. The rice grain samples were collected from a local market. Grain samples are shown in Fig. 1.

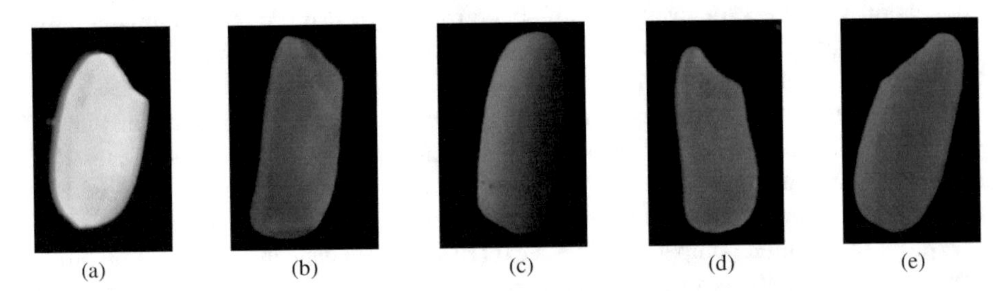

Fig. 1. Myanmar rice grains: (a) Paw San Hmwe, (b) Lone Thwe Hmwe, (c) Ayeyarmin, (d) Kauk-Nyinn-Thwe and (e) Kauk-Nyinn-Pu

4 The Proposed Method

The classification of rice grain requires several important stages of image processing, such as image acquisition; image enhancement; image segmentation; image feature extraction and classification. The stages can be seen in Fig. 2. These stages are briefly explained below.

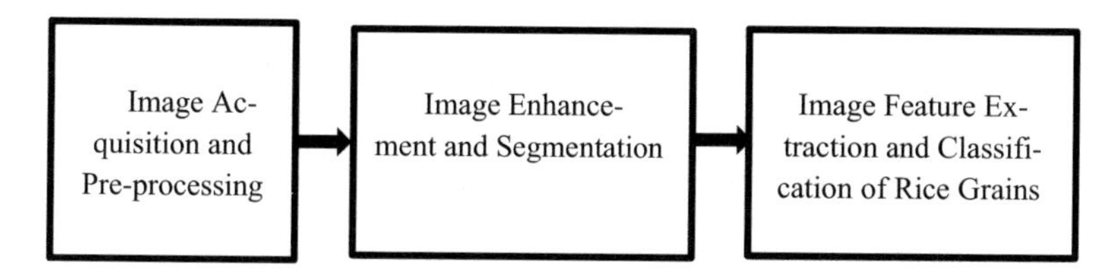

Fig. 2. System flow for rice grain classification

Image Acquisition and Pre-processing: In this study, image acquisition is the first step. The image acquisition was collected by using Canon PowerShot SX60 HS camera under uniform lighting setup. Rice grain is placed on the black sheet of paper to get the black background to the image that is used to helps in parameter extraction from the image. The camera is placed at the fixed location and mounted on stand to get grain images. Images were captured and stored in JPG format. Color representation is RGB type and horizontal and vertical resolution is 180dpi. All the grains in the sample image were arranged in arbitrary direction and position.

Image Enhancement: After acquisition step, all images will inevitably contain some amount of noise. Image enhancement step improves the visual quality and clarity of images. Firstly, the grain image is converted to gray scale image. A median filter is a non-linear digital filter and is very effective in removing salt and pepper noise. Therefore, median filter with 5×5 kernel size is used in this present work as the pre-processing to smooth and remove noise from each image. Sobel edge detection technique is also used to preserve the edges of the image during noise removal. Then binary image is produced by using convolution method with proper creation mask. Optionally, image opening operation is applied for break the touching grain images.

Image Segmentation: The subsequent step is to segment an image which is the most important stage in image analysis. Image segmentation is that the image is subdivided into different parts or objects [2]. It can also be accomplished by using three different techniques such as edge detection, region growing and threshold. In this study, threshold is used for image segmentation. It is the simplest image segmentation method. Image binarization process is performed by using threshold value. Threshold is used to segregate the region in an image with respect to the object which is to be analyzed. This separated region is based on the variation of intensity between the object pixel and the background pixel. After separating the necessary pixels by using the proper threshold value, the binary image is produced as shown in Fig. 3.

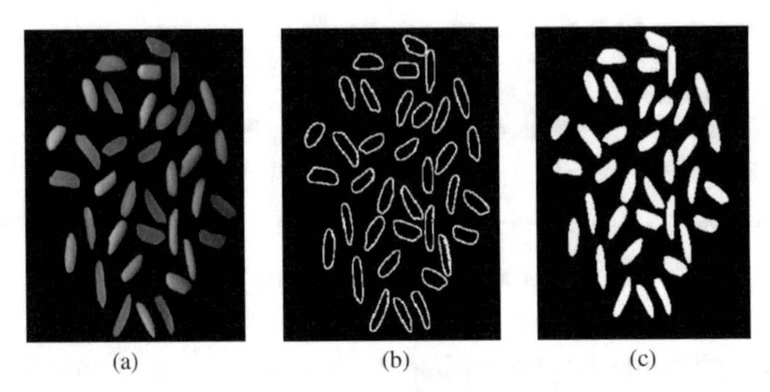

Fig. 3. Steps of rice grain classification: (a) samples of Myanmar grain that are mixed five different types of rice varieties, (b) boundary detection of rice varieties, (c) threshold image of rice varieties

Feature Extraction: To get quantitative information from segment objects, features extraction is very important. The object classification and object extraction for image processing is based on that features extraction process [1]. System considered five features to classify the type of Myanmar rice in mixture different rice. This system considered the area feature, major axis length feature, minor axis length feature, eccentricity feature and perimeter features.

Area (cm2) Feature: This system measured the number of total pixels of boundary and inside the rice grain object.

Major Axis Length Feature: That distance is assumed as a largest length of the rice grain. That length is longest length of that rice grain and is measured end point of two boundaries. Compute the pixels value along the length and that value will become as one feature to consider grains type.

Minor Axis Length Feature: That feature is perpendicular line with major axis. That line referred the width of that rice grain. The pixels count of that line will become one feature.

Eccentricity Feature: The eccentricity feature is the ratio of the distance between the foci of the rice grain ellipse shape and its major axis length. That feature values refer between 0 and 1.

Perimeter Feature: That feature is considered the total pixel count of that rice grain object. That feature can help to classify the types of different rice grains.

After the feature extraction processing the classification process is carried on digital images. Many features were highly correlated with another features and system considered some features to classify the types of grains [5]. In image processing, classification process is very important process for object and that process based on feature extraction and considering [1]. Rice grains have many difference features and system considered only five features to classify the type of Myanmar rice grain. By depending on the basic five features, Myanmar rice grains are classified exactly and clearly.

5 Results and Discussion

In this research, five varieties of rice grain were tested to achieve the good classification accuracy. The system implemented to extract the features from the mixture grains image as shown in Fig. 3(a). This testing image with the total of 38 numbers of grains is carried out. Table 1 shows the testing result on the number of rice grain varieties. It gets the good result in Paw San Hmwe. However, the measure of Lone Thwe Hmwe is slightly similar with the measure of Kauk-Nyinn-Pu. Kauk-Nyinn-Thwe and Aye-yarmin are also very similar. Therefore, the good classification accuracy is very difficult. The measurements of extracted five features of testing image are shown in Table 2. The expected classification ranges of the extracted features (Area, Major axis length, Minor axis length, Eccentricity, Perimeter) for the five varieties of rice grain are shown in Table 3. The difference of major axis length between Kauk-Nyinn-Thwe and Ayeyarmin is very small and the major axis length between Kauk-Nyinn-Pu and Lone Thwe Hmwe do not differ much. Therefore, accuracy is defined as measuring the percentage of correct classification with respect to the overall data. Therefore, the system has resulted with 80% accuracy for myanmar rice samples using the extracted features.

Table 1. Testing results on the number of rice grain varieties

Rice varieties	Actual numbers of rice grains in testing image	Testing results on the numbers of rice grains
Paw San Hmwe	6	6
Lone Thwe Hmwe	9	5
Ayeyarmin	8	12
Kauk-Nyinn-Thwe	7	3
Kauk-Nyinn-Pu	8	12
	Total = 38	Total = 38

Table 2. Measurements of extracted five features of testing image

Myanmar rice varieties		Area	Major axis length	Minor axis length	Eccentricity	Perimeter
Paw-San-Hmwe	Psm-1	0.1478	0.491	0.301	0.294	1.570
	Psm-2	0.1478	0.501	0.295	0.287	1.576
	Psm-3	0.1477	0.475	0.311	0.308	1.566
	Psm-4	0.1260	0.492	0.256	0.251	1.486
	Psm-5	0.1227	0.489	0.251	0.259	1.496
	Psm-6	0.0653	0.493	0.285	0.275	1.536

(*continued*)

Table 2. (*continued*)

Myanmar rice varieties		Area	Major axis length	Minor axis length	Eccentricity	Perimeter
Lone-Thwe-Hmwe	Ltm-1	0.037	0.172	0.217	0.195	0.734
	Ltm-2	0.143	0.709	0.202	0.191	1.800
	Ltm-3	0.158	0.722	0.219	0.208	1.860
	Ltm-4	0.161	0.730	0.220	0.210	1.880
	Ltm-5	0.155	0.719	0.215	0.205	1.848
Aye-Yar-Min	Aym-1	0.122	0.671	0.182	0.180	1.702
	Aym-2	0.122	0.659	0.185	0.179	1.676
	Aym-3	0.118	0.664	0.178	0.173	1.674
	Aym-4	0.122	0.642	0.190	0.185	1.654
	Aym-5	0.128	0.675	0.182	0.181	1.712
	Aym-6	0.124	0.672	0.185	0.183	1.710
	Aym-7	0.118	0.655	0.180	0.178	1.666
	Aym-8	0.121	0.666	0.181	0.177	1.686
	Aym-9	0.127	0.649	0.195	0.190	1.688
	Aym-10	0.131	0.647	0.202	0.193	1.698
	Aym-11	0.127	0.646	0.197	0.191	1.686
	Aym-12	0.131	0.653	0.201	0.192	1.708
Kauk-Nyinn-Thwe	Kn-1	0.126	0.652	0.193	0.185	1.690
	Kn-2	0.127	0.649	0.195	0.190	1.688
	Kn-3	0.131	0.647	0.202	0.193	1.698
Kauk-Nyinn-Pu	Knp-1	0.155	0.685	0.227	0.205	1.780
	Knp-2	0.156	0.699	0.223	0.209	1.816
	Knp-3	0.158	0.689	0.23	0.207	1.792
	Knp-4	0.147	0.685	0.215	0.203	1.776
	Knp-5	0.156	0.693	0.225	0.208	1.802
	Knp-6	0.122	0.671	0.182	0.18	1.702
	Knp-7	0.122	0.659	0.185	0.179	1.676
	Knp-8	0.118	0.664	0.178	0.173	1.674
	Knp-9	0.122	0.642	0.19	0.185	1.654
	Knp-10	0.155	0.685	0.227	0.205	1.780
	Knp-11	0.156	0.699	0.223	0.209	1.816
	Knp-12	0.158	0.689	0.23	0.207	1.792

Table 3. Results of classification ranges of the extracted features

Myanmar Rice grain	Classified range on area	Classified range on major axis length	Classified range on minor axis length	Classified range on eccentricity	Classified range on perimeter
Paw San Hmwe	0.120–0.150	0.470–0.501	0.250–0.311	0.251–0.308	1.485–1.570
Lone Thwe Hmwe	0.143–0.161	0.710–0.730	0.202–0.220	0.195–0.210	1.810–1.880
Ayeyarmin	0.115–0.122	0.650–0.675	0.175–0.190	0.173–0.185	1.670–1.715
Kauk-Nyinn-Thwe	0.126–0.131	0.645–0.655	0.193–0.202	0.180–0.193	1.680–1.690
Kauk-Nyinn-Pu	0.145–0.160	0.680–0.701	0.215–0.227	0.203–0.210	1.770–1.820

6 Conclusion

Rice grain classification is a challenge because manual classification that is being used in the industry may not be efficient. The rice classification system is implemented for Myanmar rice grain varieties. In the present research, it is tested on five varieties of rice grains such as Paw San, Lone Thwe Hmwe, Aye Yar Min, Kauk Nyinn Thwe and Kauk Nyinn Pu. Deciding of variety of rice grain is based on five features of rice grain that are different. This system can properly apply in identification and classification of varieties of Myanmar rice using image processing. As the future work we will concentrate on the optimization of classification accuracy for real-time applications and to achieve the accurate results on bulk of Myanmar rice grain more than the present testing of five varieties.

References

1. Gujjar, H.S., Siddappa, D.M.: A method for identification of basmati rice grain of india and its quality using pattern classification. Int. J. Eng. Res. **3**(1), 268–273 (2013)
2. Herath, H.M.K.K.M.B., Eng. de Mel, W.R.: Rice grains classification using image processing technics. Department of Mechanical Engineering, The Open University of Sri Lanka, Nawala Nugegoda, Sri Lanka (2016)
3. Liu, Z., Cheng, F., Ying, Y., Rao, X.: A digital image analysis algorithm based color and morphological features was developed to identify the six varieties. J. Zhejiang Univ. Sci. **6**(11), 1095–1100 (2005)
4. Aki, O., Gullu, A., Uçar, E.: Classification of rice grains using image processing and machine learning techniques. In: International Scientific Conference UNITECH (2015)
5. Ajaz, R.H., Hussain, L.: Seed classification using machine learning techniques. J. Multi. Eng. Sci. Technol. (JMEST) **2**(5), 1098–1102 (2015)
6. Birla, R., Chauhan, A.P.S.: An efficient method for quality analysis of rice using machine vision system. J. Adv. Inf. Technol. **6**(3), 140–145 (2015)
7. Rexce, J., Usha Kingsly Devi, K.: Classification of milled rice using image processing. Int. J. Sci. Eng. Res. **8**(2) (2017). ISSN 2229-5518
8. "Ricepedia: The online authority of Rice", Research Program on Global Rice Science Partnership. http://ricepedia.org/myanmar

9. Kambo, R., Yerpude, A.: Classification of basmati rice grain variety using image processing and principal component analysis. Int. J. Comput. Trends Technol. (IJCTT) **11**(2), 306–309 (2014)
10. Kaur, S., Singh, D.: Geometric feature extraction of selected rice grains using image processing techniques. Int. J. Comput. Appl. **124**(8), 0975–8887 (2015)
11. Ribeiro, S.S.: Classification of grain based on the morphology, color and texture information extracted from digital images. Int. J. Comput. Appl. Eng. Technol. **5**(3), 359–366 (2016)

Color Segmentation Based on Human Perception Using Fuzzy Logic

Tin Mar Kyi(✉) and Khin Chan Myae Zin

Myanmar Institute of Information Technology, Mandalay, Myanmar
{tin_mar_kyi, khin_chan_myae_zin}@miit.edu.mm

Abstract. Color segmentation is important in the field of remote sensing and Geographic Information System (GIS). Most of the color vision systems need to classify pixel color in a given image. Human perception-based approach to pixel color segmentation is done by fuzzy logic. Fuzzy sets are defined on the H, S and V components of the HSV color space. Three values (H, S and V), the fuzzy logic model has three antecedent variables (Hue, Saturation and Value) and one consequent variable, which is a color class ID are fuzzified with Triangular Fuzzy Numbering Method. Fuzzy Rules are constructed according to the linguistic fuzzy sets. One of Discrete Defuzzification method based on zero-order takagi-Sugeno model is used for color segmentation. To define the output color value, Fuzzy reasoning with zero order Takagi-Sugeno model is used for assigning the color of the given. There are sixteen output colors: Black, White, Red, Orange, Yellow, Dark Gray, Light Gray, Pink, Light Brown, Dark Brown, Aqua, Blue, Olive, Light Green, Dark Green and Purple.

Keywords: Color segmentation · Fuzzy logic · Takagi-Sugeno model

1 Introduction

This research is the importance of the color segmentation in vision system. It also explains general structure of Color Segmentation System. Moreover, it shows that how to convert the RGB value of color image to HSV color space for further processing. The identification of the fuzzy set ranges for inputs and outputs of the system. This system is the first part of the computer vision system. It can be extended the color object detection application. And also sixteen output colors are used to define for color classification. For getting more accurate color classification, the system can extend to define more output colors. In this system, the machines task more challenging than human eye classify colors in the spectrum visible. This research is based on fuzzy logic modeling to segment an image via color space and the human intuition of color classification using HSV color space.

T. T. Zin and J. C.-W. Lin (Eds.): ICBDL 2018, AISC 744, pp. 333–341, 2019.
https://doi.org/10.1007/978-981-13-0869-7_37

2 Color Segmentation System

Color is one of the major features in identifying objects, and color vision has been the most intensively studied sensory process in human vision. In machine vision, color has recently received attention as a useful property for image segmentation and object recognition. The purpose of color vision is to extract aspects of the spectral property of object surfaces, while at the same time discounting various illuminations in order to provide useful information in image analysis. The most notable applications include recognition and identification of colored objects and image segmentation [13].

3 Converting RGB to HSV

It is based on segmentation of the HSV color space using a fuzzy logic model that follows a human intuition of color classification. It predefines the segments using a fuzzy logic model, and divides the color space into segments based on linguistic terms. So the input RGB color space is required to convert to HSV model. RGB to HSV conversion formula is as in the following procedure: Firstly, the *R*, *G*, *B* values are divided by 255 to change the range from 0.255 to 0.1. And then *R*', *G*', *B*' are computed as following.

$$R' = R/255, G' = G/255, B' = B/255 \quad (1.1)$$

After that, maximum and minimum values among them are extracted as C_{max} and C_{min}.

$$C_{max} = \max(R', G', B') \quad (1.2)$$

$$C_{min} = \min(R', G', B') \quad (1.3)$$

The difference between C_{max} and C_{min} are computed as delta value for further processing.

$$\Delta = C_{max} - C_{min} \quad (1.4)$$

After finished calculating of these above parameters, Hue, Saturation and Value are defined as in Eqs. 1.5, 1.6, and 1.7.

Hue calculation:

$$H \begin{cases} 60^\circ \times \left(\frac{\acute{G}-\acute{B}}{\Delta} mod6\right), Cmax = \acute{R} \\ 60^\circ \times \left(\frac{\acute{B}-\acute{R}}{\Delta} + 2\right), Cmax = \acute{G} \\ 60^\circ \times \left(\frac{\acute{R}-\acute{G}}{\Delta} + 4\right), Cmax = \acute{B} \end{cases} \quad (1.5)$$

$$\text{Saturation calculation}: \; S = \Delta/C_{max} \quad (1.6)$$

$$\text{Value calculation} : V = C_{max} \tag{1.7}$$

3.1 Fuzzy Color Identification

In this system, HSV color space each color is defined by three values (H, S and V), the fuzzy logic model has three antecedent variables (Hue, Saturation and Value) and one consequent variable, which is a color class ID. The domain of the variables Hue, Saturation and Value is the interval (0,240). The domain of the consequent variables discrete, and depends on the number of the predefined color classes.

There are 10 fuzzy sets for Hue, 5 fuzzy sets for Saturation and 4 fuzzy sets for Value. All membership functions are in the form of a triangular function. The fuzzy sets of the antecedent fuzzy variable Hue are defined based on 10 basic hues distributed over the 0–240 spectrum. The hues are Red, Dark Orange, Light Orange, Yellow, Light Green, Dark Green, Aqua, Blue, Dark Purple, Light purple. Saturation is defined using the five fuzzy sets Gray, Almost Gray, Medium, Almost Clear, and Clear. Value is defined using the four fuzzy sets Dark, Medium Dark, Medium Bright and Bright.

The fuzzy rules in this model are defined based on human observations. Since this model has 10 fuzzy sets for Hue, 5 for Saturation and 4 for Value, the total number of rules required for this model is 10 5 4 = 200. The reasoning procedure is based on a zero-order Takagi-Sugeno model, so that the consequent part or the output color of each fuzzy rule is a crisp discrete value of the set Black, White, Red, Orange, Yellow, Dark Gray, Light Gray, Pink, Light Brown, Dark Brown, Aqua, Blue, Olive, Light Green, Dark Green, and Purple.

3.2 Sugeno-Type Fuzzy Inference System

The fuzzy inference process that is referring to so far is known as Mamdani's fuzzy inference method, the most common methodology. In this section, it discusses the so-called Sugeno, or Takagi-Sugeno-Kang, method of fuzzy inference. It is similar to the Mamdani method in many respects. The first two parts of the fuzzy inference process, fuzzifying the inputs and applying the fuzzy operator, are exactly the same. The main difference between Mamdani and Sugeno is that the Sugeno output membership functions are either linear or constant [19].

A typical rule in a Sugeno fuzzy model has the form of

If Input 1 = x and Input 2 = y, then Output is z = ax + by + c

For a zero-order Sugeno model, the output level z is a constant (a = b = 0).

The output level z_i of each rule is weighted by the firing strength w_i of the rule. For example, for an AND rule with Input 1 = x and Input 2 = y, the firing strength is

$$w_i = \text{And Method}\left(F_1(x), F_2(y)\right)$$

Where F1, 2 (.) are the membership functions for Inputs 1 and 2. The final output of the system is the weighted average of all rule outputs, computed as

$$FinalOutput = \frac{\sum_{i=1}^{N} w_i z_i}{\sum_{i=1}^{N} w_i}$$

4 Implementation

Input to the system is RGB image but HSV color space is used for further computing. Firstly, the RGB value from the given image is extracted pixel by pixel basic. In second stage, the extracted RGB value of each pixel is required to convert to HSV color space. In the third stage, the HSV values are fuzzified using Triangular Fuzzy Numbering for computing crisp value to fuzzy value. Fuzzy rules are predefined for further processing. According to the fuzzy values of HSV, the respective fuzzy rules for each pixel are achieved. In the fourth stage, Zero-order Takagi-Sugeno model [19, 20] is used for defuzzification. After that, the color of each pixel is assigned as last step as in Fig. 1. The procedure works iteratively and it is done for all pixel of the given image.

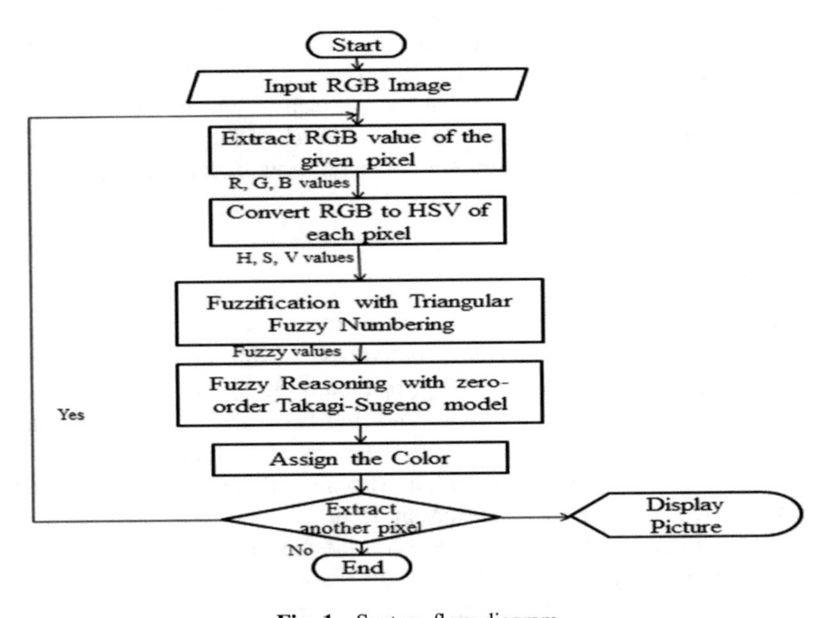

Fig. 1. System flow diagram

5 Results and Discussion

In this research, the color classification system defines the color of the input image based on pixel by pixel. The detail implementation of the system such as converting RGB to HSV color space, fuzzification of HSV value, and extraction of the color output from the consequences of fuzzy rules. The implementation is very simple and the input to the system is color image. After the input image is loaded as in Fig. 1, RGB values from each pixel are extracted [3]. The extracted RGB values are converted to HSV value as in Table 1.

Table 1. RGB value and HSV value of each pixel of the input image

Pixel	R	G	B	H	S	V
0x0	255	209	61	45.773	76.078	100
0x1	255	209	61	45.773	76.078	100
0x2	255	209	61	45.773	76.078	100
0x3	255	209	61	45.773	76.078	100
0x4	255	209	61	45.773	76.078	100
0x5	255	209	61	45.773	76.078	100
0x6	254	210	59	46.462	76.772	99.608
0x7	254	210	59	46.462	76.772	99.608
0x8	252	211	59	47.254	76.587	98.824
0x9	252	211	59	47.254	76.587	98.824
0x10	251	211	61	47.254	76.587	98.824

5.1 Fuzzy Membership Function

There are three inputs H, S and V, so it has at most six fuzzy set and at least three fuzzy set. For each fuzzy set, it has the value of fuzzy value. As an example, Table 2 mentions the 'Hue' value and its respective fuzzy values. Table 3 mentions the 'Situation' value and its respective fuzzy values. Table 4 mentions the 'Value' value and its respective fuzzy values.

Table 2. 'Hue' value and its respective fuzzy values

Pixel	Hue Fuzzy value1	Hue Fuzzy set1	Hue Fuzzy value2	Hue Fuzzy set2
0x0	0.711	Light Purple	0.218	Red
0x1	0.811	Light Purple	0.189	Red
0x2	0.968	Light Purple	0.32	Red
0x3	0.013	Dark Purple	0.987	Light Purple
0x4	0.052	Dark Purple	0.948	Light Purple
0x5	0.04	Dark Purple	0.96	Light Purple
0x6	0.973	Light Purple	0.027	Red
0x7	0.032	Dark Purple	0.968	Light Purple
0x8	1	Light Purple	0	Red
0x9	0.914	Light Purple	0.086	Red
0x10	0.043	Dark Purple	0.957	Light Purple

HSV value of each pixel may extract at most eight rules. The maximum strength of rules is the output color. The rules and the respective output color for each pixel are shown as follow.

Table 3. 'Situation' value and its respective fuzzy values

Pixel	Saturation Fuzzy value1	Saturation Fuzzy set1	Saturation Fuzzy value2	Saturation Fuzzy set2
0x0	0.25	Almost Gray	0.75	Medium
0x1	0.25	Almost Gray	0.75	Medium
0x2	0.266	Almost Gray	0.734	Medium
0x3	0.518	Almost Gray	0.482	Medium
0x4	0.492	Almost Gray	0.508	Medium
0x5	0.25	Almost Gray	0.75	Medium
0x6	0.258	Almost Gray	0.742	Medium
0x7	0.25	Almost Gray	0.75	Medium
0x8	0.25	Almost Gray	0.75	Medium
0x9	0.25	Almost Gray	0.75	Medium
0x10	0.361	Almost Gray	0.639	Medium

Table 4. 'Value' value and its respective fuzzy values

Pixel	Value Fuzzy value1	Value Fuzzy set1	Value Fuzzy set2	Value Fuzzy value2
0x0	0.513	Dark	0.487	Blue
0x1	0.516	Dark	0.484	Blue
0x2	0.974	Dark	0.026	Blue
0x3	0.719	Dark	0.281	Blue
0x4	0.752	Dark	0.248	Blue
0x5	0.51	Dark	0.49	Blue
0x6	0.513	Dark	0.487	Blue
0x7	0.993	Dark	0.007	Blue
0x8	0.5	Dark	0.5	Blue
0x9	0.507	Dark	0.493	Blue
0x10	0.824	Dark	0.176	Blue

If Hue = Red (0.019) and Saturation = Almost Gray (0.266) and Value = Dark (0.484) then Output = Black (0.019)

If Hue = Red (0.019) and Saturation = Almost Gray (0.266) and Value = Medium Dark (0.033) then Output = Dark Brown (0.019)

If Hue = Red (0.019) and Saturation = Almost Gray (0.266) and Value = Dark (0.484) then Output = Dark Brown (0.019)

If Hue = Red (0.019) and Saturation = Medium (0.734175) and Value = Medium Dark (0.033) then Output = Dark Brown(0.019)

If Hue = light purple (0.981) and Saturation = Almost Gray (0.266) and Value = Dark (0.484) then Output = Blue (0.266)

If Hue = light purple (0.981) and Saturation = Almost Gray (0.266) and Value = Medium Dark (0.033) then Output = Blue (0.033)

If Hue = light purple (0.981) and Saturation = Medium (0.734175) and Value = Dark (0.484) then Output = Blue (0.484)

If Hue = light purple (0.981) and Saturation = Medium (0.734175) and Value = Medium Dark (0.033) then Output = Blue (0.033)

The output color is Blue (0.848).

5.2 Experimental Results

The output color from fuzzy rule based system of each pixel are updated at each pixel. Experiments are performed on Intel i5 CPU at 1.80 GHz and 4 GB of RAM. The following results are obtained when the system is experimented. The output color segmented image is as shown in the figure.

No	Input Image	Output Image	Dimensions and Size(KB)	Computation Time (seconds)
1.			150x85 5.9	608
2.			66x50 10.7	87
3			150x84 6.3	446

6 Conclusion

This research is a model based on the human perception of colors by using fuzzy logic. HSV color space is used for getting near the human vision color. Fuzzy logic based color segmentation model used in many real-life application. Triangular Fuzzy Numbering and Fuzzy Rule based on Zero Order Takagi_Sugeno Method are used to classify the output color of each pixel. The output color of each pixel needs to calculate. So the computation time is relatively high according to the size of image. Getting the right human perceptual output color depend the fuzzy rule. This system is just only the first part of the computer vision system. It can extend color object detection. And also sixteen output colors (Black, White, Red, Orange, Yellow, Dark Gray, Light Gray, Pink, Light Brown, Dark Brown, Aqua, Blue, Olive, Light Green and Dark Green) are used to define for color classification. For getting more accurate color classification, the system can extend to define more output colors. Sample image and its color segmented image can be seen in Appendix.

References

1. Voloshyn, A., Gnatienko, G., Drobot, E.: Fuzzy membership functions in a fuzzy decision making problem (2002). cited by 20-21
2. Khaliq, A., Ahmad, A.: Fuzzy logic and approximate reasoning. Blekinge Institute of Technology (2010)
3. BasketBall Robot: Converting from RGB to HSV, GE 423 – Mechatronics Lab Final Project (2005)
4. Basic Mechatronics Lab Final Project (2005)
5. Healey, C.G., Enns, J.T.: A perceptual color segmentation algorithm. University of British Columbia (1996). cited by 38
6. Amante, J.C., Fonseca, M.J.: Fuzzy color space segmentation to identify the same dominant colors as users. Department of Computer Science and Engineering, INESC-ID/IST/TU Lisbon, Portugal (2010-2014)
7. Chen, J., Pappas, T.N., Mojsilović, A., Rogowitz, B.E.: Adaptive perceptual color-texture image segmentation. IEEE Trans. Image Proces. (2005)
8. Bruce, J., Balch, T., Veloso, M.: Fast and inexpensive color image segmentation for interactive robots, (IROS 2000) (2000)
9. Delon, J., Desolneux, A., Lisani, J.L., Petro, A.B.: Color image segmentation using acceptable histogram segmentation. Springer (2005)
10. Busin, L., Vandenbroucke, N., Macaire, L.: Color spaces and image segmentation. Université des Sciences et Technologies de Lille, France (2007)
11. Shamir, L.: Human Perception-based color segmentation using fuzzy logic. In: Asian Conference on Computer Vision, Taiwan, Department of Computer Science, Michigan Tech (2000). Tan, L.: Takagi-sugeno and mamdani fuzzy control of a resort management system, Blekinge Institute of Technology School of Engineering, November 2011
12. Batavia, P.H., Singh, S.: Obstacle detection using adaptive color segmentation and color stereo homography, Carnegie Mellon University, Robotics Institute, Seoul, Korea 21-26 May 2001

13. Puranik, P., Bajaj, P., Abraham, A., Palsodkar, P., Deshmukh, A.: Human perception-based color image segmentation using comprehensive learning particle swarm optimization. J. Inf. Hiding Multimedia Signal Proces. (2011)
14. Zhana, Q., Liangb, Y., Xiaoa, Y.: Color-based segmentation of point clouds. IAPRS, Paris, France (2009)
15. Bajcsy, R., Lee, S.W., Leonardis, A.: Image segmentation with detection of highlights and inter-reflections using color (1989)
16. Hanek, R., Schmitt, T., Buck, S., Beetz, M.: Fast image-based object localization in natural scenes. In: IEEE Conference on Intelligent Robots and Systems, Switzerland (2002)
17. Belongie, S., Carson, C., Greenspan, H., Malik, J.: Color- and texture-based image segmentation using em and its application to content-based image retrieval. In: IEEE ICCV (1998)
18. Sural, S., Qian, G., Pramanik, S.: Segmentation and histogram generation using the HSV color space for image retrieval. In: IEEE ICIP (2002)
19. Sugeno-Type Fuzzy Inference Tutorial
20. Takagi, T., Sugeno, M.: Fuzzy identification of systems and its applications to modeling and control. IEEE Trans. Syst. Man Cybern. (1985)
21. Skarbek, W., Koschan, A.: Colour image segmentation, Techinical University of Berlin (1994)

Key Frame Extraction Techniques

Mie Mie Khin(✉), Zin Mar Win, Phyo Phyo Wai,
and Khaing Thazin Min

Myanmar Institute of Information Technology, Mandalay, Myanmar
{mie_mie_khin, zin_mar_win, phyo_phyo_wai}@miit.edu.mm,
khaing_thazin_min@edu.mm

Abstract. Video is the most challenging filed because it combines all the form of multimedia information. Key Frame extraction plays an important role in many video processing applications. Key Frame extraction reduces the useless information of video. Key frames are essential to analysis on large amount of video frame sequences. This paper describes different key frame extraction techniques and helps to choose the best key frame method for key frame extraction.

Keywords: Key frame · Key frame extraction techniques

1 Introduction

Nowadays, video data is increasing because of the development of digital video capture and technology. Video is a collection of key frames. To discover the content of the video, the main point is to eliminate the redundant information from this video. Key frame extraction is a powerful tool that implements video content by selecting a set of summary key frames to represent video sequences. A key component of the video is the Image; the standard term for the image is the frame.

In video summarizing technique, key frame extraction is one main useful part of that process. Video summarization is to create main link of these video file and to show core information form that video. Video summarizing are used for many application area i.e. entertainments, security area, health care and etc. [7]

Finding appropriate frame from video files retrieval are used key frame extraction methods. [8] The static key-frames extraction from the video is hold the important content of the video. These frames represent that video summarization. Dynamic key-frames care for dynamic nature of video. These dynamic senses are used temporal ordered sequence of key-frames extracted [3].

Key frame extraction techniques can be roughly categorized into four types, based on shot boundary, visual information, movement analysis, and cluster method. And then sometimes it could be completed in compressed domain. Nowadays, cluster-based methods are mostly applied in video content analysis research. In these methods, key frame extraction is usually modeled as a typical clustering process that divides one video shot into several clusters and then one or several frames are extracted based on low or high level features. The methods in compressed domain usually are not suitable

T. T. Zin and J. C. Lin (Eds.): ICBDL 2018, AISC 744, pp. 342–345, 2019.
https://doi.org/10.1007/978-981-13-0869-7_38

for diverse formats of videos from the Internet. Transcoding may increase time complexity and inaccuracy.

The organization of the paper is as follows. Key frame extraction methods are presented in Sect. 2 and the conclusions are drawn in Sect. 3.

2 Key Frame Extraction Techniques

Narayanan extracted key frame by using the absolute value of histogram of consecutive frames. The experiment is tested on the KTH action database which contains six types of human actions (walking, jogging, running, boxing, hand waving and hand clapping. The performance of their technique is evaluated based on fidelity measure and compression ratio [2].

Ghatak used the threshold technique for extracting key frame. This technique consists of three modules: Module 1 inputs the video of avi format and outputs a set of frames from the input video file. After that, module 2 finds the frame difference. To find the difference, firstly the images are converted to gray scale and gray scale images are plotted to histogram. Finally, the difference matrix is calculated. In module 3, threshold value is calculated. The input frame is considered as key frame or not based on the threshold value [1].

Mentzelopoulos et al. proposed key frame extraction algorithm by using Entropy Difference [4]. Entropy gives the information that can be coded for by a compression algorithm. In this algorithm author used the entropy as local operator instead of global feature for the total image. Entropy is one of the good way of representing the impurity or unpredictability of a set of data because it is dependent on the context in which the measurement is taken. Here author said that first distribute the entropy among the image, then the higher distribution of entropy will give us the regions containing the salient objects of a video sequence. Therefore any change in the object appearance of salient regions will affect its relevant semantic information in the entire story sequence.

Rasheed and Shah [5] have proposed the color histogram based method of UCF. This algorithm uses the color histogram for measuring the intersection similarity to extract key frames. This algorithm proposes a method to select the first key frame and then based on first frame and next frame difference, key frame will be decided. There will be comparison with the next frames and new frame will be selected based on difference between frames if there is significant difference, the current frame will be the key frames.

Zhuang *et al.* proposed a GGD model for key frame extraction. GGD model, α and β, estimates the density function of coefficients. GGD estimates give a good approximation to the 2D wavelet subband histograms. In content analysis process, appropriate features extraction like motion information, color histograms or features has to be employed from a 2D-transform of the video frame. However, none of the existing methods are structurally matched with the Human Visual System (HVS), which is the main sources of the error in locating the key frames. To clear this error, features are extracted from wavelet transformed sub-bands of each frame, leading to a better match with the HVS in GGD. GGD parameters are helpful to construct the feature vectors and examine similarity between two frames in a cluster and discriminate between two frames from different clusters. These results are more accurate and more subjective to the human section as compared with the previous methods [6].

A various number of Scene Change Detection (SCD) methods are proposed by Dolly group. They discuss that is video summarization or abstraction, video classification, video annotation and content based video retrieval research. Their research used block processing scene change detection techniques & algorithms [9].

Dong, Zhang, Jia and Bao et al. present a novel keyframe selection and recognition method for robust markerless real-time camera tracking. This paper describes an offline module to select features from a group of reference images and an online module to match them to the input live video in order to quickly estimate the camera pose. Their workflow is like that. For offline space abstraction, they extract SIFT (Scale Invariant Feature Transform) features from the reference images, and recover their 3D positions by SFM (Structure-From-Motion). And, select optimal keyframes and construct a vocabulary tree for online keyframe recognition. For online real-time camera tracking, they extract SIFT features for each input frame of the incoming live video. Then, they quickly select candidate keyframes. They match the feature with candidate keyframes. And, they estimate camera pose with the matched features [10].

This proposed approach is based on correlation method for key frame extraction and parallel processing. This approach uses a correlation technique to summarize video. In this paper, they use a simple technique for comparing frames and key frame extraction. They use correlation technique to find how much two frames are similar or different from each other. When frames are similar, they will assume that these frames are duplicate frames and will remove them. Otherwise, different frames will be stored as key frames for using it in summarizing the video. They use mapping for map key frames to retrieve proper video sequence from original video. In their proposed system, they can extract frames from video, and at the same time they compare those extracted frames simultaneously and choose unique frames from them. They use keyframe extraction algorithm for choosing unique frames. Although they can apply keyframe extraction algorithms directly on this frame, it will work in linear way which will increase computational time. Therefore, they use Task function to decrease computational time. Task is mainly used for parallel processing. Task is done parallel the operations of extracting frames, creating, applying. By using parallel processing they can decrease processing speed. Parallel processing is also used to find frame difference correlation [11].

Goularte et al. propose a new keyframe extraction method based on SIFT local features. Initially, they group keyframe candidate sets (KCS) to select the best frames in the keyframes of each shot. The first frame in KCS is defined as shot first frame. The next frame in the KCS is the window rule. Widow size 25 is a good one because most of movies have 25 fps as capturing rate and there is no significant variation on consecutive frames content. Then, SIFT features are extracted from the frames in KCS. Each frame represents 128 dimensions of feature vectors. They experimented on five video segments on movies domain [12]. They evaluated their method in the scene segmentation context, with videos from movies domain, developing a comparative study with three state of the art approaches based on local features. The results show that their method overcomes those approaches. Their method is not dependent on the scene segmentation technique.

3 Conclusion

This paper describes about the different key frame extraction techniques. Each technique has its own advantages and disadvantages. Keyframe will provide more compact and meaningful video summary. So with the help of keyframes it will become easy to browse video and it will give the appropriate search result. Each technique is explained briefly along with their performance results and feedbacks by comparing with each other. This paper will definitely help the researchers to choose the best key frame extraction technique.

References

1. Ghatak, S.: Key frame extraction using threshold techniques. Int. J. Eng. Appl. Sci. Technol. **1**(8), 51–56 (2016). ISSN No. 2455–2143
2. Omidyeganeh, M., et al.: Video key frame analysis using a segment based statistical metric in a visually sensitive parametric space. IEEE (2011)
3. Mithlesh, C.S., Shukala, D.: A case study of key frame extraction techniques. Int. J. Adv. Res. Electr. Electron. Instrum. Eng. **5**(3) (2016)
4. Mentzelopoulos, M., Psarrou, A.: KeyFrame extraction algorithm using entropy difference. In: Proceedings of the 6th ACM SIGMM International Workshop on Multimedia Information Retrieval, MIR 2004, 15–16 October 2004, New York, NY, USA (2004)
5. Rasheed, Z., Shah, M.: Detection and representation of scenes videos. IEEE Trans. Multimedia **7**(6) (2005)
6. Nasreen, A., Shobha, G.: Key frame extraction from video-a survey. Int. J. Comput. Sci. Commun. Netw. **3**(3), 194–198 (2013)
7. Asadi, E., Charkari, N.M.: Video summarization using fuzzy C-means clustering. In: 20th Iranian Conference on Electrical Engineering, pp. 690–694 (2012)
8. Sheena, C.V., Narayanan, N.K.: Key-frame extraction by analysis of histograms of videoframes using statistical methods. In: 4th International Conference on Eco-friendly Computing and Communication Systems, pp. 36–40 (2015)
9. Shukla, D., Mithlesh, C.S., Sharma, M.: Design, implementation & analysis of scene change detector using block processing method. J. Appl. Sci. Res. (TJASR) **2**(2), 103–110 (2015)
10. Dong, Z., Zhang, G., Jia, J., Bao, H.: Keyframe-based real-time camera tracking. In: 2009 IEEE 12th International Conference Computer Vision, 29 September–2 October 2009
11. Satpute, A.M., Khandarkar, K.R.: Video summarization by removing duplicate frames from surveillance video using keyframe extraction. Int. J. Innovative Res. Comput. Commun. Eng. **5**(4) (2017)
12. de Souza Barbieri, T.T., Goularte, R.: KS-SIFT: a keyframe extraction method based on local features. In: 2014 IEEE International Symposium on Multimedia (2014)

A Study on Music Retrieval System Using Image Processing

Emi Takaoka(✉) and Thi Thi Zin(✉)

Graduate School of Engineering, University of Miyazaki, Miyazaki, Japan
hl14025@student.miyazaki-u.ac.jp,
thithi@cc.miyazaki-u.ac.jp

Abstract. The music retrieval system has made it possible to easily search music by simply listening to songs on machine with the spread of smartphone applications. However, it requires enormous data by voice information existing system. Therefore, we proposed new music retrieval system using images visualized by sound signal processing. In the experiment, we used 35 songs and made data for retrieval by extracting a part from them and using only melody information. As a result, the percentage of correct answers that appeared in the top 3 places was 89%, which proved that the proposed method is useful.

Keywords: Music retrieval · Spectrogram · Chroma vector · Cross-correlation

1 Introduction

Music retrieval systems are widely used in smartphone applications and the like. With the widespread use of the system, it became easier to search music by simply listening to songs on the machine. Most of existing systems read waveform data from music and judge the songs by comparing them with the waveform data that stored in the database. However, when sound data is accumulated in a database, the amount of data becomes enormous. In this research, we aim to reduce the amount of data by creating an image that visualizing the pitch by audio signal processing and image processing.

2 Proposed Methods

2.1 Create Database

The proposed music retrieval system is two components namely: database creation component and music search component as described Fig. 1(a) and (b). In this method, each song is clustered and images are created using the calculation result of the spectrogram and it is saved in database.

T. T. Zin and J. C.-W. Lin (Eds.): ICBDL 2018, AISC 744, pp. 346–354, 2019.
https://doi.org/10.1007/978-981-13-0869-7_39

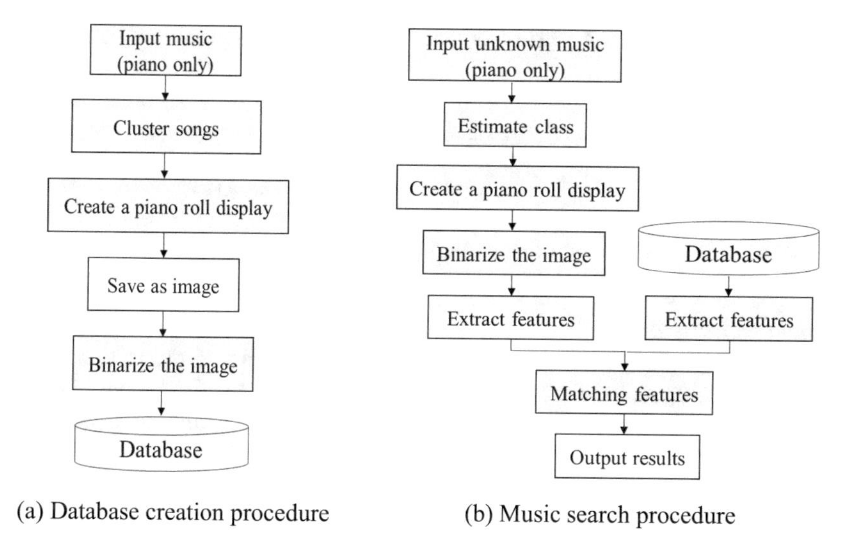

(a) Database creation procedure (b) Music search procedure

Fig. 1. Proposed system procedure

(i) Chroma Vector

The chroma vector shows the strength related to the 12 pitches, and it shows the characteristic value by the tune. Relationship between chroma vector and tone pitch can be expressed by a spiral structure (Fig. 2).

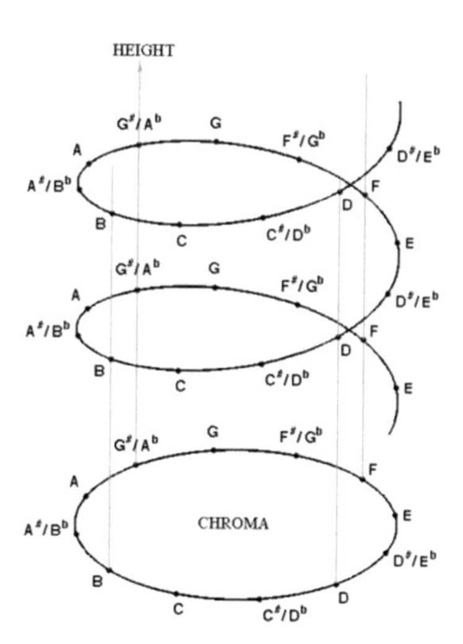

Fig. 2. Relationship between chroma vector and tone pitch (Source: http://www.nyu.edu/classes/bello/MIR_files/tonality.pdf)

The chroma vector is a sum of power spectra for every semitone between a pluralities of octaves [1, 2].

$$c(k,t) = \log\left(\sum_{i=0}^{I-1} H(12i+k,t)\right)_{k=0,\cdots,11}$$

Where H(i,t) is the power of frequency spectrogram bin i and time frame t, and I is the number of octaves [3].

(ii) Piano-roll display
The piano-roll display shows the pitch on the vertical axis and the time on the horizontal axis, and inputs the timing of sound utterance along the time series. This makes it possible to visualize performance information such as sound transitions on the screen.

(iii) Spectrogram
The spectrogram is a feature obtained from an audio signal, and it is obtained by spectrum using short-time Fourier transform and time change of the sound spectrum is expressed by color and graphic.

In order to investigate the nature of the signal at a certain time t, the wave form is cut out in the time window. Here, by using the fact that the value of the window function becomes 0 outside the not observable, and by preparing the window function with the width of the waveform to cut out, the nature of the signal at a certain time can be examined.

Using the window function centered at time t, the signal becomes as follows:

$$s_t(\tau) = s(\tau)h(\tau - t) \tag{2.1}$$

The Fourier transform reflecting the frequency distribution around time t is as follows;

$$S_t(\omega) = \frac{1}{\sqrt{2\pi}} \int e^{-j\omega\tau} s(\tau)h(\tau - t)d\tau \tag{2.2}$$

Therefore, the energy spectral density becomes as follows:

$$\begin{aligned} P_{SP}(t,\omega) &= |S_t(\omega)|^2 \\ &= \left|\frac{1}{\sqrt{2\pi}} \int e^{-j\omega\tau} s(\tau)h(\tau - t)d\tau\right|^2 \end{aligned} \tag{2.3}$$

The entire spectrum represents the time frequency distribution (spectrogram) [4].

Associating the frequency window corresponding to the time window *h(t)* with (2.4), the frequency distribution has the relationship as follows:

$$H(\omega) = \frac{1}{\sqrt{2}} \int h(t) e^{-j\omega t} dt \tag{2.4}$$

$$S_t(\omega) = e^{-j\omega t} s_\omega(t) \tag{2.5}$$

According to the frequency resolution, the spectrogram is divided into narrowband spectrogram and broadband spectrogram. In this paper, we used narrowband spectrogram, it can be calculated by widening the frequency window $H(\omega)$.

(iv) k-means clustering
k-means clustering is an iterative algorithm that divides data.
The classification procedure is as follows:

(1) Calculate the center of gravity of any initial cluster.
(2) Calculate the distance between the center of gravity of the cluster and the point to the center of gravity.
(3) Assign each point to the nearest cluster.
(4) Calculate the center of gravity of the new cluster.
(5) Repeat (2)–(4) until there is no change in assignment.

In this research, we classify songs by the k-means method using the calculation result of chroma vector of each song. By doing this, the time it takes to search is shortened.

2.2 Music Retrieval Procedure

The procedure of the proposed retrieval system is shown in the Fig. 1(b).

(i) Template matching
A method of calculating the degree of coincidence and calculating whether or not there is an image (template) of a specific part in the entire input image is called template matching.

The calculation formula of the cross-correlation at a certain coordinate (m,n) when the template is g and the input image is f is as follows:

$$R_{fg}(m,n) = \frac{\sum\sum_{(i,j)\in D} g(i,j) f(i+m, j+n)}{\sqrt{\left[\sum\sum_{(i,j)\in D} g^2(i,j)\right]\left[\sum\sum_{(i,j)\in D} f^2(i+m, j+n)\right]}} \tag{2.6}$$

The larger the cross-correlation $R_{fg}(m,n)$ is, the better matching is obtained [5].

(ii) Dynamic Time Warping
In music performance and conversation, even with the same music, the length of the spectrum series is different by the performer. Therefore, Dynamic Time Warping (DTW) is widely used as a method of calculating the degree of matching of two time series by nonlinear expansion matching of time series pattern.

Consider two K-dimension signals.

$$X = \begin{bmatrix} x_{1,1} & x_{1,2} & \cdots & x_{1,M} \\ x_{2,1} & x_{2,2} & \cdots & x_{2,M} \\ \vdots & \vdots & \ddots & \vdots \\ x_{K,1} & x_{K,2} & \cdots & x_{K,M} \end{bmatrix}$$
$$Y = \begin{bmatrix} y_{1,1} & y_{1,2} & \cdots & y_{1,N} \\ y_{2,1} & y_{2,2} & \cdots & y_{2,N} \\ \vdots & \vdots & \ddots & \vdots \\ y_{K,1} & y_{K,2} & \cdots & y_{K,N} \end{bmatrix} \quad (2.7)$$

The signals X and Y include M and N samples. Given the distance $d_{mn}(X,Y)$ between the mth sample of signal X and the nth sample of signal Y, this distance is stretched until the cumulative value of the distance scale from the start point to $d_{mn}(X,Y)$ becomes the minimum value. First, the possible values of $d_{mn}(X,Y)$ are arranged in a lattice pattern. Next, the minimum distance is calculated using the following formula.

$$d = \sum_{\substack{m \in ix \\ n \in iy}} d_{mn}(X, Y) \quad (2.8)$$

A recurrence formula based on Manhattan distance is used for distance calculation. This recurrence formula is provided with a path with tilt limitation to prevent extreme stretching and contraction, and a matching window for limiting the computation area of cumulative distance. By doing this, speech recognition performance and calculation amount reduction are realized [6, 7]. The situation in which two signals are aligned using this method is shown in the Fig. 3.

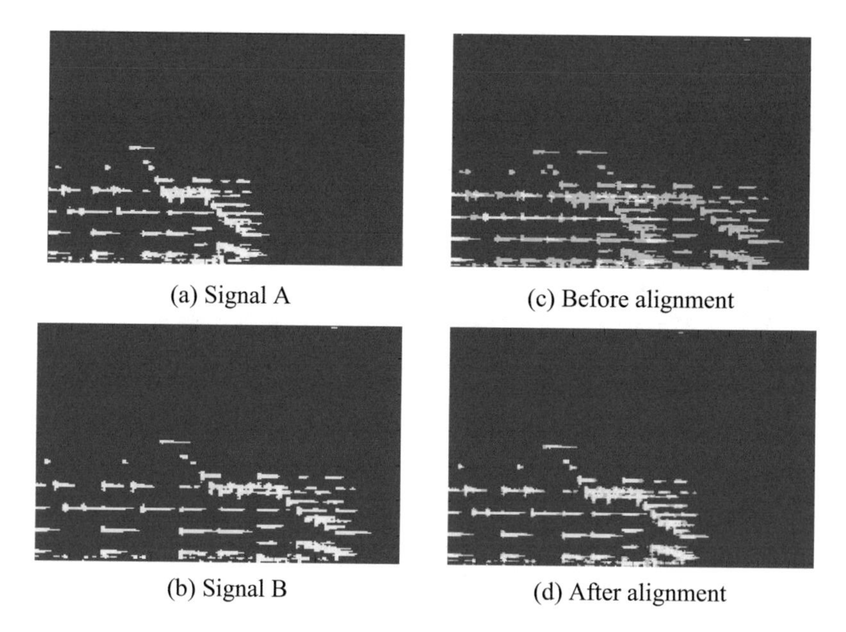

Fig. 3. DTW calculation example

(iii) Create Piano-roll display
In this research, images in which performance information is visualized are accumulated in a database. However, the piano roll of the sound source has not been automated, therefore manual input is necessary. Therefore, to automate the retrieval system, an image that similar to the piano-roll is created by using the spectrogram. After that, binarize it using image processing and save it in a database.

(iv) Feature matching
We use cross-correlation and DTW for matching songs, and evaluate whether these methods are useful.

3 Experiments

3.1 Song Data

The data used for the experiment was obtained by playing the piano with the electronic piano and the recording software connected. The list of used songs is shown in Table 1. For the experiment, 35 songs were used, among which 81 sound sources for music search and 33 sound sources for melody search were prepared.

Table 1. Music lists

Song name	Composer
Nocturne Op.9-2	F.Chopin
24 Preludes Op.28-7	F.Chopin
Etude Op.10-3	F.Chopin
Etude Op.10-12	F.Chopin
Fantasie Impromptu Op.66	F.Chopin
Valse Op.64-1	F.Chopin
Valse Op.64-2	F.Chopin
Bagatelle 'Fur Elise' WoO.59	L.Beethoven
6 Menuette WoO.10-2	L.Beethoven
Sonate fur Klavier Nr.14 Op.13-1	L.Beethoven
Sonate fur Klavier Nr.14 Op.13-2	L.Beethoven
Sonate fur Klavier Nr.14	L.Beethoven
Op.13-3	L.Beethoven
Inventions and Sinfonias BWV 772	J.s. Bach
Das Wohltemperierte Klavier BWV846	J.s. Bach
Sonate fur Klavier Nr.11 Mov.3 Alla Turca-Allegretto	W.A.Mozart
Sonate fur Klavier Nr.16 Mov.1 Allegro	W.A.Mozart
Moments Musicaux Op.94-3	F.P.Schubert
Impromptus Op.90-2 D899	F.P.Schubert
Kinderszenen Op.15-1	R.A.Schumann
Kinderszenen Op.15-7	R.A.Schumann
2 Arabesques No.1	C.A.Debussy
La fille aux cheveux de lin	C.A.Debussy
25 Etude Op.100-2	J.F.F.Burgmuller
25 Etude Op.100-25	J.F.F.Burgmuller
The Entertainer	S.Joplin
Liebestraume 3 notturnos S.541 R.211	F.Liszt
Lapriere d'une vierge Op.4	T.Badarzewska
Blumenlied (Flower Song) Op.39	G.Lange
Lieder ohne Worte Heft 5 Op.62 U161	F.Mendelssohn
Gavotte	F-J.Gossec
Csikos Post	H.Necke
3 Gymnopedies 'Lent et douloureux'	E.Satie
8 Humoresky Op.101-7 B.187	A.Dvorak
2 Rhapsodien Op.79-2	J.Brahms

3.2 Evaluation

Experiments are two types of music retrieval and melody retrieval, and evaluate with ranking using three methods.

Each method is explained below:

(1) The cross correlation between the piano-roll image (input image) stored in the database and the piano-roll image (template) of the search music is calculated, and the images of five songs with high correlation are ranked.
(2) For the five songs outputted in (1), adjust the tempo using DTW, superimpose the adjusted images and rank them in descending order of the calculation result of the overlapped area.
(3) Rank points are added to the above two methods and ranked in descending order of points earned.

The graphs of the experiment results by each method are shown in the Fig. 4. Experimental results show the correct answer rate for search songs in the top three for each method.

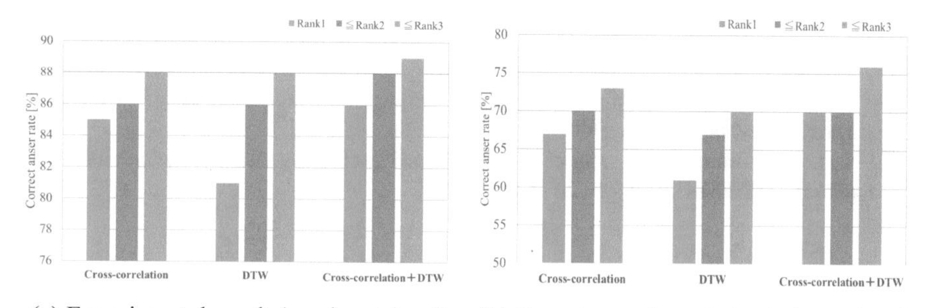

(a) Experimental result (music retrieval) (b) Experimental result (melody retrieval)

Fig. 4. DTW calculation example

The correct answer rate is as follows:

$$A = \frac{M_{collect}}{M_{all}} \tag{3.1}$$

In this case, A is the correct answer rate, M_{all} is the number of songs to be retrieved, and $M_{collect}$ is the number of correct answers.

4 Conclusions

In this paper, we proposed a new method using images created from audio signal information as music retrieval system. From the experimental results, 72 songs in the music search and 25 songs in the melody search were ranked correctly within the top 3 songs. Also, rather than ranking using only cross-correlation information and DTW, ranking the two methods combined and ranking higher ranked search songs improved the accuracy rate. Therefore, it is considered useful to combine these two for ranking.

On the other hand, songs that could not correctly output the retrieval results are commonly found to have a lot of modulation. It was excluded from the search target at the time of class estimation of the search song due to a lot of modulation. As a solution, by making songs with a lot of modulation belong to more than one class, it can be prevented from being excluded.

References

1. Bello, J.P.: Chroma and tonality, January 2018. http://www.nyu.edu/classes/bello/MIR_files/tonality.pdf
2. Giannakopoulos, T., Pikrakis, A.: Introduction to Audio Analysis: A MATLAB Approach, pp. 91–92. Academic Press (2014)
3. Fujishima, T.: Real-time chord recognition of music sounds. In: International Computer Music Conference (ICMC), pp. 464–467 (1999)
4. Cohen, L.: Time-frequency analysis (Japanese), pp. 94–96. Asakura Publishing (1998)
5. Minami, S., Nakamura, O.: Image engineering -Image electronics- (Japanese), p. 126. Corona Publishing
6. Distance between signals using dynamic time warping. MathWorks, February 2018. https://jp.mathworks.com/help/signal/ref/dtw.html
7. Shikano, K., Nakamura, S., Ise, S.: Digital signal processing of voice/sound information (Japanese). In: SHOKODO, pp. 20–22 (1997)

Analysis of Environmental Change Detection Using Satellite Images (Case Study: Irrawaddy Delta, Myanmar)

Soe Soe Khaing(✉), Su Wit Yi Aung(✉), and Shwe Thinzar Aung(✉)

University of Technology (Yatanarpon Cyber City), Pyin Oo Lwin, Myanmar
khaingss, suwityiaung123@gmail.com, shwethinzaraung@gmail.com

Abstract. Myanmar is known as a rich land and it is an agriculture-based country. Rice is the staple food of Myanmar and Irrawaddy delta is the main source of rice production in Myanmar. And there was a Nargis Cyclone hit that region in the early May, 2008 and it was extremely severe caused the worst natural disaster in the recorded history of Myanmar. So it is necessary to be a lot more attention on that region in order to analyze the change in environment to support decision making, urban development and planning. The aim of this study is to analyze the environmental changes in Irrawaddy Delta using Landsat satellite images between the years 2000 and 2017. Collection of Landsat images for the years, 2000, 2005, 2010 and 2017 is used to determine changes of environment. Images are classified into five types: water, forest, buildup, fields and others. For land cover index classification, two supervised classification algorithms: Random Forest (RF) and Support Vector Machine (SVM) are utilized and compared the results of classification using producer accuracy and user accuracy. According to the experimental result, the result of RF is more précised than that of SVM for environmental change detection.

Keywords: Random Forest · Support vector machine · Index classification

1 Introduction

The Irrawaddy Delta lies in the Irrawaddy Division, the lowest expanse of land in Myanmar. [9]. This region was the most threatened region in deforestation which is an average rate of 1.2% from 1990–2000, a rate four times the national average [1]. The population in this region is over 8 million and which is the main region for rice production. There is about 35% of rice production comes from this region. Because of the agricultural expansion, increasing population, demolished natural resources, environmental pollution and the use of wood for fuel, environmental change has been significantly increasing that leads to the effect of economically and environmentally. So, it becomes a vital for monitoring and managing natural resources and urban development.

This research is focused on developing the change detection system for Irrawaddy region. It includes detection of change in environment, remote sensing data and Geographic Information System (GIS) techniques.

T. T. Zin and J. C.-W. Lin (Eds.): ICBDL 2018, AISC 744, pp. 355–363, 2019.
https://doi.org/10.1007/978-981-13-0869-7_40

2 Study Area

The Irrawaddy Delta lies between north latitude 15° 40′ and 18° 30′ approximately and between east longitude 94° 15′ and 96° 15′ which covers about an area of 34,000 km^2. The location map of the study is shown in Fig. 1.

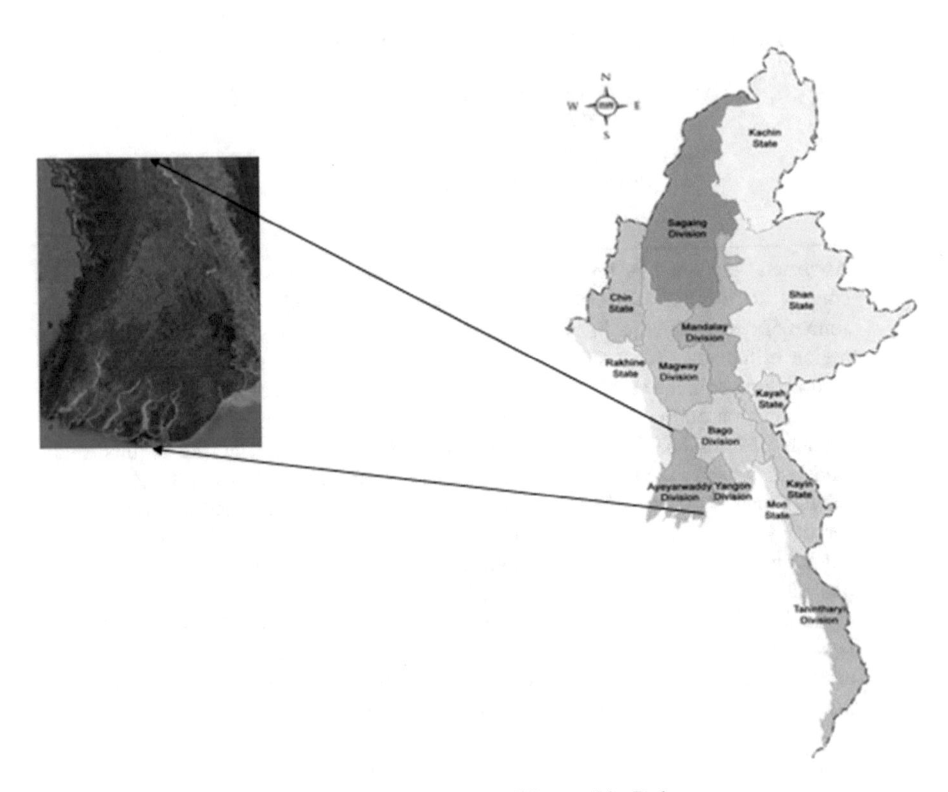

Fig. 1. Location map of Irrawaddy Delta

3 Methodology

3.1 Block Diagram of Proposed System

The step by step process shown in block diagram of the proposed system is illustrated in Fig. 2.

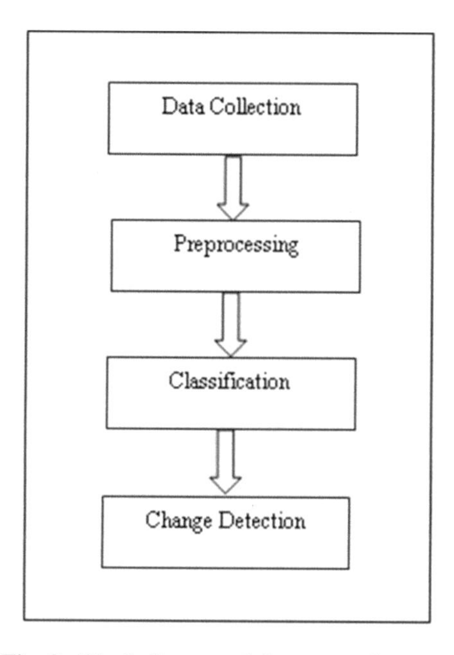

Fig. 2. Block diagram of the proposed system

Data Collection

Data Collection is done by using Landsat images for the following years. We analyzed the environment change detection from 2000 to 2017 period. The data for the year 2000 is collected by using Landsat 5 and the data for the year between 2005 to 2010 is collected from Landsat 7. For the year 2017, the data is collected by using Landsat 8.

Preprocessing

Preprocessing is done by using Google Earth Engine which is a cloud-based geospatial processing platform for planetary-scale environmental data analysis. The main components of Earth Engine includes:

- Datasets
- Compute power
- APIs
- Code Editor

The satellite images are used as input and then preprocessing is done. The classification is done by using Random Forest classifier and SVM classifier then the analysis is performed based on the results of these two classifiers.

Random Forest Classifier

The Random Forest algorithm has been used in many data mining applications, however, its potential is not fully explored for analyzing remotely sensed images [2]. Random forests are a combination of tree predictors such that each tree depends on the values of a random vector sampled independently and with the same distribution for all trees in the forest. The generalization error of a forest of tree classifiers depends on the strength of the individual trees in the forest and the correlation between them. Using a random selection of features to split each node yields error rates that compare favorably to Adaboost, and are more robust with respect to noise.

Among many advantages of Random Forest the significant ones are: accuracy among current algorithms, efficient implementation on large data sets, and an easily saved structure for future use of pre-generated trees [3]. Gislason et al. [4] have used Random Forests for classification of multisource remote sensing and geographic data. The Random Forest approach should be of great interest for multisource classification since the approach is not only nonparametric but it also provides a way of estimating the importance of the individual variables in classification. Steps of the random forest classifier are as follow:

- Firstly, the random forest classifier randomly select features and a combination of features at each node to grow a tree.
- Features are randomly selected with replacement.
- Gini index is used as an attribute selection measure. In dataset T, Gini index is defined as

$$Gini(T) = 1 - \sum_{j=1}^{n} (p_j)^2 \tag{1}$$

  where p_j is the relative frequency of class j in T.
- After the calculation of Gini index for selecting attribute, select the lowest Gini index value.
- Mid-point of every pair of consecutive value is chosen as the best split point for the attribute.
- The procedure is repeated for the remaining attribute in the dataset.
- Each tree is fully grown for the training dataset.
- Finally, the decision tree is taken the most popular voted class from all the tree predicator in the forest.

Support Vector Machine

SVM has been successfully used for data classification in the remote sensing arena [4]. Support vector machine (SVMs, also support vector networks) is supervised learning model with associated learning algorithms that analyze data used for classification and regression analysis. The goal of SVM is to produce a model (based on the training data) which predicts the target values of the test data given only the test data attributes. A Support Vector Machine (SVM) is a discriminative classifier formally defined by a separating hyper plane. In other words, given labeled training data (supervised learning), the algorithm outputs an optimal hyper plane which categorizes new examples. In two dimensional space, this hyper plane is a line dividing a plane in two parts where in each class lay in either side.

A hyper plane is a line that splits the input variable space. In SVM, a hyper plane is selected to best separate the points in the input variable space by their class, either class 0 or class 1. In two-dimensions, a line can be visualized and assume that all of the input points can be completely separated by this line. For example:

$$B0 + (B1 * X1) + (B2 * X2) = 0 \quad (2)$$

where the coefficients (B1 and B2) that determine the slope of the line and the intercept (B0) are found by the learning algorithm, and X1 and X2 are the two input variables.

Classifications can be made using this line. By plugging in input values into the line equation, whether a new point is above or below the line can be calculated.

3.2 Experimental Results

Comparison of Classification Result for the Year 2000

The classification results of the Landsat images for the year 2000 using Random Forest classifier and Support Vector Machine are shown in Fig. 3.

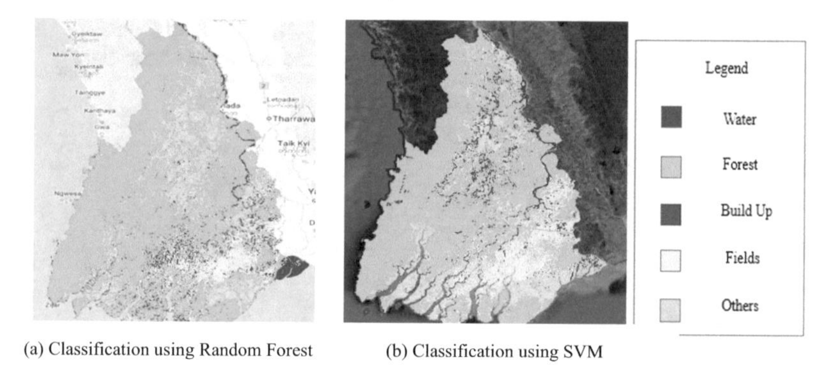

(a) Classification using Random Forest (b) Classification using SVM

Fig. 3. Classification results of Irrawaddy Delta for the year 2000

The total area calculation for each land cover indices is shown in Table 1. And it is calculated using Square-Kilometers for RF and SVM classifier.

Table 1. Total land cover area for land cover indices for the year 2000

2000 year	RF-square-km	SVM-square-km
Water	1131.94	868.63
Forest	21130.66	20149.37
Buildup	1208.22	1638.38
Others	6959.29	6360.89
Fields	3081.97	4494.82

Comparison of Classification Results for the Year 2005

The classification results of Irrawaddy Delta for the year 2005 is shown in Fig. 4. The total land cover indices area is shown in Table 2.

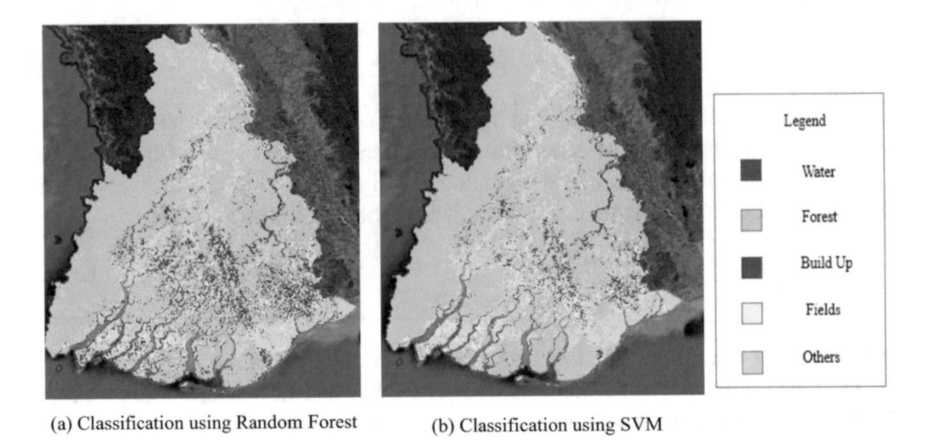

(a) Classification using Random Forest (b) Classification using SVM

Fig. 4. Classification results of Irrawaddy Delta for the year 2005

Table 2. Total land cover area for land cover indices for the year 2005

2005 year	RF-square-km	SVM-square-km
Water	790.38	802.07
Forest	19244.75	20830.71
Buildup	2298.48	1053.66
Others	7231.38	8129.89
Fields	3947.10	2695.76

Comparison of Classification Results for the Year 2010

The classification results of Irrawaddy Delta for the year 2010 is shown in Fig. 5 and the total land cover indices area is shown in Table 3.

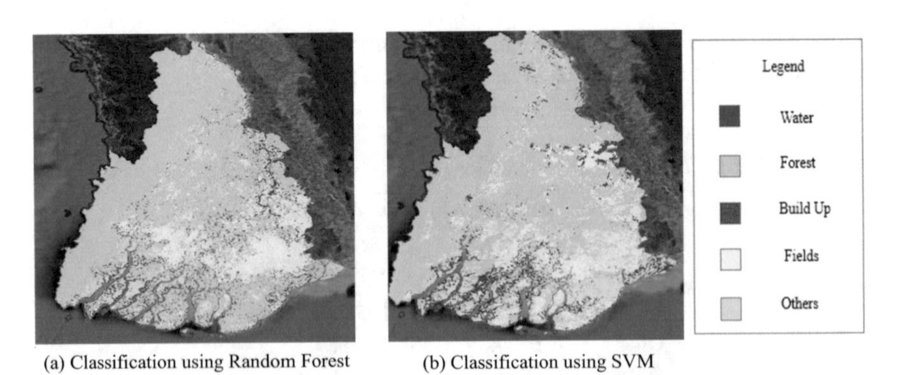

(a) Classification using Random Forest (b) Classification using SVM

Fig. 5. Classification results of Irrawaddy Delta for the year 2010

Table 3. Total land cover area for land cover indices for the year 2010

2010 year	RF-square-km	SVM-square-km
Water	649.90	1054.50
Forest	19407.39	19836.92
Buildup	1766.86	1876.16
Others	5337.64	4202.57
Fields	6350.29	6541.93

Comparison of Classification Results for the Year 2017

The classification results of Irrawaddy Delta for the year 2017 is shown in Fig. 6 and the total land cover indices area is shown in Table 4.

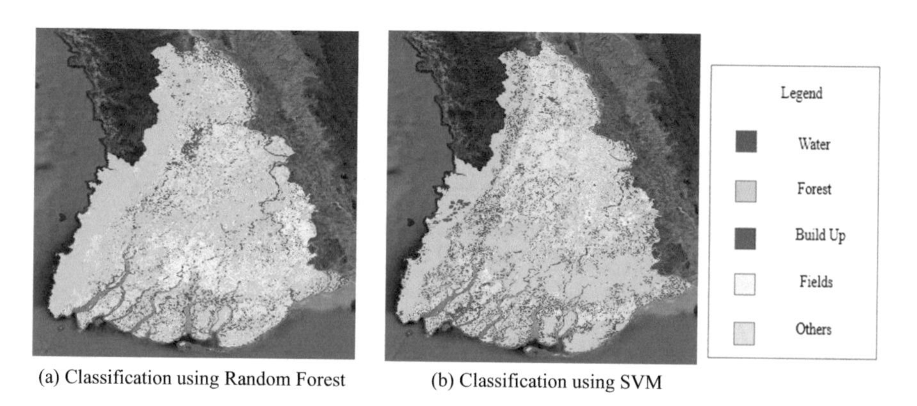

(a) Classification using Random Forest (b) Classification using SVM

Fig. 6. Classification results of Irrawaddy Delta for the year 2017

Table 4. Total land cover area for land cover indices for the year 2017

2017 year	RF-square-km	SVM-square-km
Water	688.82	658.91
Forest	16707.08	18145.42
Buildup	2365.78	3788.00
Others	9449.83	9237.10
Fields	4300.56	1682.65

Change Detection

The changes in five land cover indices between 2000 and 2017 is shown in Fig. 7. By comparing the usage of each land cover indices from Fig. 7, it can be seen that forest areas has been declined significantly due to deforestation and disaster. Buildup area is steadily increased because of population and rapid urbanization. Since Irrawaddy delta is also main rice production region, the fields area are also significantly increased (Fig. 8).

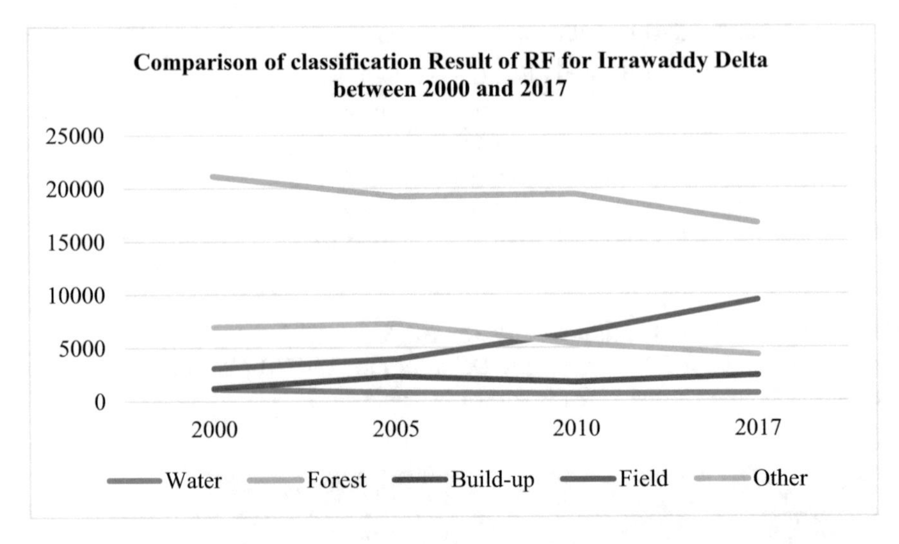

Fig. 7. Comparison of classification results between 2000 and 2017

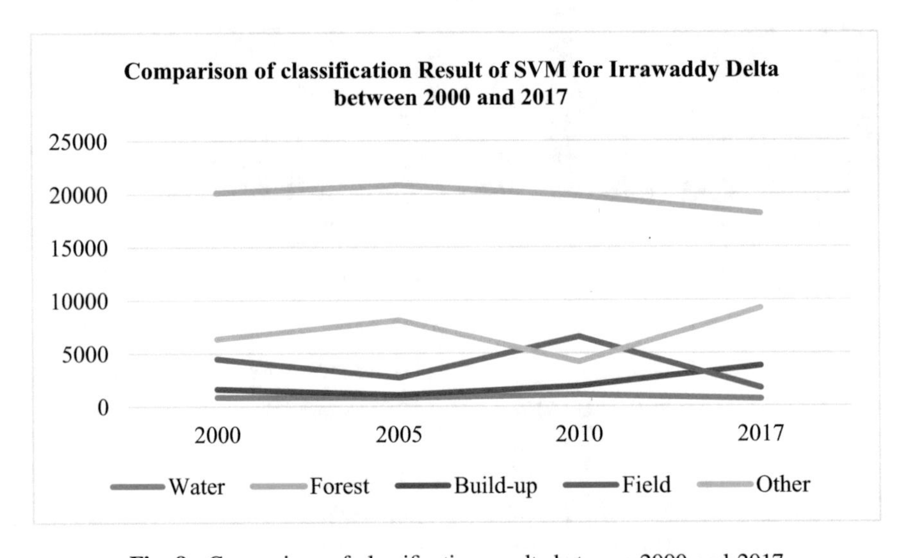

Fig. 8. Comparison of classification results between 2000 and 2017

4 Accuracy Assessment

The final stage of the image classification process usually involves an accuracy assessment step [5]. Many methods for accuracy assessment have been applied in remote sensing. In this proposed system, the performance parameters "user's accuracy", "producer's accuracy" and "overall accuracy" through the use of percentages are used to calculate for the two classification algorithms. Producer's accuracy refers to the probability that a certain land cover of an area on the ground is classified as such,

Table 5. Accuracy assessment of land cover classification

	Random forest classifier		Support vector machine	
	Producer's accuracy	User's accuracy	Producer's accuracy	User's accuracy
Water	100	100	99	99
Forest	99	99	99	99
Buildup	67	82	50	73
Others	89	88	92	97
Fields	87	82	42	97

whereas the user's accuracy refers to the probability that a pixel labeled as a certain land cover class in the map is really of this class [6]. The producer's accuracy and user's accuracy percentage for land cover indices are shown in Table 5.

The overall accuracy value for random forest classifier is 99.6% and support vector machine classifier is 99% respectively. According to the tested results, random forest gives better accuracy results than support vector machine for land cover classification.

5 Conclusion

In this research, environmental change detection system for Irrawaddy delta is proposed. For land cover index classification of Irrawaddy delta, Random Forest and Support Vector Machine algorithms are used and compared the results of these two classifiers. The classification performance of random forest classifier gives higher accuracy results than support vector machine. Therefore, Random Forest classifier has been suggested as an optimal classifier for land cover classification in this study area.

References

1. Wang, C., et al.: Preparing for Myanmar's environment-friendly reform. Environ. Sci. Pol. **25**, 229–233 (2013)
2. Kulkarni, A.D., Lowe, B.: Random forest algorithm for land cover classification. Int. J. Recent Innov. Trends Comput. Commun. **4**(3), March 2016. ISSN 2321-8169
3. Breiman, L.: Random forests. Mach. Learn. **45**(1), 5–32 (2001)
4. Gislason, P.O., Benediktsson, J.A., Sveinsson, J.R.: Random forest for land cover classification. Pattern Recogn. Lett. **27**, 294–300 (2006)
5. Huang, C., Davis, L.S., Townshend, J.R.G.: An assessment of support vector machines for land cover classification. Int. J. Remote Sens. **23**, 725–749 (2002)
6. Manisha, B.P., Chitra, G., Umrikar, N.: Image classification tool for land use/land cover analysis: a comparative study of maximum likelihood and minimum distance method. Int. J. Geo. Earth. Environ. **6**, 189–196 (2012)
7. Lambin, E.F., Turner II, B.L., Geist, H.J., Agbola, S.B., Angelsen, A., et al.: The causes of land-use and land-cover change: moving beyond themyths. Glob. Environ. Change Hum. Pol. Dimen. **11**(4), 5–13 (2001)
8. https://en.wikipedia.org/wiki/Deforestation_in_Myanmar
9. https://en.wikipedia.org/wiki/Irrawaddy_Delta

Analysis of Land Cover Change Detection Using Satellite Images in Patheingyi Township

Hnin Phyu Phyu Aung(✉) and Shwe Thinzar Aung(✉)

Department of Information Science, University of Technology (Yatanarpon Cyber City), Pyin Oo Lwin, Myanmar
hninphyuphyuaung1994@gmail.com,
shwethinzaraung@gmail.com

Abstract. Patheingyi Township is located in the eastern part of Mandalay, Myanmar. Patheingyi Township is an area that experienced a fast increase of urban population in the recent decades in Mandalay region. Rapid urbanization has significant effect on resources and urban environment. The remote sensing technique using satellite imagery has been recognized as an effective and powerful tool in monitoring and detecting land use and land cover changes. The aim of the system is to quantify changes in urban area of Patheingyi Township using satellite images. In this study, Landsat 7 and Landsat 8 images are used to study land cover change in study area. Land cover maps of 2000, 2005, 2010 and 2015 are produced using one of the supervised classification techniques. In this system, Random Forest Classifier model is used to detect the growth of urban area. The system is classified three major land cover classes: built-up, water and other. And then, ArcGIS tool is used to classify built-up area into five classes: urban built-up, suburban built-up, rural built-up, urbanized open land and rural open land. A total of 295 ground check points are used to calculate the accuracy assessment. According to the tested results, built-up area was increased by 285 ha from 2010 to 2015 in Patheingyi Township.

Keywords: Random Forest · Patheingyi Township · Google Earth Engine
Remote sensing · ArcGIS

1 Introduction

Nowadays, the world is becoming rapidly urbanized and this process needs to be monitored. One way of monitoring this process is to perform change detection in urban areas using satellite images. Change detection is the process of identifying differences in the state of an object or phenomenon by observing it at different times [3]. Change detection is not only used for urban applications but also used for detecting forest or landscape change, disaster monitoring and in many more applications. Nowadays, urban research is shifting towards the use of digital, and towards the development of remote-sensing image classification [2]. The main purpose of the system is to detect changes that have taken place particularly in the built-up land and to predict the urban area growth in four years period (2000, 2005, 2010 and 2015) in Patheingyi Township.

T. T. Zin and J. C.-W. Lin (Eds.): ICBDL 2018, AISC 744, pp. 364–373, 2019.
https://doi.org/10.1007/978-981-13-0869-7_41

Satellite images for four years period (2000, 2005, 2010, and 2015) are collected from Landsat satellites by using Google Earth Engine. Training data is collected by drawing polygons on the pixel values of Landsat images which represent land cover classes. Firstly, water index and built-up index is extracted. And then, random forest classifier is used to classify three land use areas: built-up, water and other. Among three land use areas, the system is extracted built-up class and classified it into five regions: urban built-up, suburban built-up, rural built-up, urbanized open land and rural open land by using Urban_Landscape Analysis (ULA) Model in ArcGIS. After that, changing area of built-up land is calculated and change map of Patheingyi Township is produced.

The remainder of the paper is organized as follows: Sect. 2 describes about related work, Sect. 3 details study area and Sect. 4 provides system flow. Section 5 describes about methodology and Sect. 6 explains about classification with ArcGIS. Section 7 gives about experimental result of the system. Finally, Sect. 8 concludes the system.

2 Related Work

Land use/land cover (LULC) data is data that is a result of classifying raw satellite data based on the return value of the satellite image. LULC change, including land transformation has altered a large proportion of the earth's land surface [4].

Prakasam studied LULC change over a period of 40 years in Kodaikanal, Tamil Nadu. In this study, major changes have been observed that area under built-up land and harvested land has increased whereas the area under forest and water body has decreased [5]. In Kalu District, Ethiopia, Kebrom Tekle reported the size of open areas and settlements have increased at the expense of shrub lands and forests in twenty eight years (between 1958 and 1986). Similarly, the authors reported increase of homestead made in the central highlands, during 1957 and 1986 [7].

Land Cover Classification using Random Forest Classifier is a well-studied field in [6]. The authors researched the effects of various machine learning techniques (Naïve Bayes (NB), Maximum Entropy (ME), and Support Vector Machines (SVM)) in the specific domain of change detection systems. They were able to achieve an accuracy of 82.9% using SVM.

3 Study Area

The aim of the system is to detect urban growth changes in Patheingyi Township. Patheingyi Township is located in the eastern part of Mandalay, Myanmar. Mandalay is incorporated with townships such as Patheingyi, Pyigyitagon, Chanayethazan, Aungmyethazan, Amarapua, and Chanmyathazi. It is situated between North Latitudes 21° 51′ 30″ N to 22° 09′ N and East Longitudes 96° 01′ E to 96° 22′ E. The total area of Patheingyi Township is 148192 acres. The study area is one that experienced a fast increase of urban population in the recent decades in Mandalay region. The study area of the system is shown in Fig. 1.

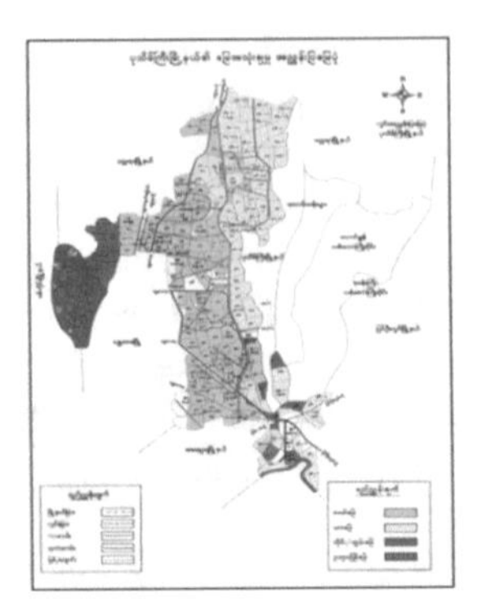

Fig. 1. Study area

4 System Flow

In this system, Landsat 7 and Landsat 8 satellites are used to get images of study area. Landsat 7 was launched on April 15, 1999 and still operational. Landsat 7 data has eight spectral bands with spatial resolution ranging from 15 to 60 m. Landsat 8 was launched on February 11, 2013. Landsat 8 data has eleven spectral bands with spatial resolution ranging from 15 to 100 m.

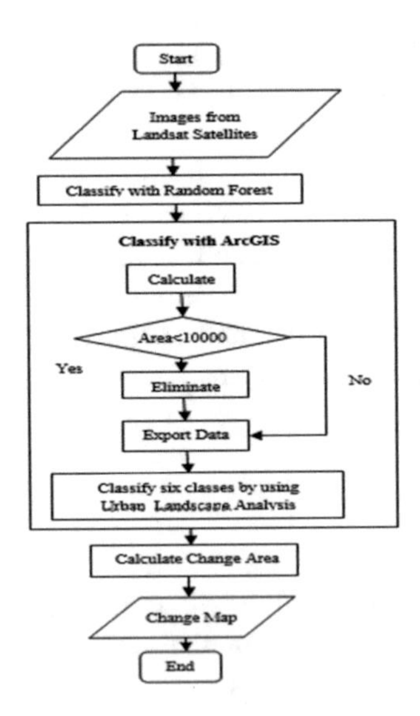

Fig. 2. System flow

The system flow is illustrated in Fig. 2. Firstly, Landsat images for 2000, 2005 and 2010 are collected from Landsat 7 and images for 2015 are collected from Landsat 8. These images are used as input data of the system. Then, Random Forest Classifier is used to classify land use areas. After that, the area of three land use classes is calculated using ArcGIS software which is explained in Sect. 6. If area is less than 10000 m^2, this area is eliminated by ArcGIS tool. After elimination, the exported map that is classified with Random Forest is used and classified it into five regions by using ULA model. Finally, resulted image is generated and changing area is represented with a map.

5 Methodology

False color composite (band 4, 3 and 2) for Landsat 7 and (band 5, 4 and 3) for Landsat 8 are used in this system. The prime step of this system is land use and land cover classification by using water index and built-up index in Google Earth Engine.

Two indices: built-up and water area are extracted using Normalized Difference Built-up Index (NDBI) and Normalized Difference Water Index (NDWI) as shown in Eqs. (1) and (2). The rest area is categorized as the 'Other' class.

For NDBI index, Band 5 (MidIR) and band 4 (NIR) of Landsat Satellite Image are employed.

$$\mathrm{NDBI} = \frac{(\mathrm{Band5} - \mathrm{Band4})}{(\mathrm{Band5} + \mathrm{Band4})} \tag{1}$$

For NDWI index, Band 2 (Green) and band 4 (NIR) of Landsat Satellite Image are used.

$$\mathrm{NDWI} = \frac{(\mathrm{Band2} - \mathrm{Band4})}{(\mathrm{Band2} + \mathrm{Band4})} \tag{2}$$

5.1 Random Forest Classifier

After getting built-up index and water index, Random Forest (RF) Classifier is used to classify three land cover classes: water, built-up and other. Random Forest Classifier is one of the supervised Machine Learning algorithms. Random Forest Classifier is a simple model which is worked well on classification of multisource remote sensing and geographic data. For many data sets, a highly accurate result is produced by RF Classifier [1]. RF is based on tree classifiers and many classification trees are generated in RF.

The flow chart of RF classifier is expressed in Fig. 3. The beginning of RF Algorithm is started with randomly selecting pixel values out of total pixel values. For each tree, the training subsets are chosen from selected pixel values with replacement. In the next stage, the best split approach is used to build a tree for each training subset. In the best split approach, Gini information gain is used to find the root node and leaf node for a tree. According to Gini values, the best split point is chosen and then the procedure is repeated for remaining trees. A classification is given by each tree. To perform prediction, the tree is "voted" for that class. Then, the forest is chosen the classification having the most votes over all the trees in the forest [4].

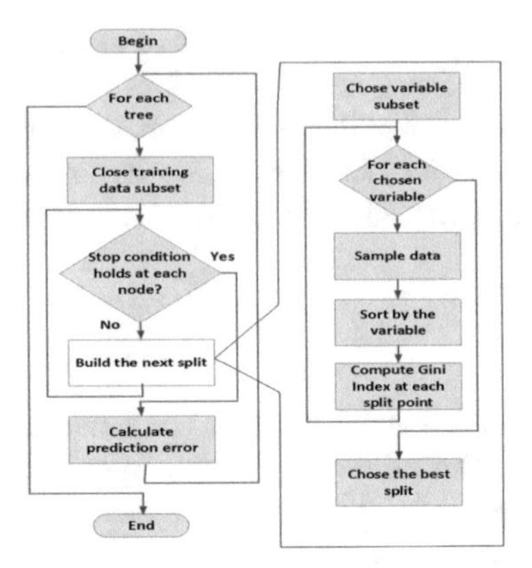

Fig. 3. Flow Chart of RF Classifier

Sample pixel values that are representing three classes of Landsat image in 2015 are described in Table 1.

Table 1. Sample pixel values

No.	B1	B2	Class
1	16.8	14.7	Water
2	16.1	14	Other
3	16.3	14.6	Other
4	17.9	16.2	Built-up
5	17.6	15.9	Built-up

A sample classification tree using Random Forest Classifier is shown in Fig. 4. Pixel values that are mentioned in Table 1 are used as input training data for Random Forest Classifier. Then, Gini index is computed for each training data. After the calculation of Gini index, the lowest Gini index value is selected. Midpoint of every pair of consecutive Gini index values is chosen as the best split point for the training subset. Then, the procedure is repeated for the remaining attribute in the training data. Finally, tree is fully grown for the training dataset [6].

The first (top) decision node is called the root node. The tree is chosen Band 1 as the root node in Fig. 4. "If band 1 is less than or equal to 16.55", it reaches to the terminal node and that node is assigned to the class "other". If band 1 is greater than 16.55, then another binary decision node is evaluated. If band 2 is less than or equal 15.3, that node is assigned to the class "water". If band 2 is greater than 15.3, the terminal node is assigned to the class "built-up".

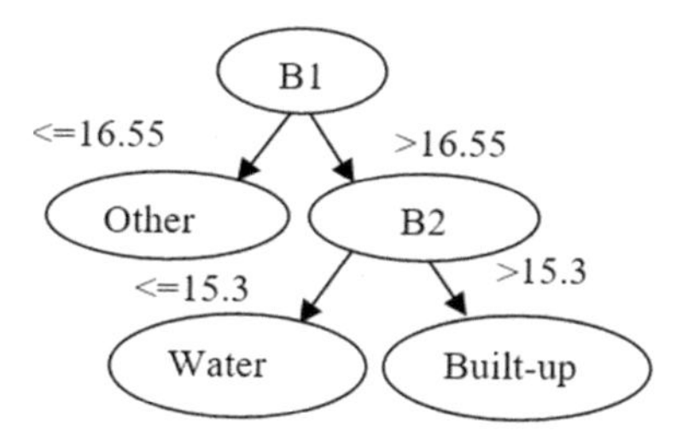

Fig. 4. Classification tree by using RF Classifier

6 Classification with ArcGIS

In the classified output, some misclassified isolated pixels or small regions of pixels is existed. The output of a "salt and pepper" or speckled appearance is occurred due to misclassification. The result of classified output is improved by the process of removing noise. In this system, ArcGIS software is used to get the best result.

ArcGIS is a geographic information system (GIS) for working with maps and geographic data. It is used for creating and using maps, compiling geographic data, analyzing mapped information, sharing, discovering, and managing geographic information in a database. Capabilities for geographic data manipulation, editing, and analysis is included in ArcGIS. It is a powerful tool used for computerized mapping and spatial analysis. ArcGIS Desktop 10 that is the newest version of popular GIS software is used in this system.

After developing three land use areas by using Random Forest Classifier, ArcGIS tool is used in system. In ArcGIS, Urban_Landscape Analysis Tool (ULAT) is used to characterize patches into five classes: urban built-up, suburban built-up, rural built-up, urbanized open land and rural open land. Furthermore, Urbanized Area (UA) and New Development (ND) are calculated based on the resulted maps covering two years.

7 Experimental Results

Analysis of the obtained land use/land cover (LULC) maps is indicated that the urban area grew 60% between 2000 and 2015. Built-up area was increased by 1143.40 ha during 2000–2015 phases. The built-up area was linearly increased over the years by extension of new township due to migration from rural to urban area and getting opportunity for employee offered by new medium industries. The area covered by surface water was obviously decreased over that period.

Classified maps of Patheingyi Township in 2000, 2005, 2010 and 2015 are shown in Fig. 5.

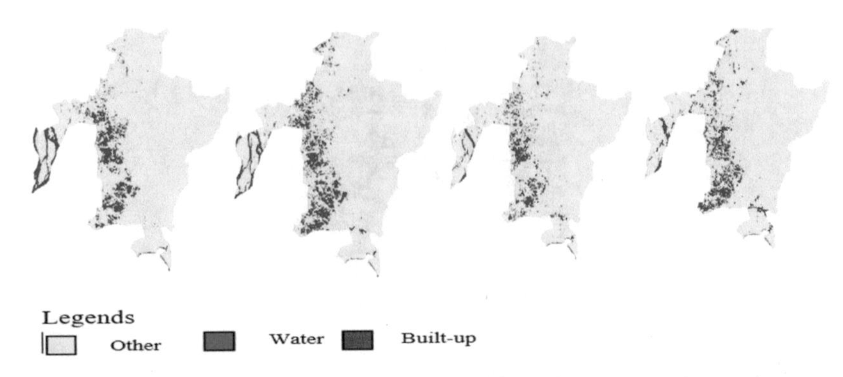

Fig. 5. Classified maps of Patheingyi Township in 2000, 2005, 2010 and 2015

7.1 Identifying Urbanized Area

The urbanized area is combination of built-up area and urbanized open space which is heavily influenced by urbanization. Five classes are classified for urbanized area: urban built-up, suburban built-up, rural built-up, rural open land and urbanized open land. It is found that suburban built-up area grew basically as a big urban cluster from 2000 to 2015.

Land use areas of Patheingyi Township in 2000–2015 periods are shown in Table 2.

Table 2. Classified areas of Patheingyi Township

Year	Urban Built-up (ha)	Suburban Built-up (ha)	Rural Built-up (ha)	Urbanized Open Land (ha)	Rural Open Land (ha)
2000	10.5	528.8	282.8	3.5	53366.2
2005	17.4	700.7	270.6	6.4	52925.2
2010	28.3	1086.0	265.3	6.4	50009
2015	362.3	1371.6	254.5	126.5	48000.6

Urbanized area maps of Patheingyi Township in 2000, 2005, 2010 and 2015 are shown in Fig. 6.

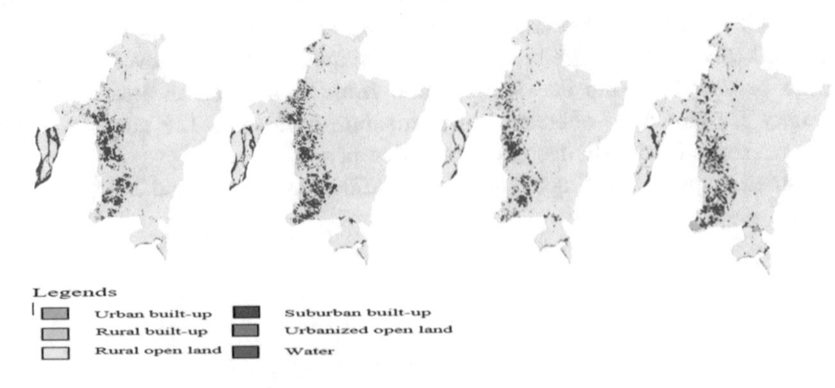

Fig. 6. Urbanized area maps of Patheingyi Township in 2000, 2005, 2010 and 2015

7.2 Land Cover Changes

By land cover change detection, it is obvious that the new development area was attributed to extension that accounts for more than 66.61% during 2000–2015 periods. In the same period, leapfrog 33% and infill 0.39% were contributed to new development area. New development areas of Patheingyi Township in 2000–2015 periods are shown in Fig. 7.

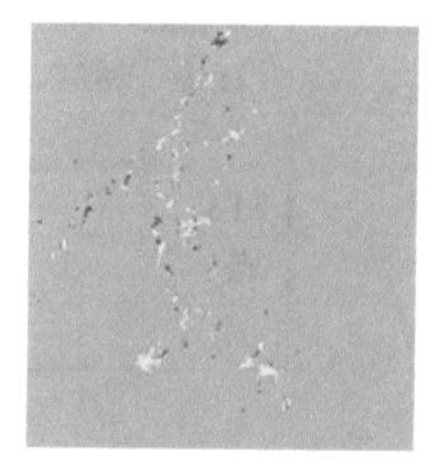

Fig. 7. New development areas of Patheingyi Township during 2000–2015 period

Growth of urban areas of Patheingyi Township during 2000–2015 periods is expressed in Fig. 8.

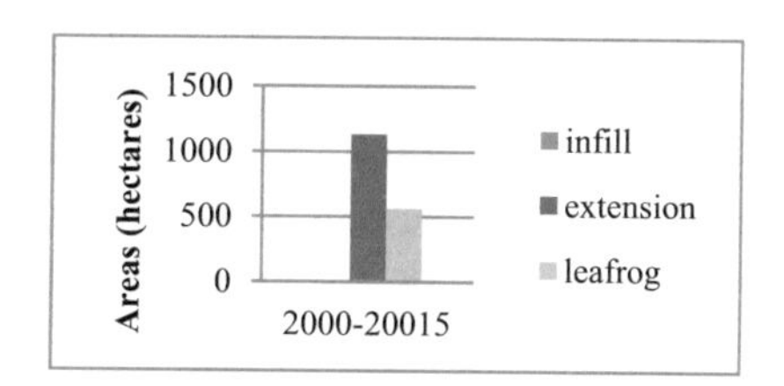

Fig. 8. Urban growth areas of Patheingyi Township

7.3 Accuracy Assessment of Land Cover Classification

Accuracy assessment is very important to understand the output results and making decisions. The measurement of accuracy is expressed as "user's accuracy", "producer's accuracy" and "overall accuracy" through the use of percentages.

Producer's accuracy is the map accuracy from the point of view of the map maker (the producer). It is calculated by taking the total number of correct classifications for a particular class and divided it by the column total. The user's accuracy is the accuracy from the point of view of a map user, not the mark maker. It is calculated by taking the total number of correct classifications for a particular class and divided it by the row total [8].

The accuracy assessment of land cover classification is shown in Table 3. A total of 295 ground check points are used to calculate the accuracy percentage. These check points are reference pixel points of Landsat image in 2015 that are representing three classes mapped with the ground truth point. In the second row of table, the total number of ground truth points is 119 and the correct classification as built-up is 95. The incorrect classification as built-up is 24 where 6 points are wrongly classified as water and 18 points as other. Producer's accuracy for built-up, water and other are 96.94%, 93.88% and 80.81. User's accuracy for built-up, water and other are 79.83, 96.84, and 98.77. Overall accuracy for classification is 90.5%.

Table 3. Accuracy assessment of land cover classification

	Built-up	Water	Other	Ground Truth Points	Producer's Accuracy (%)	User's Accuracy (%)
Built-up	95	6	18	119	96.94	79.83
Water	2	92	1	95	93.88	96.84
Other	1	0	80	81	80.81	98.77

8 Conclusion

The purpose of this study island cover classification analyzed by the remote sensing technology using Landsat satellite images. In combination with GIS, the total urban area and its change were easily assessed. The overall accuracy for classification is 90.5%. In this research, rural open land area had changed mostly to urban built-up and suburban built-up. This is implied that an increase in population due to migration from rural to urban area is occurred in the study area. Suburban built-up area was increased by 285 ha from 2010 to 2015. During the same period, water covered area was obviously changed. According to this research, Patheingyi Township is an area that was experienced a fast increase of urban population in the recent decades in Mandalay region.

References

1. John, K.M., Stuart, E.M., William, G.K., Curtis, M.E.: An Accuracy Assessment of 1992 Landsat-MSS Derived Land Cover for the Upper San Watershed (U.S./Mexico). U.S. Environmental Protection Agency (2002)
2. Pol, R.C., Marvine, B.: Change detection in forest ecosystems with remote sensing digital imagery. Remote Sens. Rev. **13**, 207–234 (1996)
3. Sader, S.A., Hayes, D.J., Hepinstall, J.A., Coan, M., Soza, C.: Forest change detection with remote sensing image. Int. J. Remote Sens. **22**, 1937–1950 (2001)
4. Giles, M.F.: Status of land cover classification accuracy assessment. Remote Sens. Environ. **80**, 185–201 (2001)
5. Jenness, J., Wynne, J.J.: Cohen's Kappa and Classification Table Metrics 2.0: An ArcView 3x Extension for Accuracy Assessment of Spatially Explicit Models. U.S. Geological Survey, Reston, Virginia (2005)
6. Eric, K.F., Adubofour, F.: Analysis of forest covers change detection. Int. J. Remote Sens. Appl. **2**, 82–92 (2012)

7. Congalton, R.G.: A review of assessing the accuracy of classification of remotely sensed data. Remote Sens. Environ. **37**, 35–46 (1991)
8. Lambin, E.F., Turner II, B.L., Geist, H.J., Agbola, S.B., Angelsen, A., et al.: The causes of land-use and land-cover change: moving beyond the myths. Glob. Environ. Change Hum. Policy Dimensions **11**(4), 5–13 (2001)

Environmental Change Detection Analysis in Magway Division, Myanmar

Ei Moh Moh Aung and Thu Zar Tint(✉)

Faculty of Information and Communication Technology,
University of Technology (Yatanarpon Cyber City), Pyin Oo Lwin, Myanmar
eimohmohaung@gmail.com, thuzartint1984@gmail.com

Abstract. Extracting environmental information accurately from remotely sensed images plays a vital role in policy planning, hydrological planning and environmental planning. Environmental change is facing a critical problem due to several factors such as increasing population, demolished natural resources, environmental pollution, land use planning, and others. Moreover, land use and land cover change are considered one of the central components in current strategies for managing natural resources and monitoring environmental changes. The main objective of this study is to assess the environmental changes in Magway Division using Landsat satellite images in the period from 2000 to 2017. In this study, the land cover change detection has been performed based on the analysis of the digital data of Landsat 5 (TM) for the year 2000, 2005, and 2010 and then Landsat 8 (OLI) for 2017. The collected images are classified into water, forest, buildup, vegetation and others for the representative Division in Myanmar. After that, Random Forest (RF) and Support Vector Machine (SVM) classifiers are applied for land cover index classification. The user's accuracy, producer's accuracy, overall accuracy and kappa coefficient are calculated to assess the method's performance.

Keywords: Random Forest · Support Vector Machine · Landsat images

1 Introduction

Magway Region is the second largest of Myanmar's seven divisions. It is made up of the districts of Pakokku, Magway, Minbu, Thayet, and Gangaw involving 25 townships and 1,696 ward village-tracts. And, this region falls in the Dry Zone of Central Myanmar, bordering Mandalay Region on the east, Sagaing Region on the north, Chin State and Rakhine State on the west and Bago Region on the south [7]. Decreasing the forest range will affect the global warming. Moreover, increasing the temperature will make the water shortage, ice melting. The major crop of the Magway Region are sesame and groundnut. Other crops such beans, potatoes, sunflower, millet, tobacco and onions are also produced. Crops can't grow well because of the lack of water, and then cattle don't get fed.

As the agriculture is the one of the main part of the civil economy, the water resources are most effect on people's livelihood in this region. Changing rivers and water surface can impact the loss of residential and farm land, loss of agricultural productivity and other valuable properties. Land use and land cover are dynamic and is an important component in understanding the interactions of the human activities with

T. T. Zin and J. C.-W. Lin (Eds.): ICBDL 2018, AISC 744, pp. 374–383, 2019.
https://doi.org/10.1007/978-981-13-0869-7_42

the environment and thus it is necessary to simulate environmental changes. The remotely sensed data enable researchers and decision makers to monitor changes in an area of interest in the long term without observation in the field. For this reason, it is less costly and less labor to collect data and analyze the change detections, and it also serves to explain the reasons for these changes and linkages with the socioeconomic data [1].

The land cover monitoring is an important part of environmental management system, which requires much attention to protect the natural resources in our environment. Due to this motivated facts, this environmental change detection system for Magway Region is implemented using remote sensing data and Geographic Information System (GIS) techniques. This system employed the Google Earth Engine (GEE) code editor which is a cloud-based geospatial processing platform for multi-temporal satellite imagery classification.

The paper is organized in five sections. Section 1 introduces the system. In Sect. 2, the study area of the system is presented and Sect. 3 describes the system methodology. Section 4 presents the experimental result of the system and the system is concluded in Sect. 5.

2 Study Area

Magway Division is the second largest division in Myanmar. It lies between north latitude 18° 50′ and 22° 47′ approximately and between east longitude 93° 47′ and 95° 55′. It covers about an area of 44818.96 km^2. The location map of study area is as shown in Fig. 1.

Fig. 1. Location map of Magway Division

3 Methodology

3.1 Block Diagram of Proposed System

The block diagram of the proposed system is illustrated in Fig. 2.

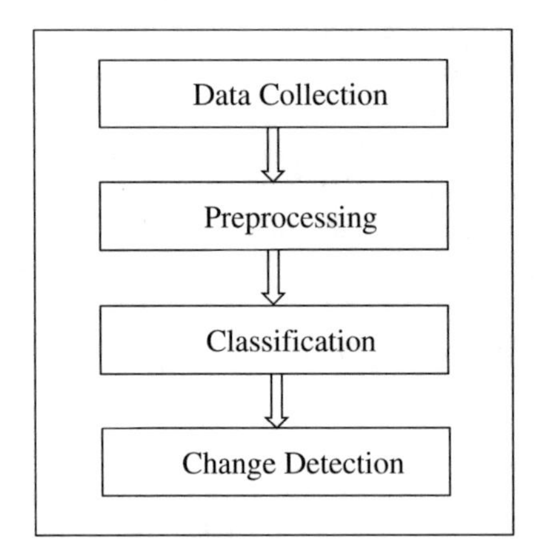

Fig. 2. Block diagram of the proposed system

3.2 Data Collection

Landsat data have become exceedingly integrated into Earth observation and monitoring applications, particularly within the last decade. Moreover, Landsat data can be freely available from U.S. Geological Survey (USGS) and these Landsat data are delivered at a pixel size of 30 m. This system analyzes the environmental change detection from the period 2000 to 2017. For the year 2000, 2005 and 2010, Landsat 5 TM (Thematic Mapper) are collected. And, Landsat 8-OLI (Operational Land Imager) are collected for the year 2017.

3.3 Preprocessing

Images acquired by Landsat sensors are subject to distortion as a result of sensor, solar, atmospheric, and topographic effects. Preprocessing attempts to minimize these effects to the extent desired for a particular application. Firstly, the reflected solar energy captured by sensors are converted to radiance and then rescale this data into digital number (DN). After that, the digital numbers (DNs) values are converted into the TOA reflectance values using conversion coefficients in the metadata file.

These pre-processing steps were performed outside GEE platform. After these products were generated they were uploaded in the GEE cloud platform for the further classification using classification algorithms [2]. Moreover, spectral band combination and gamma correction are performed to the preprocessed Landsat images in the GEE platform.

Gamma Correction

Gamma correction is defined by the following power-law expression where each input pixel value is raised to a fixed power:

$$I_{output}(i,j) = c(I_{input}(i,j))^{\gamma} \tag{1}$$

Where the non-negative input image is raised to the power γ and multiplied by the constant c, to get the output image. In the common case of c = 1, input and outputs are typically in the range 0–1. In general, a value of $\gamma > 1$ enhances the contrast of high-value portions of the image at the expense of low-value regions, whilst the reverse for $\gamma < 1$ can be seen [3].

3.4 Classification

Traditionally, classification methods in remote sensing determine how radiance values recorded by a sensor should be grouped into a number of categories or classes (e.g., land cover type). Random Forest (RF) and Support Vector Machine (SVM) are applied for land cover classification.

Random Forest Classifier

Random Forests, one of ensemble classification family that are trained and their results combined through a voting process, can be considered as an improved version of bagging, a widely used ensemble classifier [4].There are two steps involving random selection that are used when forming the trees in the forest. The first step involves randomly selecting, with replacement, data from supplied training areas to build each tree. The second random sampling step is used to determine the split conditions for each node in the tree. At each node in the tree a subset of the predictor variables is randomly selected to create the binary rule [5]. Gini index is used as an attribute selection measure. In dataset T, Gini index is defined as following equation, where p_j is the relative frequency of class j in T.

$$Gini(T) = 1 - \sum\nolimits_{j=1}^{n} (p_j)^2 \tag{2}$$

After the calculation of Gini index for selecting attribute, select the lowest Gini index value. Mid-point of every pair of consecutive value is chosen as the best split point for the attribute. This procedure is repeated for the remaining attribute in the dataset. Each tree is fully grown for the training dataset. Finally, the output of the classifier is determined by a majority vote of all individually trained trees.

Support Vector Machine

Support vector machine (SVM) is supervised learning model with associated learning algorithms that analyze data used for classification and regression analysis. A Support Vector Machine (SVM) is a discriminative classifier formally defined by a separating hyper plane. In other words, given labeled training data (supervised learning), the algorithm outputs an optimal hyper plane which categorizes new examples. In two dimensional space, this hyper plane is a line dividing a plane in two parts where in each class lay in either side [6]. A hyper plane is a line that splits the input variable space. In SVM, a hyper plane is selected to best separate the points in the input variable space by their class, either class 0 or class 1. In two-dimensions, a line can be visualized and assume that all of the input points can be completely separated by this line. For example:

$$B0 + (B1 * X1) + (B2 * X2) = 0 \tag{3}$$

where the coefficients (B1 and B2) that determine the slope of the line and the intercept (B0) are found by the learning algorithm, and X1 and X2 are the two input variables. Classifications can be made using this line. By plugging in input values into the line equation, whether a new point is above or below the line can be calculated.

3.5 Experimental Result

The classification results of the specified year are presented in this section. Firstly, create the Fusion table to get Myanmar Township Boundary. And then, collect the Landsat images for the specified year and extract the Region of Interest (ROI) from the collected images. After that, mosaicking the image collections. Finally, the resultant images are classified by using Random Forest classifier and Support Vector Machine classifier.

The classification results of year 2000 Landsat image using Random Forest classifier and Support Vector Machine classifier are shown in Fig. 3.

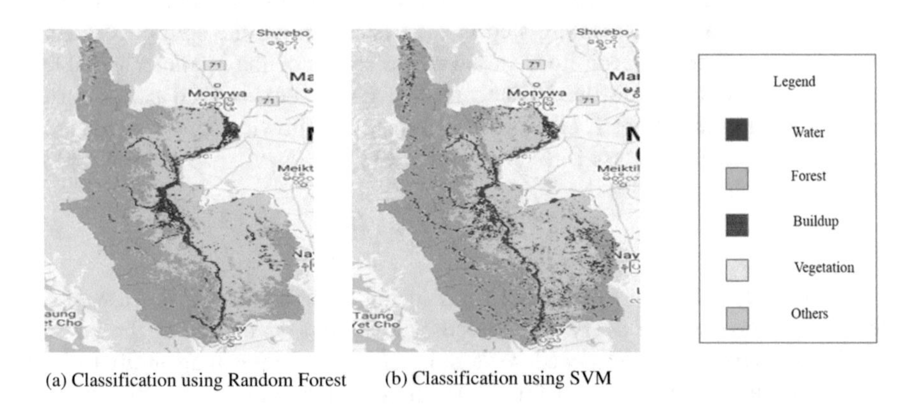

(a) Classification using Random Forest (b) Classification using SVM

Fig. 3. Classification results of Magway Division for the year 2000

The land cover index, the water area, forest area, buildup area are described in blue color, green color, red color respectively. And, vegetation area are shown in yellow color and the remaining area (others) are described in cyan color. The total area calculation for each land cover indices are shown in Table 1. In this testing, the total area of each land cover index is calculated using Square-Kilometers for RF and SVM classifier.

Table 1. Total land cover area for land cover indices for the year 2000

2000 Year	RF-Square-km	SVM-Square-km
Water	2193.935338	1026.069742
Forest	25787.675	25583.46285
Buildup	1929.904294	1255.374844
Vegetation	987.3912897	367.9820167
Others	13878.88804	16544.90452

The classification results of Magway Division for the year 2005 are shown in Fig. 4. The total land cover indices area are shown in Table 2.

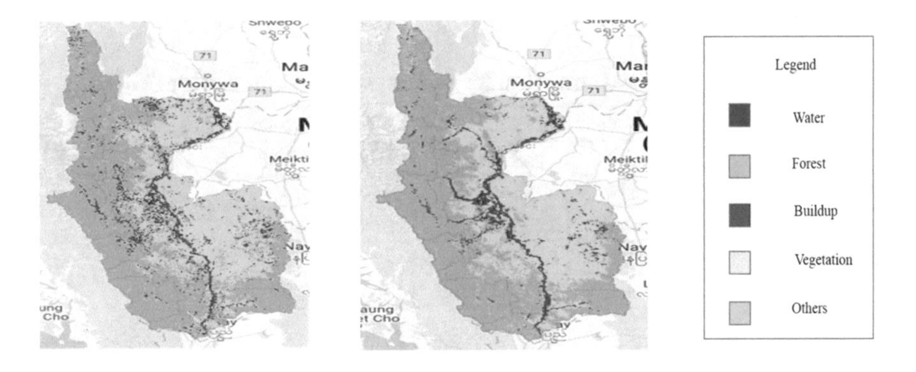

(a) Classification using Random Forest (b) Classification using SVM

Fig. 4. Classification results of Magway Division for the year 2005

Table 2. Total land cover area for land cover indices for the year 2005

2005 Year	RF-Square-km	SVM-Square-km
Water	1706.844613	1013.489776
Forest	24420.03729	24330.54175
Buildup	2309.861295	2398.935508
Vegetation	1442.80694	1005.977668
Others	14898.24382	16028.84926

The classification results of Magway Division for the year 2010 are shown in Fig. 5 and the total land cover indices area are shown in Table 3.

The classification results of Magway Division for the year 2017 are shown in Fig. 6 and the total land cover indices area are shown in Table 4.

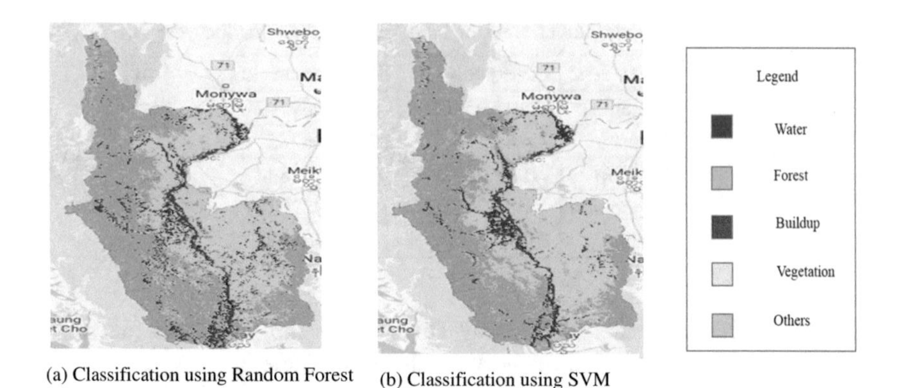

(a) Classification using Random Forest (b) Classification using SVM

Fig. 5. Classification results of Magway Division for the year 2010

Table 3. Total land cover area for land cover indices for the year 2010

2010 Year	RF-Square-km	SVM-Square-km
Water	1480.003137	984.7079755
Forest	22864.03483	22549.43304
Buildup	22864.03483	2444.778048
Vegetation	1299.970626	1278.949912
Others	15950.60569	17519.92499

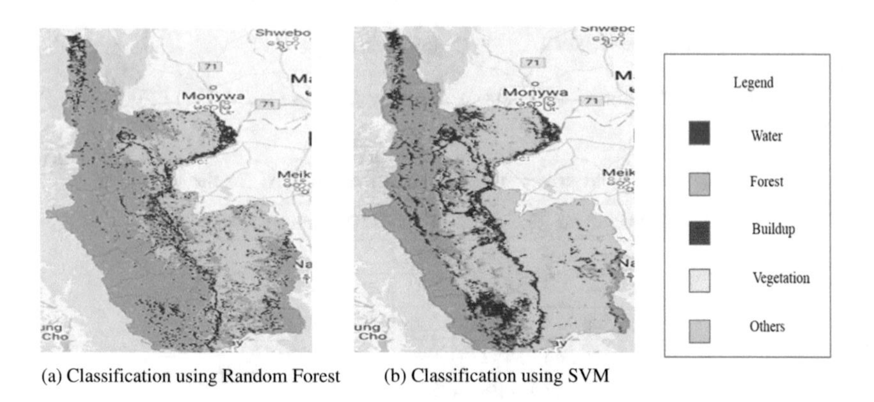

(a) Classification using Random Forest (b) Classification using SVM

Fig. 6. Classification results of Magway Division for the year 2017

3.6 Change Detection

Change detection is the use of remotely sensed imagery of a single region, acquired on at least two dates, to identify changes that might have occurred in the interval between the two dates. The changes of five land cover indices between 2000 and 2017 using Random Forest classifier are shown in Fig. 7. And, the five land cover indices changes between 2000 and 2017 using Support Vector Machine are also described in Fig. 8.

Table 4. Total land cover area for land cover indices for the year 2017

2017 Year	RF-Square-km	SVM-Square-km
Water	855.5453137	920.8156
Forest	17892.13395	16776.74
Buildup	3681.842619	4724.944
Vegetation	538.7718626	423.6252
Others	21809.50022	21931.67

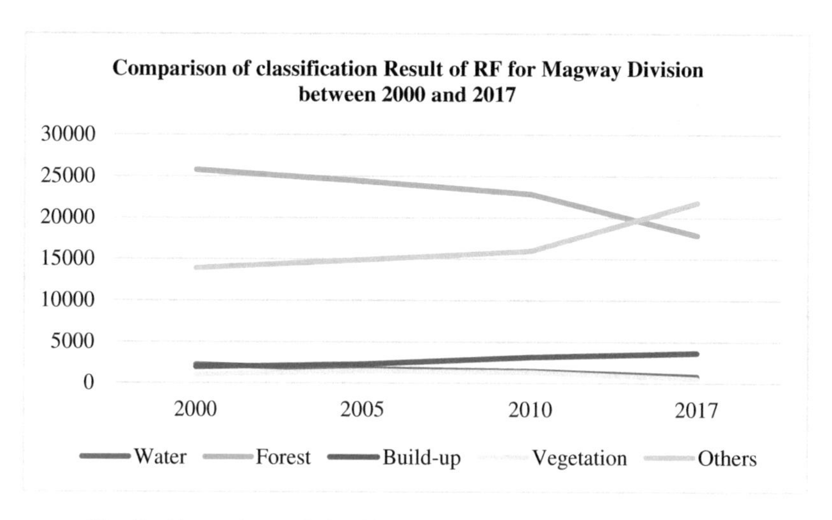

Fig. 7. Comparison of classification results between 2000 and 2017

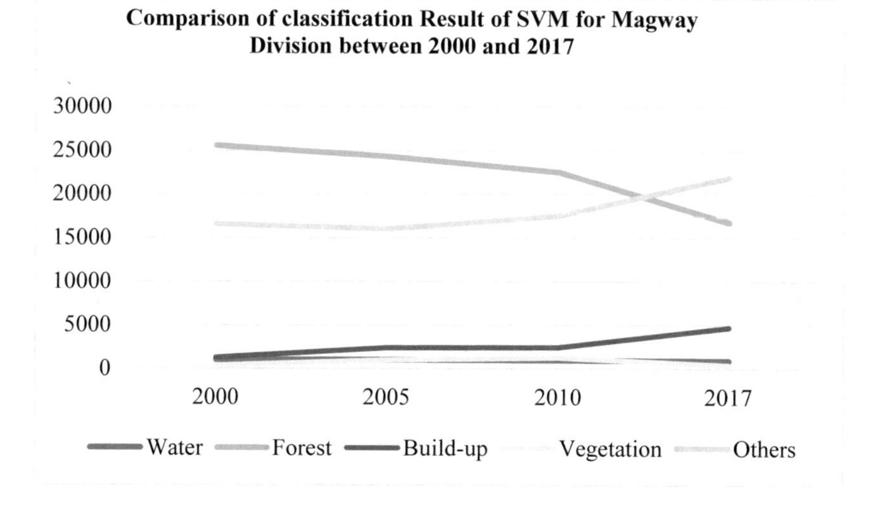

Fig. 8. Comparison of classification results between 2000 and 2017

According to testing result, the water areas are slightly decrease during seventeen years. And, the forest areas have been significantly reduced from 2000 to 2017, it may be seen that the deforestation and urbanization. Moreover, the buildup area are steadily increase because of increase population. The vegetation areas are slightly decrease in 2017, it may be seen that the water shortage has become in these year.

4 Accuracy Assessment

Performance measures are evaluated to prove the quality of the proposed work. To test the performance of the system using remote sensing techniques, accuracy assessment methods such as user's accuracy, producer's accuracy, overall accuracy and kappa coefficient are applied. The comparison of producer's accuracy and user's accuracy percentage for land cover indices are shown in Table 5.

Table 5. Comparison of accuracy assessment for land cover classification

	Random Forest Classifier		Support Vector Machine	
	Producer's accuracy	User's accuracy	Producer's accuracy	User's accuracy
Water	98	91	97	95
Forest	98	98	97	96
Buildup	59	66	58	78
Vegetation	80	89	82	86
Others	98	95	97	87

According to testing result, the overall accuracy for the Random Forest classifier is about 97% and Support Vector Machine is 95%. The kappa coefficient value for the Random Forest classifier is 0.91 and the kappa coefficient value for Support Vector Machine is 0.88. The overall accuracy and kappa coefficient values of the system are calculated as the average value from the results of the year 2000, 2005, 2010 and 2017 for land cover classification.

5 Conclusion

Information gathering about the land cover changes is the fundamental for a better understanding of the natural environment and human beings. In this paper, the environmental change detection system for Magway Division is proposed. To compare the land cover index classification result, Random Forest and Support Vector Machine are applied. The system is implemented with the Google Earth Engine code editor which is the cloud-based geospatial processing platform. According to the experimental result, water area, forest area and vegetation area rates are slightly decrease in the period from 2000 to 2017. And, buildup area are steadily increase in these period. The performance of the index classification is evaluated using user's accuracy, producer's accuracy, overall accuracy and kappa coefficient. According to testing result, the classification performance of random forest classifier provides the better accuracy results than support vector machine for land cover classification.

References

1. Canaz, S., Aliefendioğlu, Y., Tanrıvermiş, H.: Change detection using Landsat images and an analysis of the linkages between the change and property tax values in the Istanbul Province of Turkey. J. Environ. Manage. **200**, 446–455 (2017)
2. Shelestov, A., Lavreniuk, M., Kussul, N., Novikov, A., Skakun, S.: Exploring Google Earth Engine platform for big data processing: classification of multi-temporal satellite imagery for crop mapping. Frontiers Earth Sci. **5**, Article 17 (2017). https://doi.org/10.3389/feart.2017.00017
3. Solomon, C., Breckon, T.: Fundamental of Digital Image Processing: A Practical Approach with Examples in Matlab. Wiley-BlackWell, Chichester (2011)
4. Guan, H., Yu, J., Li, J., Luo, L.: Random forest-based feature selection for land-use classification using LIDAR data and orthoimagery. Int. Arch. Photogrammetry Remote Sens. Spat. Inf. Sci. **XXXIX-B7**, 203–208 (2012)
5. Horning, N.: Random forests: an algorithm for image classification and generation of continuous fields data sets. In: International Conference on Geoinformatics for Spatial Infrastructure Development in Earth and Allied Sciences (2010)
6. https://medium.com/machine-learning-101/chapter-2-svm-support-vector-machine-theory-f0812effc72
7. https://en.wikipedia.org/wiki/Magway_Region

Author Index

T. T. Zin and J. C.-W. Lin (Eds.): ICBDL 2018, AISC 744, pp. 385–386, 2019.
https://doi.org/10.1007/978-981-13-0869-7

Printed in the United States
By Bookmasters